Nelson Maths

4

Teacher's Book

Karen Morrison
Lisa Greenstein

OXFORD
UNIVERSITY PRESS

Great Clarendon Street, Oxford, OX2 6DP, United Kingdom

Oxford University Press is a department of the University of Oxford.

It furthers the University's objective of excellence in research, scholarship, and education by publishing worldwide. Oxford is a registered trade mark of Oxford University Press in the UK and in certain other countries.

British Library Cataloguing in Publication Data

Data available

ISBN: 978-1-382-01016-0

1 3 5 7 9 10 8 6 4 2

Paper used in the production of this book is a natural, recyclable product made from wood grown in sustainable forests. The manufacturing process conforms to the environmental regulations of the country of origin.

Printed and bound by CPI Group UK Ltd, Croydon, CR0 4YY

Acknowledgements

The publisher and authors would like to thank the following for permission to use photographs and other copyright material:

Cover: Matthieu Nivesse. **Photos: p9:** ann_isme/Shutterstock; **p21:** Oxford University Press.

Artwork by Aviel Basil, John Haslam, Q2A Media, Integra Software Services, Pantek Media, and Oxford University Press.

Every effort has been made to contact copyright holders of material reproduced in this book. Any omissions will be rectified in subsequent printings if notice is given to the publisher.

Cover activities

The following activities are based on the Level 4 Pupil Book and Workbook cover image. You can use these stimulus questions according to children's learning to date.

Number

Multiplication problems: Ask: *How many tentacles does one octopus have?* Ask the children to write down a multiplication sentence to find the total number of octopus tentacles. Challenge them to write other multiplication sentences based on the cover image. For example, the orange coral has 3 lots of 3 branches, so a total of 9 branches. It has three leaves on each branch, so a total of 27 leaves.

Geometry

Spotting shapes and angles: How many different 2D and 3D shapes can the children see? Can they identify 2D faces on the 3D shapes? Encourage the children to describe the properties of the shapes. Are the 2D shapes regular or irregular? Can the children identify any right, acute and obtuse angles in the picture? How many parallel, perpendicular, horizontal and vertical lines can they find? Are there any lines of symmetry? Give the children small mirrors for support and encourage them to explore the coral.

Contents

How Nelson Maths works

LEVEL	PUPIL BOOK	WORKBOOKS	TEACHING SUPPORT	DIGITAL CONTENT ON OXFORD OWL
STARTER LEVEL				For all levels: • Digital versions of the Pupil Books, Workbooks and Teacher's Books • Assessment support • Parent notes • Curriculum mapping and planning guides • Vocabulary support
1				
2				
3				
4				
5				
6				

How to use Nelson Maths

Nelson Maths is a comprehensive maths programme for children aged 4–11. The course covers the following five strands: Number, Measure, Geometry, Statistics and Algebra to ensure full coverage of your chosen curriculum. The course is also full of opportunities to develop children's problem-solving skills through carefully designed questions and activities.

Nelson Maths made up of seven levels (one for each year group) and Level 5 contains 18 units, which should ideally be taught in order. The units have been arranged in a careful progression, designed to support children to build their skills and knowledge and make connections across the different strands. Your children may progress through some units more quickly than others, but it is recommended that you spend approximately two weeks on each unit.

Children each have a **Pupil Book** and write-in **Workbook**. The main lesson activities are included in the **Pupil Book**, and the **Workbook** includes extra practice and consolidation activities. The **Teacher's Book** contains detailed teaching support for each unit, which will help you to introduce new concepts, revise prior learning, deepen children's understanding of a topic and encourage mathematical conversations and discussion.

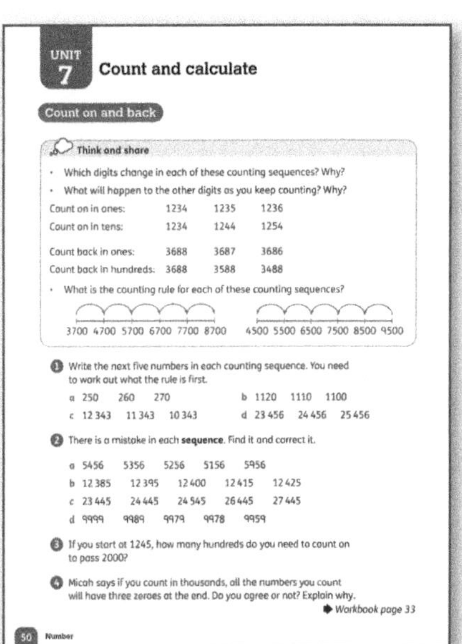

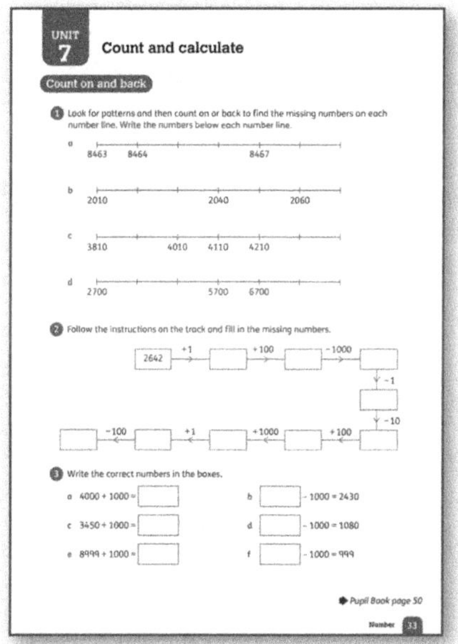

Each unit in the **Pupil Book** is supported by a corresponding unit in the **Workbook**.

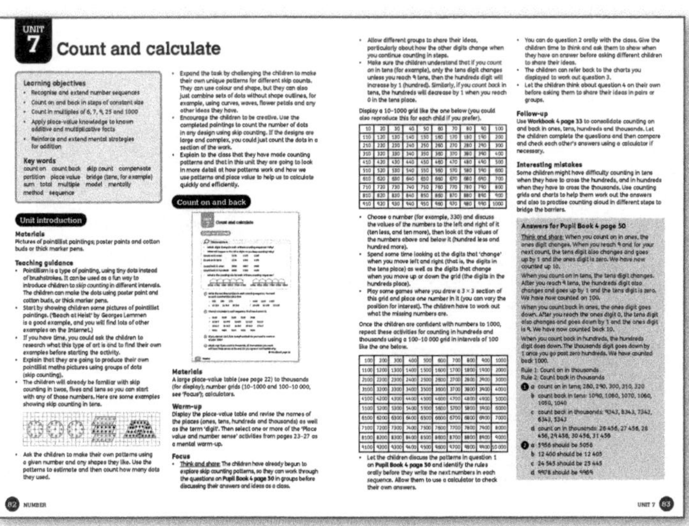

The **Teacher's Book** provides detailed teaching guidance for each unit and answers to the questions and activities.

Read the relevant unit of the Teacher's Book before you begin teaching it. It's important to familiarise yourself with the learning objectives, key vocabulary and resources you need before each lesson. The following pages will give you a good understanding of how to use the Pupil Book, Workbook and Teacher's Book together.

In addition to the printed materials, you can also find lots of supplementary digital resources online on Oxford Owl. Please see page 9 for more information.

Using the Pupil Books

The **Pupil Book** units are designed to be worked through in order, following the discussion prompts and activities suggested in the **Teacher's Book**, and using the linked **Workbook** pages to help children to consolidate their understanding through independent practice. The following features of the **Pupil Books** are designed to help you get the most out of your lessons.

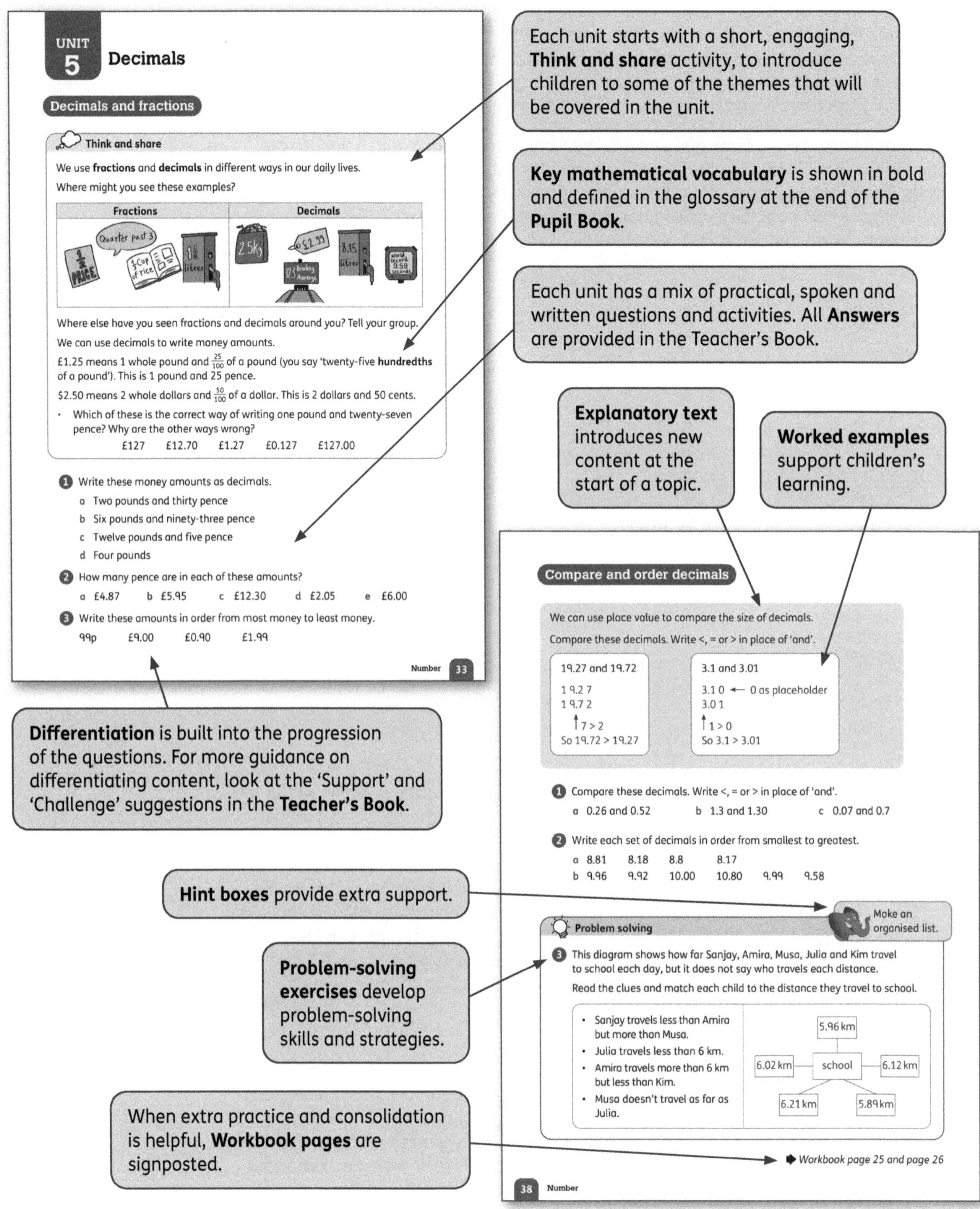

Each unit starts with a short, engaging, **Think and share** activity, to introduce children to some of the themes that will be covered in the unit.

Key mathematical vocabulary is shown in bold and defined in the glossary at the end of the **Pupil Book.**

Each unit has a mix of practical, spoken and written questions and activities. All **Answers** are provided in the Teacher's Book.

Explanatory text introduces new content at the start of a topic.

Worked examples support children's learning.

Differentiation is built into the progression of the questions. For more guidance on differentiating content, look at the 'Support' and 'Challenge' suggestions in the **Teacher's Book.**

Hint boxes provide extra support.

Problem-solving exercises develop problem-solving skills and strategies.

When extra practice and consolidation is helpful, **Workbook pages** are signposted.

The first unit in each Level is called **Think maths**. It encourages a growth mindset and builds resilience by teaching children that mistakes are a positive part of their learning journey. It prepares children to think mathematically and make connections across maths.

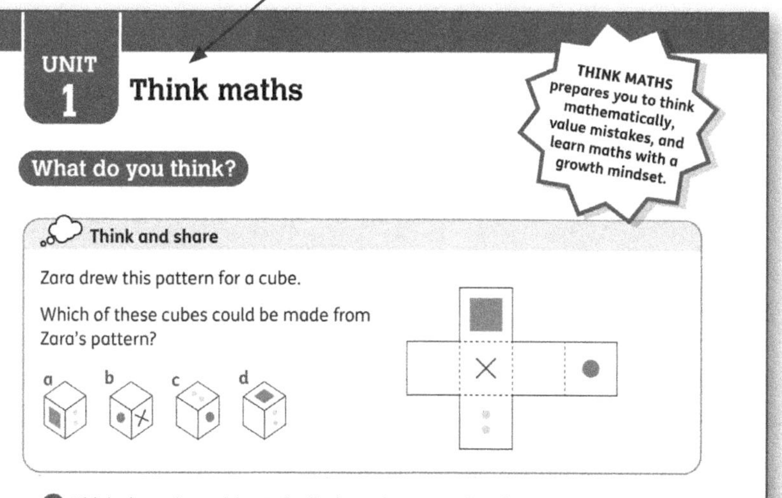

1 Think about the activity and talk about these questions in your group.

a What did you have to think about to solve the problem?

b What did you try to help you solve the problem?

c What was the most challenging part of this activity?

d Did you ever feel you didn't know what to do? What did you try to move past that?

e Did you make any mistakes? How did these help you learn?

2 Choose three of these statements about learning maths.

Everyone can learn maths.

Mistakes are valuable fo

Questions are really important.

Maths is about bein and making sense

Maths class is a place to learn, not a place to perform.

Understanding things prop is more important than doi things quickly.

Take turns with your partner. Try to persuade each other statements you chose are true.

Each book has three **Mixed practice** reviews – one per 'term'. These formative assessment questions help you to assess children's understanding. You may choose to revisit some concepts after the practice.

Mixed practice **Answers** are provided in the **Teacher's Book**.

Each book has three **Mixed practice** reviews – one per 'term'. These formative assessment questions help you to assess children's understanding. You may choose to revisit some concepts after the practice.

Mixed practice 1

1 Look at this set of digit cards.

0   0   3   3   4   4

a Write the greatest 4-digit number you can make using the cards.

b What is the smallest 4-digit number you can make using the cards?

c Copy this number line and show the approximate position of the two numbers you made.

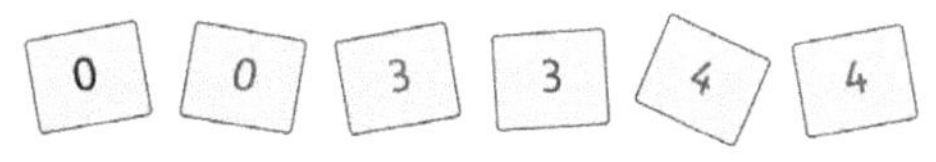

d Round each number you made to the nearest 100.

2 Cara is making a set of Roman numeral cards. She wants to be able to make all the numbers from 1 to 100.

Which of these two sets of cards will allow her to do this?

Set A			Set B		
I	I	I	V	X	I
V	V	X	X	I	X
X	L	L	I	L	C

3 What is the correct mathematical name for:

a a rectangle with four sides the same length

b a shape with five sides

c a quadrilateral that has four equal sides but no right angles

d a regular six-sided polygon?

4 The table shows the times that a bus arrives at five stops. The times are in 24-hour notation.

Stop 132	Stop 133	Stop 134	Stop 135	Stop 136	Stop 137
11:30	11:50	12:05	12:40	12:55	13:00

Where is the bus at these times?

a  b 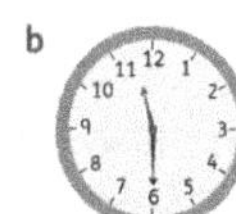 c 

Using the Workbooks

The **Workbooks** provide essential practice to help children consolidate their learning. They provide new questions and activities linked to the concepts introduced in the **Pupil Books**. They are designed for independent use, making them ideal for homework or additional classwork. We recommend that children complete each page in the **Workbook** after they have completed the corresponding lesson in the **Pupil Book**. When extra practice and consolidation is helpful, **Workbook** pages are signposted at the end of the Pupil Book lesson.

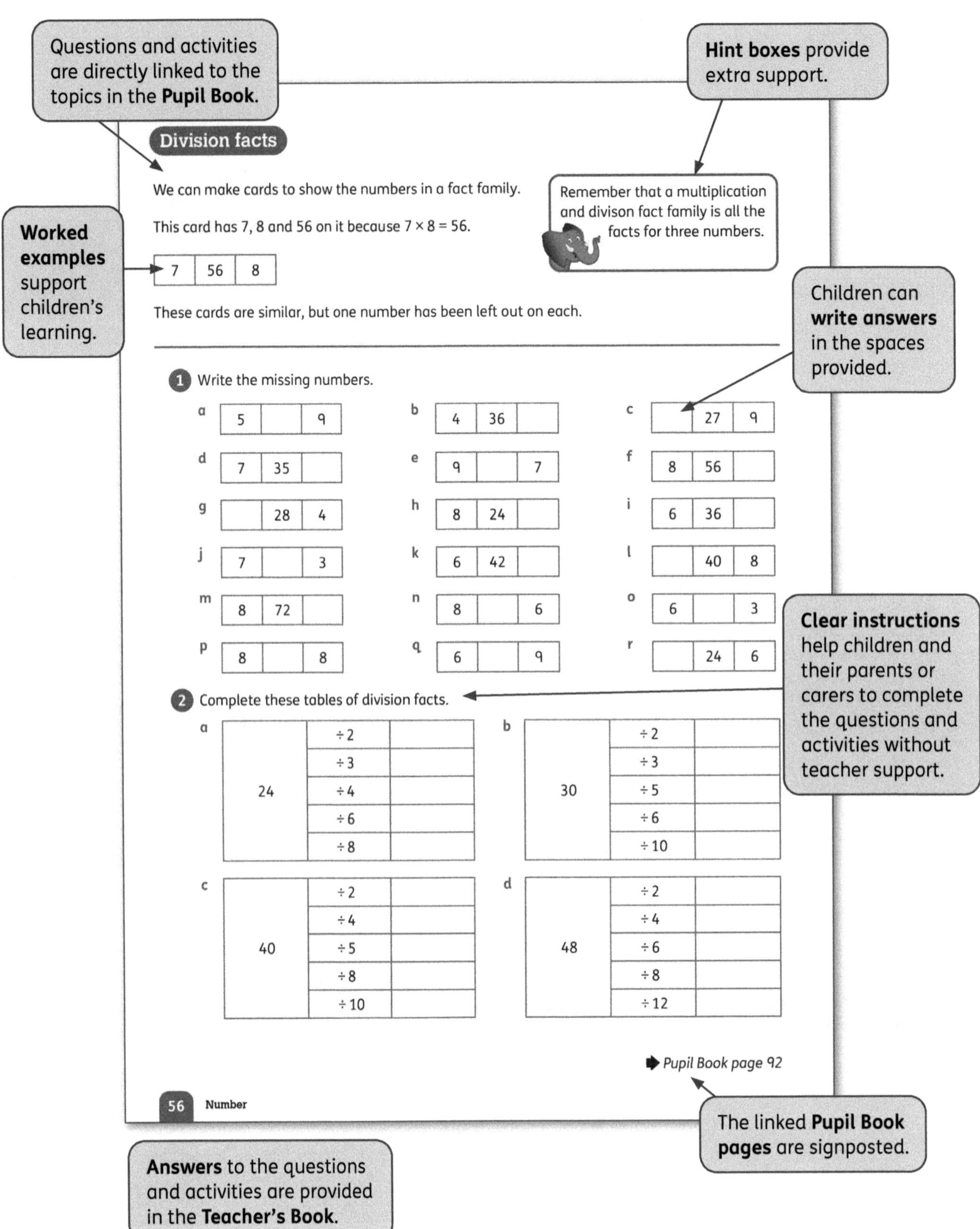

Using the Digital Content

This edition of *Nelson Maths* is supported by additional digital resources available online on **Oxford Owl**. These include:

- Digital versions of the Pupil Books, Workbooks and Teacher's Books to support planning and for front of class display
- A set of printable assessments with a progress tracking tool and mark scheme
- Guidance for you to share with children's parents or carers about their child's learning, including support for homework and ideas for incorporating maths into everyday life
- Vocabulary support
- Curriculum correlation charts
- Planning support

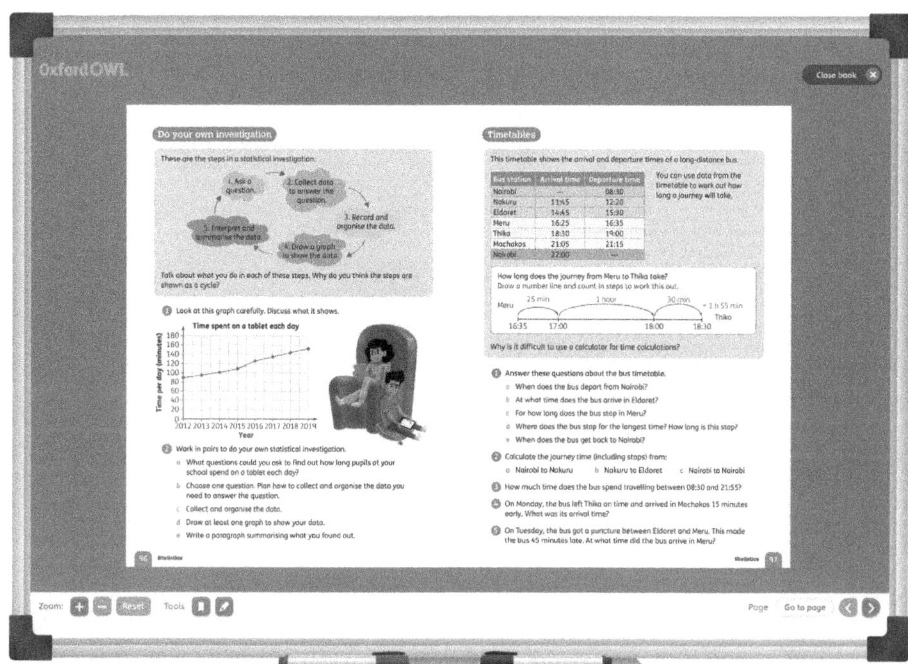

Using the Teacher's Book

The **Teacher's Book** provides step-by-step lesson notes for each unit in the **Pupil Book** and **Workbook**, including learning objectives, answers to the questions activities and suggestions for additional challenge and support.

Before you start teaching the units, take some time to familiarise yourself with these useful sections at the start of the **Teacher's Book**. These four sections are designed to help you get the most out of the Nelson Maths teaching materials:

1. **Introduction** – find out more about the mathematical strands, skills and fundamental principles that underpin *Nelson Maths*. Get ideas and support for teaching maths in an inclusive way that acknowledges individual differences and human diversity.

2. **Teaching approach** – guidance for teaching *Nelson Maths* effectively, including:

- how to develop robust learners with a growth mindset and an understanding of the importance of making mistakes
- the importance of the do-talk-record model that underpins the lessons
- using engaging teaching methods, from investigations and number talks, to exploring maths in real life contexts.

3. **An environment for exploration and play** – practical support on organising your classroom space; approaching whole-class, group and individual activities; and using manipulatives and games to deepen children's learning.

4. **Activity bank of warm-ups and mental maths** – a bank of engaging mental maths, support, and consolidation activities to use with your class.

Teacher's Book: using the unit pages

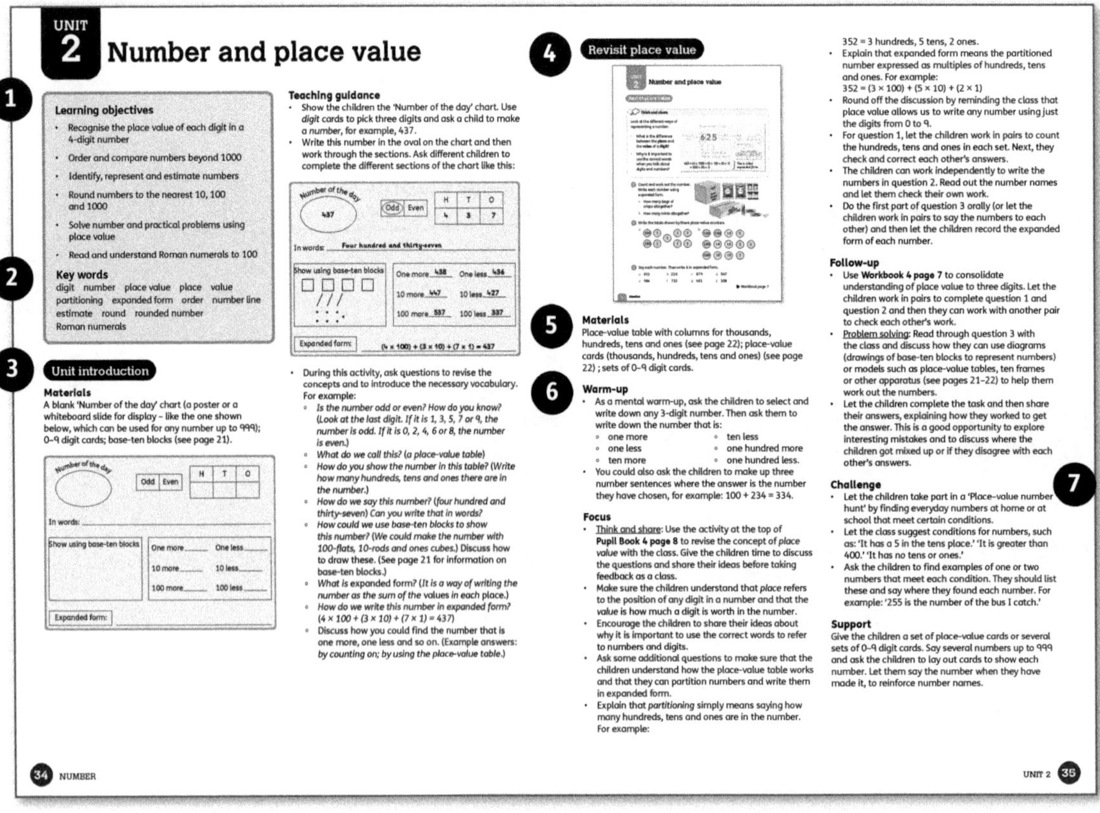

UNIT 2 — Number and place value

Learning objectives

- Recognise the place value of each digit in a 4-digit number
- Order and compare numbers beyond 1000
- Identify, represent and estimate numbers
- Round numbers to the nearest 10, 100 and 1000
- Solve number and practical problems using place value
- Read and understand Roman numerals to 100

Key words

digit number place value place value partitioning expanded form order number line estimate round rounded number Roman numerals

Unit introduction

Materials

A blank 'Number of the day' chart (a poster or a whiteboard slide for display – like the one shown below, which can be used for any number up to 999); 0–9 digit cards; base-ten blocks (see page 21).

Teaching guidance

- Show the children the 'Number of the day' chart. Use digit cards to pick three digits and ask a child to make a number, for example, 437.
- Write this number in the oval on the chart and then work through the sections. Ask different children to complete the different sections of the chart like this:

Number of the day 437 — Odd Even — H T O — 4 3 7

In words: **Four hundred and thirty-seven**

Show using base-ten blocks — One more 438 One less 436 ; 10 more 447 10 less 427 ; 100 more 537 100 less 337

Expanded form: (4 × 100) + (3 × 10) + (7 × 1) = 437

- During this activity, ask questions to revise the concepts and to introduce the necessary vocabulary. For example:
  - *Is the number odd or even? How do you know?* (Look at the last digit. If it is 1, 3, 5, 7 or 9, the number is odd. If it is 0, 2, 4, 6 or 8, the number is even.)
  - *What do we call this?* (a place-value table)
  - *How do you show the number in this table?* (Write how many hundreds, tens and ones there are in the number.)
  - *How do we say this number?* (four hundred and thirty-seven) Can you write that in words?
  - *How could we use base-ten blocks to show this number?* (We could make the number with 100-flats, 10-rods and ones cubes.) Discuss how to draw these. (See page 21 for information on base-ten blocks.)
  - *What is expanded form?* (It is a way of writing the number as the sum of the values in each place.)
  - *How do we write this number in expanded form?* (4 × 100) + (3 × 10) + (7 × 1) = 437)
- Discuss how you could find the number that is one more, one less and so on. (Example answers: by counting on; by using the place-value table.)

Revisit place value

Materials

Place-value table with columns for thousands, hundreds, tens and ones (see page 22); place-value cards (thousands, hundreds, tens and ones) (see page 22); sets of 0–9 digit cards.

Warm-up

- As a mental warm-up, ask the children to select and write down any 3-digit number. Then ask them to write down the number that is:
  - one more
  - one less
  - ten more
  - ten less
  - one hundred more
  - one hundred less.
- You could also ask the children to make up three number sentences where the answer is the number they have chosen, for example: 100 + 234 = 334.

Focus

- Think and share: Use the activity at the top of **Pupil Book 4 page 8** to revise the concept of place value with the class. Give the children time to discuss the questions and share their ideas before taking feedback as a class.
- Make sure the children understand that place refers to the position of any digit in a number and that the value is how much a digit is worth in the number.
- Encourage the children to share their ideas about why it is important to use the correct words to refer to numbers and digits.
- Ask some additional questions to make sure that the children understand how the place-value table works and that they can partition numbers and write them in expanded form.
- Explain that partitioning simply means saying how many hundreds, tens and ones are in the number. For example:

352 = 3 hundreds, 5 tens, 2 ones.

- Explain that expanded form means the partitioned number expressed as multiples of hundreds, tens and ones. For example: 352 = (3 × 100) + (5 × 10) + (2 × 1)
- Round off the discussion by reminding the class that place value allows us to write any number using just the digits from 0 to 9.
- For question 1, let the children work in pairs to count the hundreds, tens and ones in each set. Next, they check and correct each other's answers.
- The children can work independently to write the numbers in question 2. Read out the number names and let them check their own work.
- Do the first part of question 3 orally (or let the children work in pairs to say the numbers to each other) and then let the children record the expanded form of each number.

Follow-up

- Use **Workbook 4 page 7** to consolidate understanding of place value to three digits. Let the children work in pairs to complete question 1 and question 2 and then they can work with another pair to check each other's work.
- Problem solving: Read through question 3 with the class and discuss how they can use diagrams (drawings of base-ten blocks to represent numbers) or models such as place-value tables, ten frames or other apparatus (see pages 21–22) to help them work out the numbers.
- Let the children complete the task and then share their answers, explaining how they worked to get the answer. This is a good opportunity to explore interesting mistakes and to discuss where the children got mixed up or if they disagree with each other's answers.

Challenge

- Let the children take part in a 'Place-value number hunt' by finding everyday numbers at home or at school that meet certain conditions.
- Let the class suggest conditions for numbers, such as: 'It has a 5 in the tens place.' 'It is greater than 400.' 'It has no tens or ones.'
- Ask the children to find examples of one or two numbers that meet each condition. They should list these and say where they found each number. For example: '255 is the number of the bus I catch.'

Support

Give the children a set of place-value cards or several sets of 0–9 digit cards. Say several numbers up to 999 and ask the children to lay out cards to show each number. Let them say the number when they have made it, to reinforce number names.

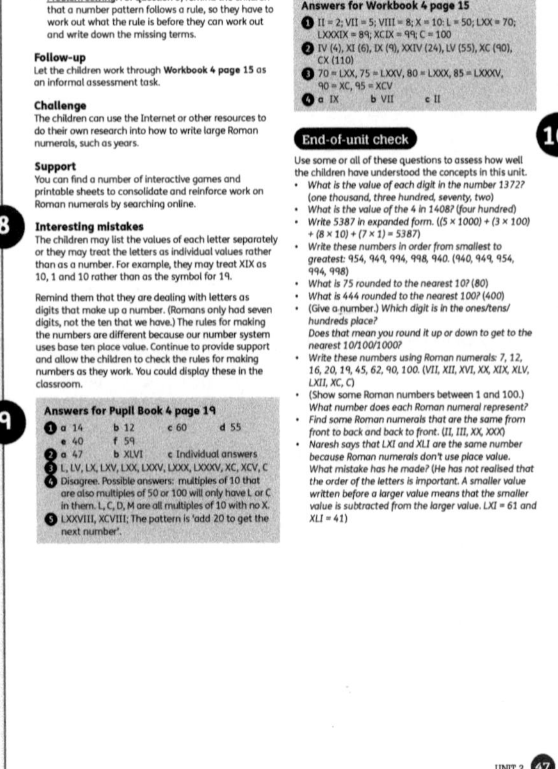

- Problem solving: For question 5, remind the children that a number pattern follows a rule, so they have to work out what the rule is before they can work out and write down the missing terms.

Follow-up

Let the children work through **Workbook 4 page 15** as an informal assessment task.

Challenge

The children can use the Internet or other resources to do their own research into how to write large Roman numerals, such as years.

Support

You can find a number of interactive games and printable sheets to consolidate and reinforce work on Roman numerals by searching online.

Interesting mistakes

The children may list the values of each letter separately or they may treat the letters as individual values rather than as a number. For example, they may treat XIX as 10, 1 and 10 rather than as the symbol for 19.

Remind them that they are dealing with letters as digits that make up a number. (Romans only had seven digits, not the ten that we have.) The rules for making the numbers are different because our number system uses base ten place value. Continue to provide support and allow the children to check the rules for making numbers as they work. You could display these in the classroom.

Answers for Pupil Book 4 page 19

1. a 14 b 12 c 60 d 55 e 40 f 59
2. a 47 b XLVI c Individual answers
3. L, LV, LX, LXV, LXX, LXXV, LXXX, LXXXV, XC, XCV, C
4. Disagree. Possible answers: multiples of 10 that are also multiples of 50 or 100 will only have L or C in them. L, C, D, M are all multiples of 10 with no X.
5. LXXVIII, XCVIII; The pattern is 'add 20 to get the next number'.

Answers for Workbook 4 page 15

1. II = 2; VII = 5; VIII = 8; X = 10; L = 50; LXX = 70; LXXXIX = 89; XCIX = 99; C = 100
2. IV (4), XI (6), IX (9), XXIV (24), LV (55), XC (90), CX (110)
3. 70 = LXX, 75 = LXXV, 80 = LXXX, 85 = LXXXV, 90 = XC, 95 = XCV
4. a IX b VII c II

End-of-unit check

Use some or all of these questions to assess how well the children have understood the concepts in this unit.
- What is the value of each digit in the number 1372? (one thousand, three hundred, seventy, two)
- What is the value of the 4 in 1408? (four hundred)
- Write 5387 in expanded form. ((5 × 1000) + (3 × 100) + (8 × 10) + (7 × 1) = 5387)
- Write these numbers in order from smallest to greatest: 954, 949, 994, 998, 940. (940, 949, 954, 994, 998)
- What is 75 rounded to the nearest 10? (80)
- What is 444 rounded to the nearest 100? (400)
- (Give a number.) Which digit is in the ones/tens/hundreds place? Does that mean you round it up or down to get to the nearest 10/100/1000?
- Write these numbers using Roman numerals: 7, 12, 16, 20, 19, 45, 62, 90, 100. (VII, XII, XVI, XX, XIX, XLV, LXII, XC, C)
- (Show some Roman numbers between 1 and 100.) What number does each Roman numeral represent?
- Find some Roman numerals that are the same from front to back and back to front. (II, III, XX, XXX)
- Naresh says that LXI and XLI are the same number because Roman numerals don't use place value. What mistake has he made? (He has not realised that the order of the letters is important. A smaller value written before a larger value means that the smaller value is subtracted from the larger value. LXI = 61 and XLI = 41)

---

1 **Learning objectives** are listed at the beginning of each unit.

2 **Key words** match the important words highlighted in bold in the **Pupil Book**.

3 **Unit introduction** activities introduce the topics in the unit, engage children and find out what they already know.

4 The **linked Pupil Book and Workbook pages** are signposted.

5 **Materials** needed for the lesson are listed.

6 Most lessons are broken down into three stages:

- **Warm-up** – introduce the topic and unlock prior learning
- **Focus** – develop a firm understanding of concepts and skills by working through the **Pupil Book**
- **Follow-up** – recap the key learning points and extend learning with the activity suggestions in the **Workbook**.

7 Where relevant, **Challenge** and **Support** sections give guidance on deepening the learning or making it more accessible, so that every child is catered for.

8 It is important for children to see mistakes as part of their learning journey. The **Interesting mistakes** section highlights common errors and misconceptions that children may have, along with what to look for and advice on how to support children.

9 **Answers** for **Pupil Book** and **Workbook** questions are provided to make marking easy.

10 An **End-of-unit check** allows you to assess how well children have understood the concepts in each unit, so you can monitor children's understanding throughout the course. A set of formative assessment questions or a closing activity are provided with answers.

Level 4 Scope and sequence

Unit	Unit title	Strand	Learning objectives	Key words
1	Think maths	Think maths	• Establish a classroom environment conducive to thinking and working mathematically • Set up positive norms • Establish key messages of growth mindset mathematics	cube, cuboid, cone, cylinder, prism, pyramid
2	Number and place value	Number	• Recognise the place value of each digit in a 4-digit number • Order and compare numbers beyond 1000 • Identify, represent and estimate numbers • Round numbers to the nearest 10, 100 and 1000 • Solve number and practical problems using place value • Read and understand Roman numerals to 100	digit, number, place value, place value partitioning, expanded form, order, number line, estimate, round, rounded number, Roman numerals
3	2D shapes	Geometry	• Compare and classify 2D shapes based on their properties • Identify and name regular and irregular polygons • Recognise and name different types of quadrilaterals • Use Carroll diagrams to sort and classify quadrilaterals • Investigate properties of squares and rectangles • Solve problems involving 2D shapes and their properties	square, rectangle, pentagon, hexagon, octagon, triangle, circle, sides, regular, irregular, polygon, heptagon, properties, length, angle, right angle, vertex/ vertices, quadrilateral, rhombus, trapezium, parallelogram, kite, Carroll diagram
4	Time	Measure	• Read and record time accurately using both analogue and digital clocks • Write times in different formats, including 24-hour notation • Understand that units of time are not decimal and know the relationship between them • Convert hours to minutes, minutes to seconds, years to months and weeks to days • Solve problems involving times and conversion of time units	second, minute, hour, analogue, digital, a.m., p.m., before noon, after noon, 24-hour notation, timetable, day, week, month, year, leap year
5	Decimals	Number	• Use decimal notation for tenths and hundredths • Extend place-value tables to include tenths and hundredths and recognise the value of digits in decimal fractions • Locate decimals on a number line • Count forward and backwards in tenths and hundredths • Compare and order decimals • Round decimals to the nearest whole number (ones) • Solve a range of problems involving decimal amounts, including money and measurements	fraction, tenths, hundredths, decimal, decimal point, counting sequence, round, whole number, kilometre, approximately
6	Measures and money	Measure	• Estimate, compare and calculate lengths, mass and capacity • Use knowledge of fractions to read and interpret measuring scales • Use decimal notation for measurements and money amounts • Convert between different units of measure • Solve problems involving units of measurement • Estimate, compare and calculate amounts of money	length, units, millimetre (mm), centimetre (cm), metre (m), kilometre (km), height, capacity, volume, millilitre (ml), litre (ℓ), mass, gram (g), kilogram (kg), estimate, measure, scale, notes, coins, currency, bar model, remainder

Unit	Unit title	Strand	Learning objectives	Key words
7	Count and calculate	Number	• Recognise and extend number sequences • Count on and back in steps of constant size • Count in multiples of 6, 7, 9, 25 and 1000 • Apply place-value knowledge to known additive and multiplicative facts • Reinforce and extend mental strategies for addition	count on, count back, skip count, compensate, partition, place value, bridge (tens, for example), sum, total, multiple, model, mentally, method, sequence
8	Symmetry	Geometry	• Identify lines of symmetry in different shapes • Complete a symmetrical shape using the line of symmetry • Identify lines of symmetry on shapes and patterns • Complete a symmetrical shape using the line of symmetry • Identify lines of symmetry on shapes and patterns	symmetrical, symmetry, line of symmetry, horizontal, vertical, diagonal, regular
9	Data and charts	Statistics	• Plan and carry out an investigation to answer a statistical question • Collect, record and organise data using tallies and frequency tables • Draw bar charts, pictograms and Venn diagrams to display data • Solve comparison, sum and different problems using data presented graphically	data cycle, tally, tally mark, tally chart, frequency table, frequency, bar chart, axis/axes, pictogram, symbol, key, Venn diagram, properties
10	Addition and subtraction	Number	• Estimate and use inverse operations to check calculations • Add and subtract numbers with up to 4 digits using formal written methods • Decide which operation and methods to use to solve problems and explain choices • Solve two-step addition and subtraction problems in context	estimate, round, inverse operation, column method, column addition, difference, column subtraction, two-step problem, area
11	Angles and triangles	Geometry	• Revisit right angles and understand that an angle is a measure of turn • Identify and classify angles as acute, right or obtuse • Compare and classify triangles based on angle size and side lengths • Understand that equilateral triangles are regular polygons	angle, turn, degree, right angle, acute angle, obtuse angle, polygon, classify, acute-angled triangle, right-angled triangle, obtuse-angled triangle, scalene, isosceles, equilateral
12	Multiplication and division facts	Number	• Recall multiplication and division facts for multiplication tables up to 12 × 12 • Understand properties of multiplication (and division), including multiplying by 1 and 0 and dividing by 1 • Multiply three numbers, changing the order as needed • Manipulate multiplication and division equations and understand and use the commutative property • Understand and use the distributive property • Use factor pairs and commutativity in mental calculations	factor, array, multiplication fact, product, times table, division fact, division, multiplication, inverse, multiply, fact family
13	Negative numbers	Number	• Use negative numbers in context • Count backwards through zero to include negative numbers • Understand that temperature can be measured in degrees and that it can drop below 0 to give a negative temperature	negative number, temperature, minus sign, degrees Celsius, Fahrenheit, thermometer
14	Perimeter and area	Geometry	• Estimate, measure and calculate the perimeter of different shapes in centimetres and metres • Find the area of shapes by counting squares • Begin to apply formulae to find the perimeter and area of rectangles (including squares)	perimeter, rectangle, area, length, width, square centimetre

Unit	Unit title	Strand	Learning objectives	Key words
15	Fractions	Number	• Understand that fractions are equal parts of a whole and that as the whole is divided into more parts, the parts become smaller • Recognise and show families of equivalent fractions using diagrams • Know that two proper fractions can be equivalent in value • Recognise and write decimal equivalents to 1/4, 1/2 and 3/4 • Use knowledge of equivalence to compare and order fractions • Reason about the location of mixed numbers on a number line • Convert mixed numbers to improper fractions and vice versa • Add and subtract proper fractions, improper fractions and mixed numbers with the same denominators • Solve a range of problems involving fractions	fraction, half/halves, quarter, third, eighth fifth, tenth, numerator, denominator, equivalent fraction, proper fraction, improper fraction, mixed number
16	Position and movement	Geometry	• Understand that position can be described using coordinates • Read and plot points in the first quadrant of a coordinate grid • Use given coordinates to draw polygons in the first quadrant • Describe movement in a straight line as a translation of any given unit up/down and left/right • Translate points and shapes in the first quadrant • Reason and solve problems related to position, movement and symmetry properties (reflect shapes) on the coordinate grid	coordinates, point, coordinate grid, horizontal, vertical, axis/axes x-axis, y-axis, vertex/vertices, translate, translation
17	Multiplication	Number	• Use place value and known facts to multiply mentally • Revise multiplication and division facts to 12 × 12 • Multiply any number by 10 and 100 and recognise that this makes the numbers 10 or 100 times greater • Convert measurements by multiplying by 10 or 100 • Use the distributive property to multiply 2-digit numbers by 1-digit numbers • Multiply 2-digit and 3-digit numbers by a 1-digit number using formal written layout • Solve problems involving multiplication, including scaling problems	times, multiply, product, array, multiplication fact, division fact, partitioning, grid method, column method, estimate
18	Work with line graphs	Statistics	• Interpret and present data using time graphs • Solve comparison, sum and difference problems using data presented on time graphs	line graph, axis/axes, horizontal, vertical, value
19	Patterns	Algebra	• Count on and back in given steps, and extend beyond zero to include negative numbers • Use objects to represent numbers • Recognise the term-to-term rule and extend linear and non-linear number sequences • Recognise and extend the spatial pattern of square numbers	sequence, term-to-term rule, term, square number
20	Division	Number	• Recall multiplication and division facts for tables up to 12 × 12 • Use place value, known and derived facts to divide mentally • Divide whole numbers by 10 and 100 • Solve scaling problems using multiplication and division • Estimate and divide whole numbers, including division with a remainder	divide, equal groups, share, inverse, fact family, remainder, multiple, inverse operation

Section 1 Introduction

This Teacher's Book is designed to support the component parts of Nelson Maths Level 4.

Sections 1–4 of this Teacher's Book (pages 14–30) provide background and reference information, tips on classroom organisation, and a bank of activities and games to use for teaching practically. Section 5 provides lesson notes for the Student Book and Workbook unit-by-unit and page-by-page.

Strands and skills

In Levels 1 to 6, the content is divided into strands:
- Number – numbers and the number system, and calculating
- Measure – money, length, mass and capacity, and time
- Geometry – shape, position and movement
- Statistics – organising, categorising and representing data
- Algebra – use of formulae, variables and equations

In this course, problem solving is not treated as a separate strand but is integrated into the materials at all levels.

Fundamental principles

This series makes the following assumptions about the teaching of mathematics:
- Children need concrete (practical, hands-on) experiences in order to acquire sound mathematical understanding. Like adults, children learn best by investigating and making discoveries for themselves.
- Children refine their understanding and develop conceptual structures by talking about their own thinking and what they have done.
- Individual children develop at different rates. While some will find certain elements of mathematics difficult, others will understand them quickly.
- Children learn in a variety of ways, so mathematics teaching should provide a rich and wide variety of experiences. Children need plenty of opportunities to apply what they have learnt and to relate their mathematics work to other areas of the curriculum and their daily lives.
- Children will become more mathematically able if they are allowed to develop reliable personal ways of working. The formal recording used by mathematicians comes with time and experience, and we cannot expect young children to understand it or use it themselves before they are ready to do so. The conventions (formal methods) should be taught only once children are confident in their own knowledge, concepts and skills.

- Children learn mathematics most effectively when they enjoy what they are doing and when they can see the relevance of what they are learning.

This course reflects current thinking about the most effective ways of teaching and learning mathematics at primary level. It recognises the professionalism of the teacher, and acknowledges that teachers are the best judges of which experience(s) are most appropriate at different stages for the children in their classes. For this reason, this course does not impose a rigid, inflexible structure. Instead, it provides a wide variety of practical activities and games linked to clearly defined purposes and objectives. The teacher selects according to the needs of their classes, groups and individuals.

Individual differences and inclusivity

Everyone learns at their own pace, and in different ways. We also need to recognise that each child enters the classroom with their own experience, their own identity, and their own hopes, fears and curiosities.

This course recognises individual differences and aims to give children the chance to explore the world of mathematics and solve problems in their own way. To achieve this, we believe that each child needs to feel that maths is for them too. Therefore, examples and illustrations need to include a wide range of children of different genders and ethnic, cultural and linguistic backgrounds and should not promote stereotypes.
- When giving children additional maths problems to solve:
  - Use examples of girls doing things that are traditionally considered 'for boys'.
  - Use examples of dads as the primary carers as well as working mothers and female leaders.
- Be aware of children's differences and acknowledge different bodies: not all children have ten fingers and ten toes, and some may not be able to walk or run as easily as others.
- Avoid comparisons of height or weight that some children may experience as body-shaming.
- Celebrate slower workers in your classroom as much as faster workers. Praise reasoning and focused working rather than speed. Never focus on speed-testing or 'who can find the answer the fastest'. Inclusivity means acknowledging that slower, deeper thinkers are some of our best mathematical minds. Maths is not about speed. It is about reasoning.
- Praise different ways of working. Some children need to move, fidget and play in order to solve problems. Allow those who think visually or rely on practical manipulatives and drawings to solve problems using their preferred methods.

- Struggle and confusion are key parts of the process of maths learning. Never dismiss a child's struggle as evidence that they cannot do maths or that they are not mathematically minded.
- When you are planning classroom activities, bear in mind the abilities and needs of your group and change or adapt activities as necessary. For example, it may be more appropriate to count sets of ten counters, or use ten frames (see page 21) than to use fingers for counting. When activities require children to identify, sort or match colours, adapt these activities for children who have colour-blindness, by focusing on pattern rather than colour.

Section 2 Teaching approach

A growth mindset

In recent years, scientists have made important discoveries about how the brain learns and grows. A key discovery is the notion of neuroplasticity or brain plasticity. Brain plasticity is the brain's ability to change connections and neural pathways. Put simply, this means that the brain can learn and grow throughout a person's life.

These discoveries have led many educators to talk about 'growth' versus 'fixed mindset'. People with a fixed mindset believe their abilities are determined and fixed, for example by genetics or by factors outside their control. A fixed mindset can lead to fears of making mistakes. Children with fixed mindsets may believe that they are either 'good at' or 'bad at' maths, and may easily get frustrated when they find problems challenging. We need to help children to understand that our maths ability changes all the time as we learn and grow. By getting children to understand and believe that growth, development and improvement are always possible, we can foster a 'growth mindset', in which children are not afraid of tackling difficult problems.

Getting rid of maths myths

Every child has the potential to learn maths, irrespective of their existing level of mathematical ability. The ability to learn maths is also not related to other traits such as gender, race or ethnicity. There is no such thing as a 'maths person'. It is extremely important that teachers understand that all children are able to learn to think mathematically. Maths is not a special subject and does not require any special natural 'talent'. Some children may grasp concepts more quickly than others – just like in other subjects – but giving them the time to develop their understanding, allows all children to achieve highly in mathematics.

The importance of mistakes

Scientists have also studied the brains of people attempting to solve problems. They found something that might seem surprising – that more neural connections and pathways form when we make mistakes than when we get things right! Although getting correct answers is usually praised by teachers, it is key that we also start understanding the importance of mistakes. Mistakes offer a valuable opportunity to explore why and how a concept makes sense. You can explore mistakes in a productive way with your class by:

- frequently reminding the children that making mistakes shows we are learning something new
- ensuring that the children know that mistakes (and questions) are welcome in your classroom
- discussing mistakes with interest and curiosity
- instead of always asking a question that invites a correct answer, sometimes phrasing questions to ask the children to identify the *incorrect* answers and then explain why they don't work.

You can read more about growth mindset in books by Carol Dweck and Jo Boaler, or search for these online.

The do–talk–record model

The learning framework for this course can be summarised as: *do–talk–record*. In other words, the starting point is always to explore and investigate concepts. Talking, discussing and explaining follows. Finally, written work can record and represent the concepts that the child has already explored practically and in discussion.

Doing

The first step in developing and understanding concepts is to let children:
- handle and manipulate apparatus
- play games
- investigate patterns and rules using real objects
- model situations and problems using real objects.

Children need time to play and explore in this way before they are expected to communicate about their work.

Talking

Children can make sense of what they have been doing by discussing:
- what they have done
- why they have done it
- what they have found out.

This allows them to generalise concepts and ideas from their particular experiences. The teacher's role is to create situations for discussion and to ask open-ended but directed questions. Most of the activities in this

Teacher's Book will help you to facilitate discussion and will encourage the children to listen to each other and experiment with different ways of thinking about and solving problems.

Recording

By Level 4, children are likely to have refined their skills and knowledge to some degree and they will have developed strategies that they find easy and useful for solving problems. They may still need to use informal and very personal methods (jottings) of recording steps in a process, or keeping track of what they have done.

Jottings are an important step in moving towards non-standard methods of calculation (such as diagrams and jumps on a number line). These will give the children a foundation for more concise standard written methods of recording. It is very important that you allow, and in fact encourage, the children to make use of jottings as they work. Here are some possible ways of doing this in the classroom:

- Do jottings of your own as you work out solutions. For example, if you are demonstrating how to calculate 144×5 you might jot the following on the board to show your thinking:
  1440
  720
- Talk through the jottings as you make them. For example: *144 times 10 is 1440, half of that is 720.* This modelling process helps the children to see that jottings are important and useful.
- Make space for jottings in the children's exercise books. You can reinforce the importance of jottings as a means of showing your working by encouraging the children to jot as they work. If you only allow jotting on scrap paper, the children may think it is not as important or valuable as the 'real' work in their book.
- Do activities where jotting is the point of the activity. For example, ask the children to represent 'three quarters' visually in as many ways as possible, or ask them to work out problems where they will need to jot down steps to keep track of the process (such as *How many ways can you find of making one dollar using any combination of 50 cent and 10 cent coins?*).
- Ask the children to share their jottings and compare them to show that there are different methods of working. This can help the children to see that some strategies are more efficient than others, which can refine their own thinking. For example, in the coin task above you may find that some children draw coin combinations, others list them and those who are more able and confident may make a table and work more systematically. All of these methods may provide the correct answers, but obviously some will take longer than others.
- In the early stages of using apparatus in a new way, recording may take the form of drawings or words and drawings. Some children will gradually find this time-consuming and will simplify their recording independently. Others may need your suggestions and encouragement. As a teacher, you will need to work out carefully when a child is ready to use a standard mathematical symbol or format, so that recording is based on full comprehension.
- Although in Level 4 you will teach children some standard written methods for operations on larger numbers, it remains crucial that you do not force the children into formal and standard methods of recording calculations before they have fully grasped the process and are confident in the methods.

Exploring and investigating

Traditionally, primary mathematics has almost always involved short, directed tasks that result in 'right' or 'wrong' answers. This course provides a balance between short, fairly self-contained activities and open-ended investigations that sometimes require the children to collect information, make their own observations, ask questions or complete activities at home. Most of the activities are designed to develop children's awareness of the range of mathematical possibilities open to them when tackling a mathematical task.

As much as possible, allow the children to take control, make decisions and explore the many avenues that can arise from a simple starting point. Always encourage children to ask 'What if … ?' and 'Why?' when investigating. These questions may lead to new challenges, fresh understanding and the development of new skills.

Many investigations have no final 'answer' or easily accessible generalisation for the children. Some have a simple pattern or rule that may be discovered and explained. However, many children will want to know why certain patterns repeat, and offer explanations about the rules that govern them. This is the first step towards generalisation; encourage this by asking questions, for example: *Why is the same number added each time?* or *Can you guess what will happen next?*

The value in investigations is in children following them as far as they can, and in the new skills that they acquire on the way. For some children, the early, often concrete, experimentation is enough to give them confidence, and increase their enjoyment of using already acquired skills.

Many everyday objects can provide rich sources of investigative work. The 100 chart, addition square and multiplication square all contain many fascinating patterns. Children can also explore patterns in solid and flat shapes, such as the relationships between faces, edges and vertices of 3D shapes, and the relationships between sides, corners and angles of 2D shapes.

Use investigations to introduce new concepts. For example, introduce number patterns by presenting number chains (sequences), and letting the children work out what comes next or what is missing. Introduce geometric patterns through explorations of colour arrangements using pattern blocks, tiles or geoboards. The children can explore the relationship between area and perimeter, and between volume and the dimensions of cuboids.

As the children develop an investigative approach, help them to work systematically, and remind them of the question they are working to answer. Encourage them to make new observations and discoveries along the way, but ensure that these do not eventually distract from the question the children are working on. Ask questions such as: *What does this show us? What do you notice? What do you wonder? How can you use this to work out the answer? What other information do you need? How can you find it?* and so on.

Number talks

Number talks are open-ended, meaning they present a question that has multiple possible answers, and multiple possible strategies for solving.

Children can use agreed hand signals as shown below, instead of calling out or putting their hands up. However, these may not be culturally appropriate in all countries, and you may want to come up with alternative signals that are suitable for your class.

- **Step 1: Present the problem and give time for independent thinking**
  First present the problem, then ask the children to think to themselves quietly how they could solve it. They can use hand signals to show when they have an idea.

I have an idea, and one way to solve it.

Note that when children signal that they have an idea, you can encourage them to think of more ways to solve it. They can indicate the number of ways by raising more fingers, as shown.

I have two ways to do it.

- **Step 2: Share in pairs**
  In pairs, children share their suggestions for how to work it out. Note that explaining how they solved a problem is as important as (or more important than) the final answer. They can use a hand signal to indicate they are finished, such as a fist to the chest.

I have finished.

- **Step 3: Children share their strategies**
  Call on individual children to share their strategies with the class. Discuss each strategy as the child demonstrates for the class. Continue using hand signals so that the children can engage with the

discussion and take turns to share their own working. The children can hold their thumb and little finger out for 'I agree' or 'Me too'; hold their hands with palms down and move them in a criss-crossing motion for 'I respectfully disagree' and raise a finger for 'I have another way to do it'.

I agree or Me too.

- **Step 4: Discussion**
  The discussion will take place as other children show that they agree or disagree, and offer alternatives strategies.

I respectfully disagree.

Principles of the number talk strategy

The number talk strategy aims to get all children to engage actively and creatively with mathematics. To achieve this engagement, it is important to remember the 'big ideas' behind the number talk:

- *Value the contributions of all children.* Throughout the discussion help them to articulate their thoughts and reach a clearer understanding of how they are working things out. The emphasis is on the process and understanding, not on reaching the right answer fast.
- Children often answer in a roundabout way, with a very general idea or 'sense' of what they want to say. Use questions to *clarify, refine and explore* the meaning behind each response.
- Encourage children to explore *conceptual explanations*, rather than procedures.
- Gently *explore interesting mistakes* because these provide opportunities for deeper understanding. We can often better understand how something does work by reaching a clearer understanding of how it doesn't work.
- Praise children for effort, for making sense of a problem and for *persistence in the face of difficulty*, rather than for efficiency. We like to tell children that if it is difficult, that tells us we are in the place of learning. If it is easy, that means we've already learnt it and it's time to find more challenge.
- Create a *safe community space* where children feel they can share their ideas without criticism or judgement.
- Encourage *sharing ideas*, rather than competition. Children should understand that learning comes from connecting with other people's ideas and building on them.

Finding out more about number talks

It is important to remember that number talk is a

format or technique, not a specific lesson or set content.

- View number talks in action, shared by teachers, on the Internet.
- Find posters and strategies for number talks through an online image search.
- Books on number talks include: *Number Talks: Whole Number Computation, Grades K-5: A Multimedia Professional Learning Resource* by Sherry Parrish; *Classroom-Ready Number Talks for Kindergarten, First and Second Grade Teachers* by Nancy Hughes

Numberless word problems

Many teachers fear problem solving because they present a problem, and then watch as children try to guess which operation to use, or which numbers to arrange into a number sentence. Numberless word problems offer a useful strategy for encouraging children to think and understand questions before looking for the 'maths answers'.

The numberless word problem is presented with no numbers at first, and the numbers are gradually added. The numberless word problem does two things:

- It removes the numbers.
- It removes the question.

This forces the children to slow down and think about the problem before they can attempt to solve it.

To present a numberless word problem, follow these steps:

- **Step 1:** Present the word problem you want to use, but remove the numbers and the question. Use words such as *some, more, joined, went away, took away, fewer* and so on. Ask the children to tell you what is happening in the story. Here is an example to practise addition:

> There are some children in the playground. Some more children come out to play. Now there are many children in the playground.

- **Step 2:** Ask the children to think about what the question could be in this story. Allow plenty of thinking time and invite the children to be 'question detectives', guessing from the clues what the question is going to be. They may have a variety of suggestions, such as: *How many were in the playground to start with? How many children came out? How many were there altogether?* Once they have identified the possible questions, you can reveal the question, but still without numbers:

> There are some children in the playground. Some more children come out to play. Now there are many children in the playground.
> How many children came out to play?

- **Step 3:** Ask the children to identify what information

they need in order to solve the problem. First, ask the children if they can answer the question. They cannot – because they don't have the information they need. Let them suggest what they need to find out in order to solve it. Give them the first piece of information they need, by crossing out 'some' and writing a number:

> There are ~~some~~ 15 children in the playground. Some more children come out to play. Now there are many children in the playground.
> How many children came out to play?

Again, ask if they can answer the question. When they say no, let them work out what other information they need to solve it. Let them work out what else they need to know. Finally, cross out 'many' and change it to a number (24).

- **Step 4:** As with number talks (see pages 17–18), give the children time to solve the problem using their own strategies. Follow the steps of a number talk as you let them share and discuss their strategies.

Mathematics in real life

Some children may struggle to understand the relevance of mathematics in their everyday lives. This course places great emphasis on making children aware of the relevance of mathematics in real life.

In this Teacher's Book, you will find ideas for using the child's own environment as a stimulus for mathematical activities. The Pupil Book and Workbook frequently require the children to look at the mathematics in the classroom, the playground and their own homes. Each set of activities and problems requires new skills and fresh understanding. Many questions are open-ended or have no exact solution, and the children are asked to make predictions, generalisations and estimates, and to evaluate their own answers. Encourage these skills in all areas of the curriculum. Children use their understanding of mathematics at home and at school, in situations such as sorting toys or other items, telling the time, looking for and making patterns, helping to prepare food, and playing board and other games.

In school

In school, there are many opportunities for you to teach mathematics through familiar situations, so that the children understand why it is useful and appreciate the order and sense that mathematics gives to life. For example, the children can identify the date each day, as well as the time at various points throughout the lesson. Registration, dinner money, timetables, sorting and putting away equipment will provide a range of relevant experience in statistics, measure and geometry as well as number.

In play

Children of all ages should have opportunities to play both in and out of school. This offers them the freedom to explore new situations, to make discoveries for themselves and to be creative. Unfamiliar mathematics equipment should be introduced through play, with the children exploring the functions and possibilities of the materials. A good example is to experiment with pairs of compasses by drawing patterns and pictures before using them as mathematical instruments.

Construction kits offer children the opportunity to explore shapes and inverse operations, through building and dismantling. Note that for any materials that are unavailable at your school or in your area, you may be able to find an online version.

At home

Part of the teacher's role is to involve parents and guardians in the children's learning. Parents can do more than supervise their children's homework. Many activities can involve the parent in the child's learning and provoke mathematical discussion and language at home.

Encourage parents to extend their children's mathematical understanding through playing board and card games, and by encouraging them to help with normal home activities such as cooking, gardening, cleaning and organising the home, drawing up plans and measuring when redecorating, and estimating how many or how much when shopping.

Many children will voluntarily help and encourage younger brothers and sisters in games and getting organised. Family visits and holidays give children opportunities to see environments different from their own, and to experience time and distance. They are also likely to be budgeting pocket money, saving for special things and predicting how long it will take to afford treats.

Children may play with computer games that require a variety of mathematical skills. They might see and use a wide range of other electronic equipment at home, too, which also demands mathematical skills. Many children will be responsible for their own timekeeping and have a degree of responsibility for others.

Some homes will not actively encourage girls to use construction kits, computers or calculators, and some parents will not be confident of their own mathematical skills or understanding. As a teacher, you can help a great deal by making explicit the mathematical content of everyday experiences and activities.

You can find additional parent notes for each level on Oxford Owl. See page 9 for more details.

Section 3 An environment for exploration and play

Organising the space

All teachers have their own preferences about how best to organise the available space. However, here are some useful guidelines for any classroom.

Storage

Always store equipment so that the children have easy access to it and you can check it periodically. Label all items clearly and encourage the children to make their own decisions about what they need. From the very beginning, insist that the children pack up and return equipment to the correct storage containers or shelves.

A mathematics centre

Create a mathematics centre in your classroom. This does not have to be where the equipment is stored, but it should be a bright and attractive part of the classroom with displays of children's work and other mathematical stimuli. The centre is a place for children to go at odd moments in the day, to be challenged with mathematics-related stories, questions and activities.

Questions and activities should be provided by both teachers and children for interactive problem solving, for example: *The answer is 3. What was the question?*, inviting children to draw or write out their suggestions.

A number pattern or sequence, on a series of cards organised by the children, may be 'secretly' altered by the teacher, and children have to discover what has changed, and put it right.

Recycled resources

Many household items can be repurposed to make mathematical equipment. Invite the children and their families to collect these materials for the classroom:

- bottle caps
- egg boxes
- clean glass jars
- buttons and beads
- plastic caps from juice or milk cartons
- plastic yogurt pots
- natural objects such as shells, seed pods, stones for the children to arrange in size order.

Store the materials neatly in plastic containers or jars.

Whole class teaching

Often you will work with your whole class, especially at the beginning and end of a lesson. The course offers plenty of ideas for this kind of approach. The children may be seated together on chairs or on the floor, arranged in a circle for a discussion, or clustered around you for a demonstration. If you are calling on individuals to show things to the group, make sure there is enough space at the front for them to come up, and make sure everyone can see and hear the volunteer child's answer.

Set up some classroom rules at the beginning of the year and go over these at the beginning of each term or even each week if necessary. Classroom rules might include listening when others are speaking and waiting turns. Many teachers find it useful to use signs for *agree*, *disagree* and *I know the answer* rather than having the children raise their hands or shout out. The traditional approach of raising hands can be intimidating for quieter children or those who take a little longer to think about their answers. Hand signals allow teachers to give more time to those in the group who need it. The illustrations on page 17 show some commonly used hand signals.

Group work

You can group the children in similar or mixed-ability groups, to suit the purpose of the work. This gives the children opportunities to collaborate and to discuss their work with each other and with you. It allows peer teaching to take place and for the work to be matched to their needs. It also allows you to work simultaneously with a number of children and this minimises the need for repeated explanations to individuals. Group teaching is an effective form of classroom organisation for both teacher and children.

Working individually or in pairs

At times it may be appropriate for the children to work as individuals or in pairs, to provide extra help to children who need it, or to stimulate and challenge children who have grasped a concept quickly. Working individually gives the children the opportunity to concentrate on their own thinking, to develop this through investigations and problem solving, and to experiment with materials. Children working in pairs have the opportunity to develop collaborative skills, to play games together and to share ideas in an investigation.

Manipulatives give children an opportunity to experience the use of mathematics through concrete materials and objects. There are suggestions for different kinds of manipulatives in the unit-by-unit teaching guidance section of this book.

> With all activities using colour, if you have colour-blind children in your class, adapt the activity to use patterns, or combinations of colours that are within their visual range.

Counting apparatus
Interlocking cubes

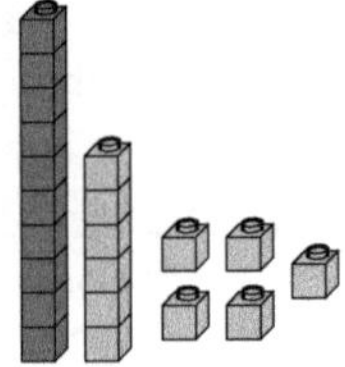

Interlocking or 'snap' cubes

Interlocking cubes are an invaluable resource for learning to count, especially as the children learn their number bonds up to ten, and then begin to work with tens and ones as they work with numbers up to 100.

Rekenreks

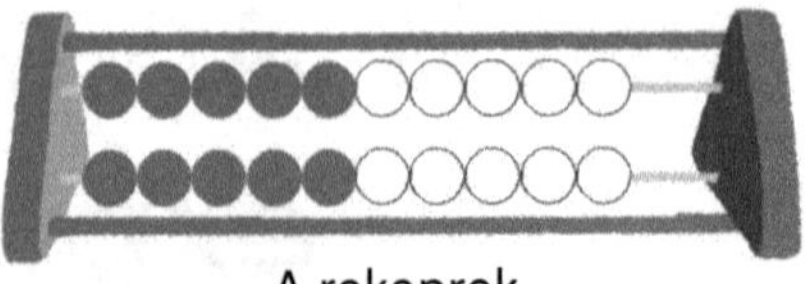

A rekenrek

The rekenrek was designed by a Dutch mathematics researcher to support children in developing their number sense in a natural, intuitive way. 'Rekenrek' means counting rack or counting frame. At the primary level, the device consists of two horizontal bars, each with ten beads. The beads are in two sets of five, shown by different colours (usually red and white). The rekenrek looks a little like an abacus, but the bars do not represent place values.

When we work with a rekenrek, we start with all the beads pushed over to the right, in the 'resting' position. To count or represent a number, we move beads over to the left. The rekenrek can be used for a range of skills, including:

* developing subitising skills within five and within ten
* counting to ten
* addition and subtraction
* composing and decomposing numbers up to 20
* recognising doubles and near doubles
* skip counting, and so on.

The beauty of working with a rekenrek is that it allows children to develop a range of strategies intuitively.

If your school does not have access to commercially available rekenreks, you may be able to make some simple ones with dowel sticks and beads. You can also search for an online interactive rekenrek app.

Ten frames

A ten frame is a 2 × 5 rectangular grid, used most commonly with counters of one more colours.

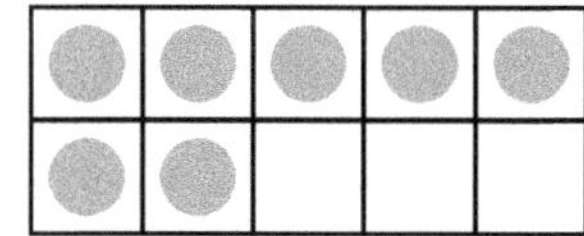

Using a ten frame to represent 7

Ten frames help children build and visualise numbers to 5, 10, 20, and 100. They can use the frames to count, represent, compare and calculate, working within a specific number range.

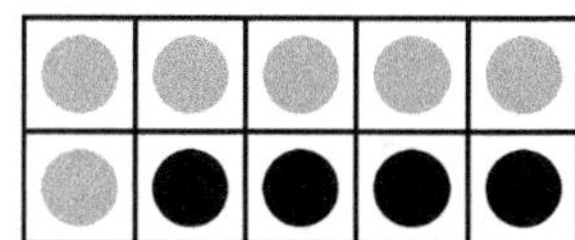

Using a ten frame to represent 6 + 4 = 10

Ten frames help children to see equivalences between quantities. So, for example, they will quickly see that 10 is equivalent to two fives, and that 20 is equivalent to two tens, or to four fives. This helps them move away from linear, one-by-one counting and develop a range of counting strategies.

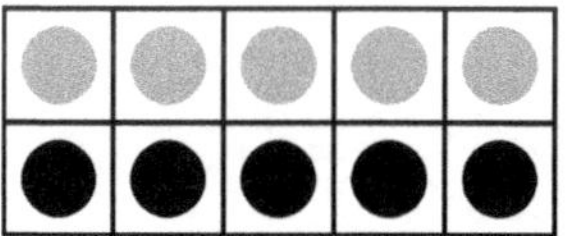 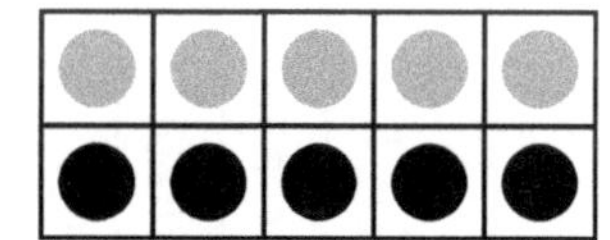

Using ten frames to represent 20 as
two tens or four fives

Your school may have access to wooden or plastic physical frames. If you don't have these at your school, you can easily make your own ten frames and use sets of counters (five of one colour and five of another).

You can also search for interactive ten frames online.

Numicon (number frames)

Numicon

Numicon is another powerful and versatile concrete resource that supports children to visualise and work with numbers. The holes in each Numicon number frame represent a number.

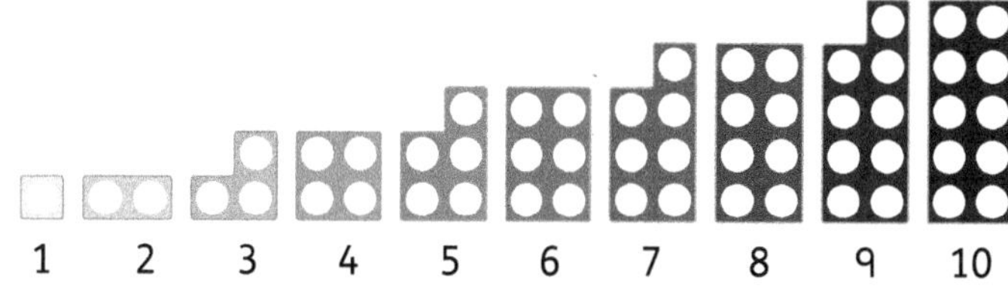

As with ten frames, the children can also use the Numicon number frames to count, represent, compare and calculate. When thinking about 5, for example, they can quickly see that 5 is 1 less than 6, but 1 more than 4. If they compare a 3- and a 6-number frame to a 9-number frame, they can see that 6 and 3 make 9.

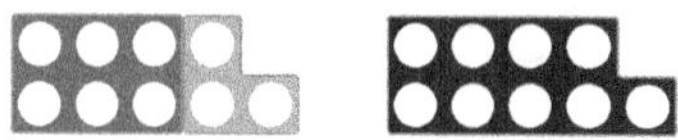

Using Numicon to represent 6 + 3 = 9

For more information on using Numicon number frames to teach mathematics, and the full range of Numicon products, visit https://global.oup.com/education/content/primary/series/numicon

Base-ten blocks

Base-ten blocks are a concrete resource made up of blocks of varying sizes. Each block can represent a different value. For whole numbers, they represent:

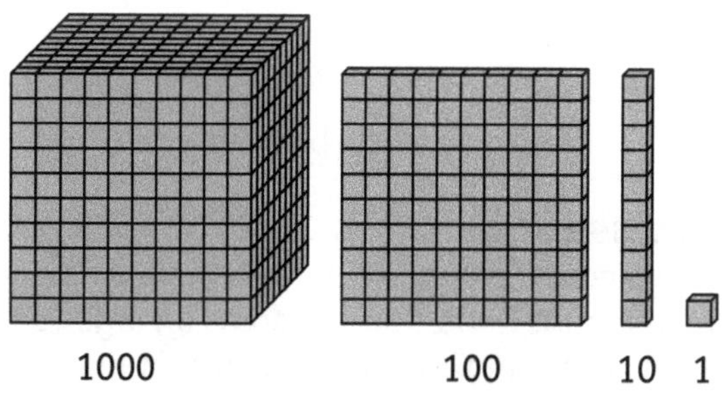

1000   100   10 1

They can also be used to introduce decimals, representing ones, tenths, hundredths and thousandths respectively.

Like ten frames and Numicon, base-ten blocks help children to represent numbers, compare and calculate. They are particularly effective when used to support partitioning, for example, into tens and ones.

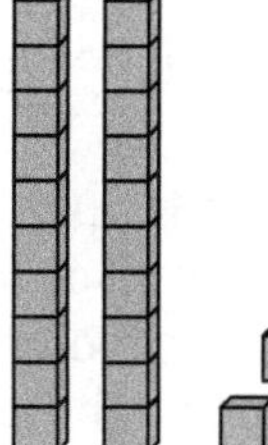 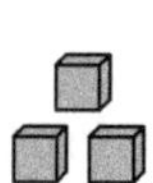

Representing 23 as two
tens and three ones
using two 10-rods and
three ones cubes

Using base-ten blocks and place-value tables helps children see how numbers change when multiplied and divided by powers of ten. Subtracting using base-ten blocks is a great, 'hands-on' way to introduce and support exchanging.

Bar models

Drawing bar models is an excellent problem-solving strategy that children can use in both primary and secondary school. A problem that may otherwise seem very challenging can become more accessible when you 'draw' it.

Children should be encouraged to both interpret bar models as well as draw their own when solving word problems.

There are two main types of bar models: the part-part-whole model and the comparison model. Which you choose to use will depend on the problem you are trying to solve.

For example, Tyra has 32 and she gives 13 to Esme. How many marbles does she have left?

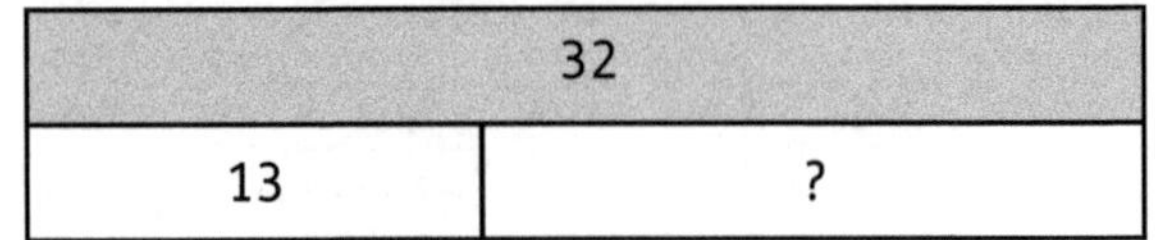

A part-part-whole bar model

Lewis completed 142 laps of the running track in one week. Joshua completed 52 laps in the same week.
How many fewer laps did Joshua complete than Lewis?

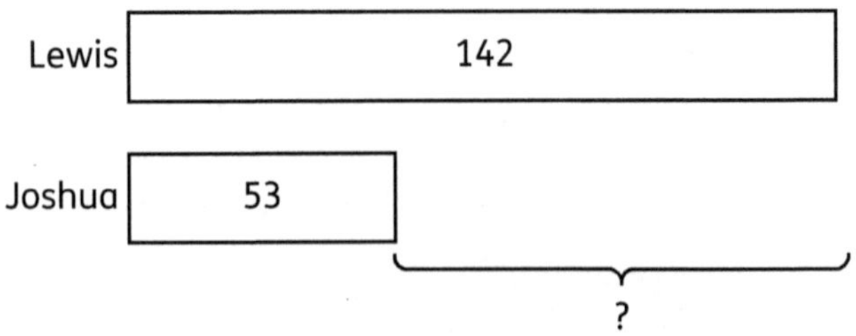

A comparison bar model

Place-value cards

Place-value cards for whole numbers have an 'arrow' or point on the right-hand side. Children can organise the cards horizontally or vertically to represent numbers in expanded notation. They can overlap cards and line up the arrows to form multi-digit numbers.

If any children have not previously worked with place-value cards, you will need to teach them how to use them. Begin by pointing out the arrows on the cards. Explain that these arrows always go on top of each other when you are making a number.

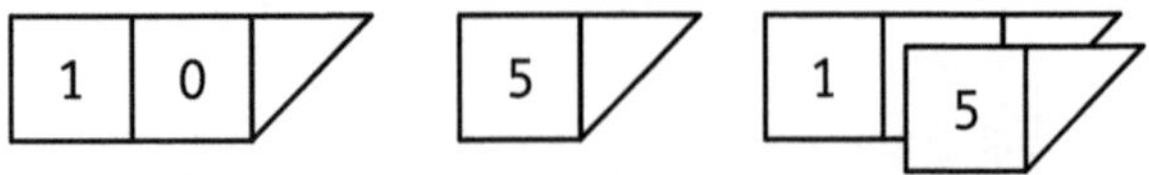

At Level 4, the place-value cards need to include thousands, hundreds, tens and ones, as well as tenths and hundredths. Place-value cards for decimals have a decimal point and the 'arrow' or point is on the left-hand side. For example:

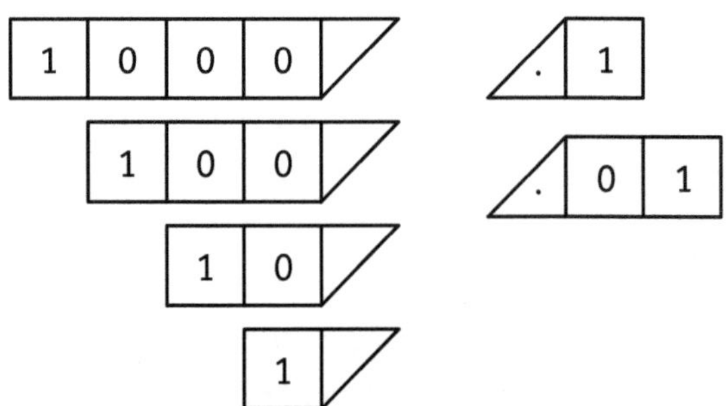

Combining cards to make a decimal value:

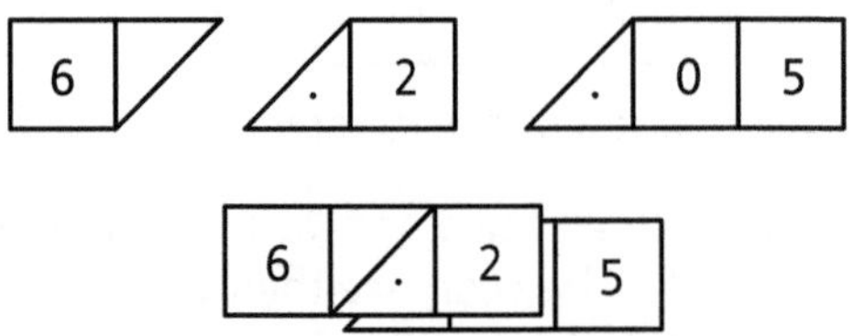

Place-value cards are an important teaching and learning resource and it would be useful to have a set available for each child. If possible, laminate the cards to make them more durable. (If you are making a set for each child, you may like to send the cards home for parents or carers to cut out.)

Place-value tables

Children can use place-value tables to build numbers using counters or base-ten blocks. If possible, make place-value tables on card and laminate them to make them more durable. This table shows the places needed for work at Level 4:

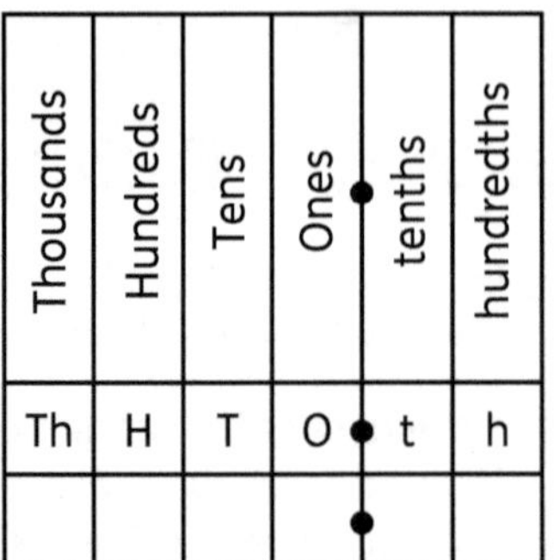

Thousands	Hundreds	Tens	Ones	tenths	hundredths
Th	H	T	O	t	h

Games

A game is different from a teaching activity because there is an element of luck and chance in a game, sometimes leading to a winner. Some games are played alone, and children try to improve on their own results, while other games are played in pairs or groups.

Make sure that the children understand the rules of the game before they start playing. If they don't, they may not play the game properly and this can result in conflict with others. The children should learn to:

- take turns and wait for their turn
- move their own game pieces only
- help each other if they can.

The children also need to learn that in games involving luck, players don't win or lose on the basis of ability. Explain that those players who get the best luck in a particular game win; those who lose were just less lucky. Winning graciously, as well as losing and playing fairly, are important life skills.

When the children play games in class, do not keep a record of who wins and who loses. If the children like a game, encourage them to play it in their free time and to take it home and play with family members if possible.

Section 4 Activity bank of warm-ups and mental maths

This activity bank includes a range of mental maths activities, as well as some suggestions for support and consolidation activities. Most of the ideas are for number work and operations, although there are also some suggestions for the other strands. You can use this as a resource for activity ideas for your class. The unit-by-unit section in the next chapter refers back to specific activities in this section as unit openers, suggestions for extra support and consolidation.

Try to include 10 minutes of mental maths activities each day. This section includes many examples of activities that you can use as they are or adapt to suit your own classroom. We have tried to provide a range of different types of activities (factual recall, games, grids, tables, problem solving and puzzles) to show some of the ways in which you can approach the mental maths part of the lesson. However, this is not a definitive list and some activities will appeal more to some classes and teachers than others.

If you need additional ideas and suggestions, there are thousands of websites and social-media platforms with creative suggestions for every maths topic. Choose the one that suits you and explore!

> Please avoid any games or activities that use countdown clocks or timers to turn mental maths into a race or speed test. Speed testing is very damaging for children's mathematical development.

Place value and number sense

Counting backwards and forwards
Counting in given steps: Ask the children to count back and forward in steps of 1, 2, 10 and 100. Start the year by revising place value to 1000, then move onto numbers to 10 000. For example:
- Count from 999 to 1025.
- Count back in twos from 2500 to 2450.

- Count in tens from 4150 to 5300.
- Count back in tens from 6500 to 6200.
- Count in hundreds from 3245 to 5245.
- Count in thousands from 1200 to 8200.
- Count forwards in thousands from 4000 to 10 000.
- Count forwards in tens from 5890 to 6020.
- Count backwards in hundreds from 6300 to 5200.

Adding and subtracting tens and hundreds: Ask questions based on counting back or forwards in tens and hundreds.
- *What is 10 more than 450? (460)*
- *What is 10 less than 900? (890)*
- *What is 100 more than 6000? (6100)*
- *What is 100 less than 1500? (1400)*
- *What is 100 less than 10 000? (9900)*
- *What is 100 more than 3500? (3600)*
- *What is 100 more than 4900? (5000)*

Find the numbers: Make a set of number cards so that you have sets of three numbers that are 100 greater and 100 less than each other. You will need enough cards to give the children one each. For example:

1324	1424	1224
8766	8866	8666
3455	3355	3555
4099	4199	3999

- Shuffle the cards and hand them out.
- Ask the children to find the other two numbers in their set by asking questions.
- When they find each other, ask them to display or write their numbers. If some children finish quickly, let them extend their number sequence in both directions.

More and fewer: Make up a set of questions based on a 4-digit number. For example, display the number 4235 and ask the children to write or say the number that has:
- five more tens (*4285*)
- three fewer ones (*4232*)
- two fewer hundreds (*4035*)
- three more thousands (*7235*).

Place value with place-value cards

Materials: place-value cards (see page 22).

Make a given number: Write a number on the board, demonstrate how to use the cards and then ask the children to make that number. For example:

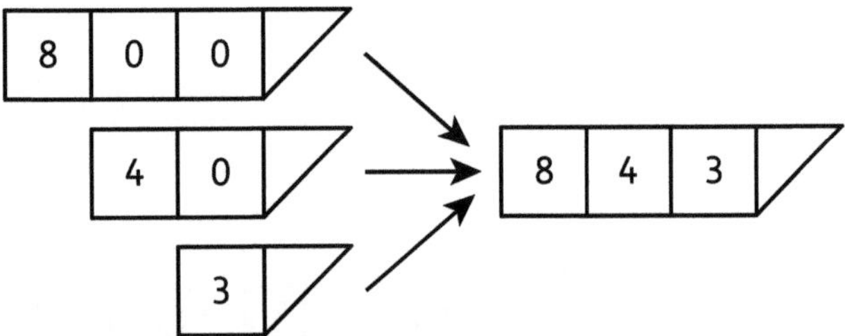

Use zero as a place holder: Use two cards to illustrate how a 3-digit number can be made. Explain that the zero is a place holder.

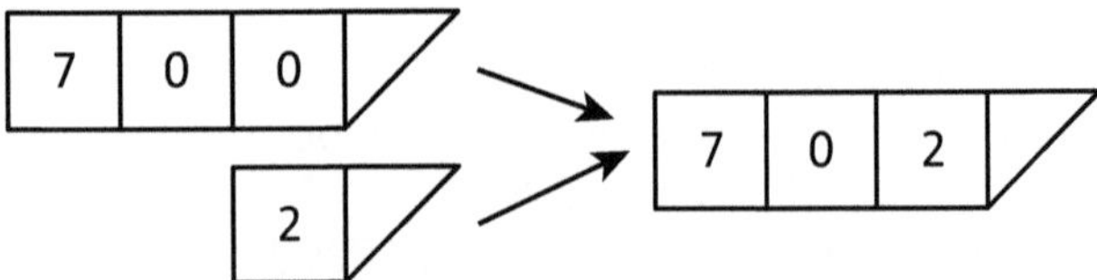

Make numbers from expanded notation:
60 + 4

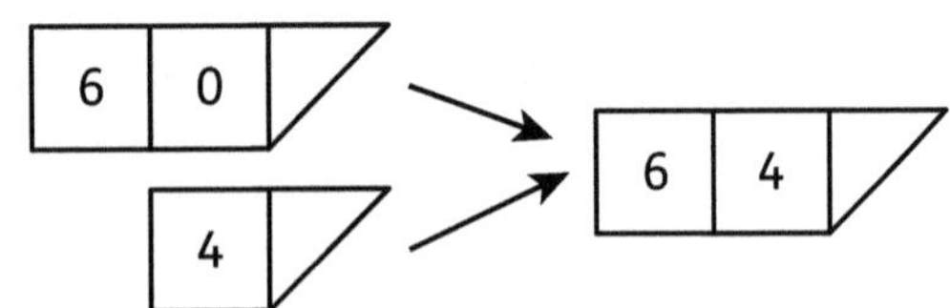

200 + 80 + 7

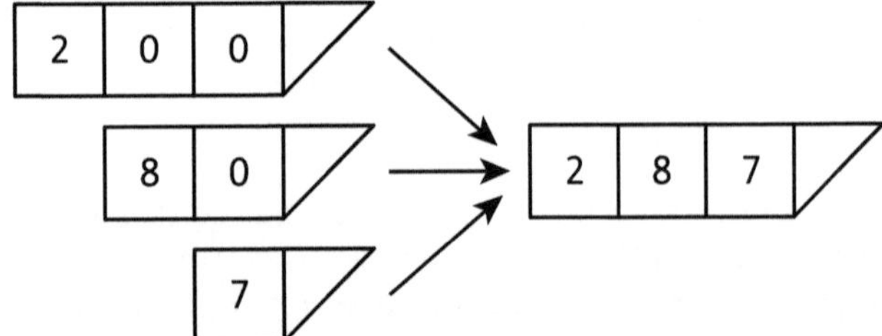

Make a number from words: Say a number, for example, *two hundred and sixty-three* and ask the children to make this number with place-value cards.

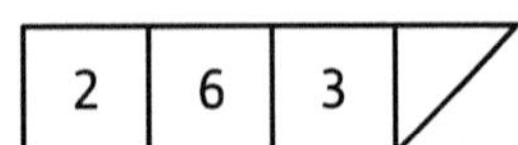

- *Show me how you would make four hundred and twenty-two.*
- *How would you make three hundred and one?*

Practise place-value vocabulary: Ask questions that use different words, for example, *Show me a 3-digit number in which all the digits are the same.*

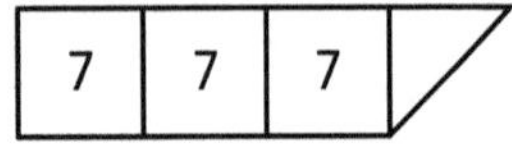

- *Show me a number that does not have any hundreds.*
- *What about a number that does not have any tens?*

Place-value activities

Number charts: Create number charts like the ones shown below. You could make large, laminated versions of the charts or create electronic versions.

Circle one number in each place on a laminated chart or highlight the blocks on an electronic chart.

Ask the children: *What number is shown on this chart?* (85)

10	20	30	40	50	60	70	80	90
1	2	3	4	5	6	7	8	9

What number is shown on this chart? (1011)

1000	2000	3000	4000	5000	6000	7000	8000	9000
100	200	300	400	500	600	700	800	900
10	20	30	40	50	60	70	80	90
1	2	3	4	5	6	7	8	9

Then ask the children: *Say and write the number in words and in numerals. (one thousand and eleven, 1011)*

Listen carefully: Ask the children to listen carefully as you say some numbers, and then write them down. For example, read out:
- *two thousand, four hundred and thirty-five (2435)*
- *one thousand, nine hundred and three (1903)*
- *five thousand and ninety-nine (5099)*
- *three thousand and seven (3007)*
- *five thousand, seven hundred (5700).*

Digit values: Read out some numbers and ask the children to write down the value of one particular digit. For example, *What is the value of 3 in each number?*

1366 (300)	1493 (3)	3678 (3000)	3908 (3000)
435 (30)	398 (300)	4213 (3)	9308 (300)

Make sure that each number has only one of the digit that the children need to spot.

Numbers on the board: Write a 4-digit number on the board such as 3029.
- Ask the children to say the number aloud.
- Ask different children to say how many thousands, hundreds, tens or ones there are.
- Point to a digit and ask the children to say its value.
- Ask the children to reverse the digits and say the new number.
- Ask the children to make five different numbers using digits from the given number in any order. They can then swap their numbers with a partner and say each other's numbers aloud.

Making 4-digit numbers: You will need several sets of 1–9 digit cards. Give each group of children four digit cards. Ask the groups to make:
- the smallest possible number
- the greatest possible number
- the smallest possible odd number or even number
- the greatest possible odd number or even number.

Find my number: Display a selection of 4-digit numbers, for example:

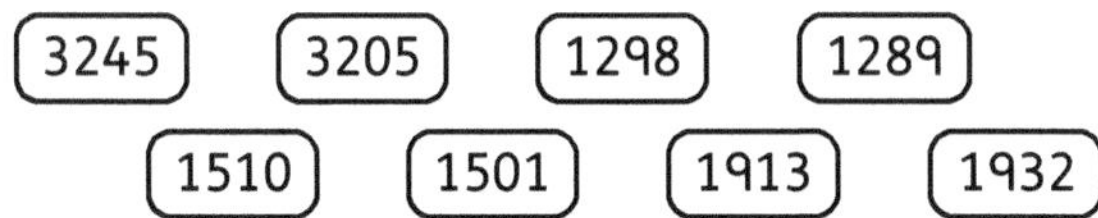

Using numbers with similar digits makes the activity more challenging.

- Say each number in words (for example, *three thousand two hundred and forty-five*) and ask children to come up and identify the number you have said.
- Ask the children to reverse the digits of a number and say the new number aloud. (Be clear about what they should do with a 0 in the ones place or avoid using any numbers with 0 in the ones place.)

The 10 questions game: Ask the children to play this game in pairs. Each child writes down a number without showing it to their partner. You could give them a range, for example, any number less than 5000.

The children then ask each other up to ten questions that can only be answered by 'yes' or 'no' to identify the number. For example:

- 'Is it odd?'
- 'Does it have three digits?'
- 'Is the digit in the ones place greater than the digit in the tens place?'

Once they have asked ten questions, they should try to guess the number. If they can't, their partner can give them a clue, for example:

- 'My number is between 300 and 350. It has no tens.'

They can then ask two more questions before guessing again.

Comparing numbers

Compare using signs: Ask the children to write down any 3-digit number. Write a random set of 3-digit numbers of your own on the board.

Let the children write number sentences using your numbers and their own number using the <, > or = signs.

Let the children use mental strategies and jottings to find the sum of or difference between the numbers in their number sentences. Then discuss their strategies.

Greater or less?: Ask the children to write a 3- or 4-digit number on paper. Choose one child to come to the front of the class and display his or her number. Ask:

- *Whose numbers are greater than this?* (The children display their 'greater' numbers.)
- *Whose numbers are smaller than this?* (The children display their 'smaller' numbers.)

Choose one child to display their number standing to the left or right of the first child (depending on whether their number is less or greater). Choose other children to position themselves correctly relative to the numbers displayed.

Ordering numbers: Write a list of positive and negative temperatures on the board. For example:

12   −3   0   −5   −4   9   5   −7   −11   −9

Draw a straight line on the board to represent a thermometer scale, including °C at one end to show the unit of measure.

Ask the children to find the lowest temperature. (*−11 °C*) Write this on the left-hand end of the line. Repeat for the highest temperature. (*12 °C*) Write this on the right-hand end of the line.

Ask for volunteers to choose one of the temperatures and position it on the line as accurately as they can. Once all the numbers have been placed, discuss whether any numbers are inaccurate. Let the children decide and suggest how to move the numbers if necessary.

You can adapt this activity to work with decimals, fractions, mixed numbers and whole numbers.

Bingo: Play 'Bingo' with the class. You will need three or four sets of 0–9 digit cards. First, the children need to write five 3- or 4-digit numbers in a simple grid like this one. They can work in pairs to do this.

456	321	999	408	765

Shuffle all the digit cards together. Take three cards at random (or four if you are making 4-digit numbers) and call out or display the digits, for example, 5, 6 and 7.

The pairs of children see whether they can reorder the digits to make any of the numbers on their grid. If they can make one of their numbers, they cross it out on their grid. (So, the pair who made the example grid above would be able to make 765 and cross it out.)

The first pair of children to cross out all their numbers can call out 'Bingo!' and they win the game.

You will need to record the digits that you call out each time so you can check the pairs' claims of 'Bingo!'

Fractions and decimals

Fraction and decimal equivalents: Display a grid of 1- or 2-place decimals. For example:

0.3	0.45	0.5	0.01	0.10
0.75	0.25	0.4	0.9	0.8

Ask the children to draw their own blank grid and rewrite each decimal as an equivalent fraction using tenths and hundredths.

Answers:

$\frac{3}{10}$	$\frac{45}{100}$	$\frac{5}{10}$	$\frac{1}{100}$	$\frac{10}{100}$
$\frac{75}{100}$	$\frac{25}{100}$	$\frac{4}{10}$	$\frac{9}{10}$	$\frac{8}{10}$

(Some children may simplify the fractions and write, for example: $\frac{1}{4}$ for 0.25 and $\frac{3}{4}$ for 0.75. Encourage them to explain their thinking and reasoning if they do this.)

You can also reverse this activity by displaying fractions as tenths and hundredths and asking the children to write the decimal equivalents.

Ordering mixed numbers: Display a set of mixed numbers and ask the children to write them in order from greatest to smallest or vice versa. Include mixed numbers with the same whole number and different fractional parts to encourage the children to compare the whole number and the fraction parts, for example, $5\frac{1}{2}$ and $5\frac{3}{4}$.

Equivalent fractions: Display pairs of equivalent fractions, including some errors. Ask the children to find, write down and correct the incorrect number sentences. For example:

$\frac{1}{2} = \frac{4}{8}$ $\frac{1}{2} = \frac{6}{10} \left(\frac{1}{2} = \frac{5}{10} \text{ or } \frac{6}{12}\right)$ $\frac{1}{5} = \frac{2}{8} \left(\frac{1}{5} = \frac{2}{10}\right)$ $\frac{2}{10} = \frac{20}{100}$

You can adapt this activity to work with fractions and their decimal equivalents

Rounding

Spot the rounding mistakes: Write several 4-digit numbers on the board. Round each to the nearest 10 or 100 (choose one place value to round to per activity). Make sure that some of the rounded values are incorrect. Ask the children to find the incorrectly rounded numbers and correct them. For example (rounding to the nearest 100):

2345 → 2300	2662 → 2660 (*2700*)
4129 → 4200 (*4100*)	3888 → 4000 (*3900*)
3999 → 4000	

Mental rounding: Draw a grid like this one on the board. If you are going to reinforce rounding to the nearest ten, make sure that the numbers all have a value other than 0 in the ones place.

456	1275	499	109
3245	6501	1295	1082
3509	8024	8019	876
103	562	901	1052

Ask the children to copy the grid and rewrite the numbers, rounding them to a given place value (for example, the nearest 10, 100 or 1000).

Alternatively, work through the grid, pointing at the numbers and asking children to round it to the nearest 10 or 100 or 1000.

Dartboard rounding: Prepare some rounding 'dartboards' like this one.

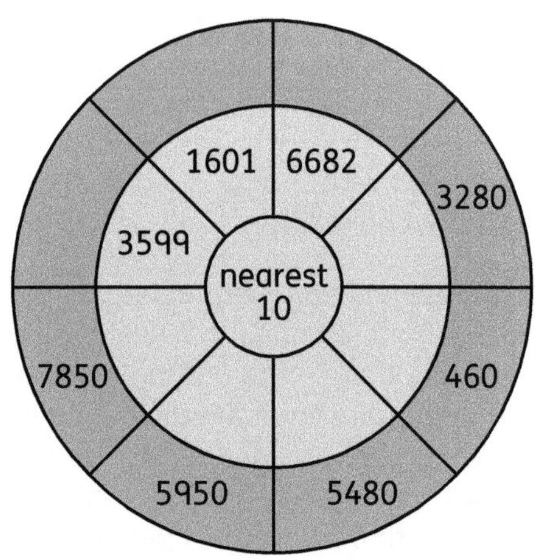

Point to a space on the dartboard and choose a child to give the answer.

Vary the game by changing what you are rounding to (centre circle), by giving the inner circle of numbers, and by giving the outer circle of numbers (remember when you do this that there are several options for the inner-circle answers.)

Matching game: Give each child a number rounded to the nearest 10 or 100, depending on what you are teaching, for example, multiples of 10 from 90 to 450. The children should write their numbers down.

Say a number, for example, 423 and ask *Who has this number rounded to the nearest ten?* The child should display the rounded number and the class can decide whether it is correct. Repeat, choosing different children to say a number between 90 and 450.

Rounding in a table: Ask the children to draw a three-column table like this one in their notebooks:

1000	1100	1200

Read out 20 numbers in the range from 950 to 1249. As you say each number, the children should write it in the correct column to show the nearest 100. For example, if you say 1050, the children write 1050 in the 1100 column.

Rounding and estimating

Which number?: Display some 3-digit numbers, for example:

342   291   450   405   325   251   320

Ask questions that involve rounding and estimating and basic operations. For example:
- *When you multiply by 20, which of these numbers give a product that is:*
  - *less than 6000 (291, 251)*
  - *greater than 6000 (342, 450, 405, 325, 320)*
  - *between 6500 and 7000 (342, 325, 320)*
  - *greater than 8000? (450, 405)*

Repeat this for different multiples of ten and a hundred and for 4-digit numbers.

Adapt the numbers and the 'answers' for other operations. For example:
- *When you add 520, which number gives a sum that is . . . ?*
- *When you subtract 90 from each number, which give an answer that is . . . ?*

Money problems: Prepare a set of money problems that involve estimating to work out whether you have enough to pay for a given number of items. For example, *Priya has $50. She wants to buy 4 sets of pens at $8.50 each and 2 pencil cases at $8.99 each. Does she have enough?*

Vary the cost of items, using local prices and currencies where possible and also vary the amount available to spend.

Estimating answers: Prepare a set of addition and subtraction problems. For example:
379 + 341   532 + 288   209 + 399
418 + 299   502 – 199

Ask questions that can be answered by rounding and estimating. For example:
- *Which calculations have an answer close to 700? (379 + 341, 418 + 299)*
- *Which calculations have an answer greater than 600? (379 + 341, 532 + 288, 209 + 399, 418 + 299)*
- *What is a good estimate for the answer to this calculation? (pointing to a calculation)*

Mental maths

Mental problem solving
Word problems: Test understanding of mathematical terms and vocabulary by posing word problems to be solved mentally. For example:
- *Find the number that can be increased by 30 to make 81. (51)*
- *What is the product of 4 and the number that is 3 greater than 4? (28)*
- *What is the difference between 16 and double 48? (80)*
- *What number do you get if you halve the product of 98 and 100? (490)*
- *What is the sum of 345 and double 90? (525)*
- *Give me a pair of 3-digit numbers with a difference of 120. (example answer: 590 – 470)*
- *The sum of two numbers is 280 and one of the numbers is half of 180. What is the other number? (190)*
- *Give me three numbers that have a total of 124. (example answer: 100 + 14 + 10) Are there any others? (yes, many others)*
- *Find four pairs of numbers with a difference of 54. (example answer: 100 – 46, 60 – 6, 74 – 20, 55 – 1)*
- *How many lengths of 10 cm can you cut from a ribbon that is 186 cm long? (18)*

Greatest and smallest totals: Give the children sets of 0-9 digit cards. Ask the children to make three 2-digit numbers using their cards, and then calculate the total mentally.

Ask them to find the three 2-digit numbers that make the smallest possible total and the greatest possible total. Spend some time discussing how they worked this out.

Logic puzzles: Write puzzles appropriate to the ability levels in your class. For example:
- *Jess is 7 years older than her sister. Their combined age is 25. How old is Jess and how old is her sister? (Jess is 16 years old; her sister is 9 years old.)*
- *The sum of two numbers is 140 and there is a difference of 6 between them. What are the two numbers? (67 and 73)*
- *A farmer has some chickens and goats in a yard. There are 15 heads and 48 legs. How many goats are there? (9)*
- *Busi and Rob have six children. Each of these children is married and has three children of their own. How many people are in this family? (32)*
- *It takes Mr Jones 4 minutes to cut a piece of pipe into two parts. How long would it take him to cut it into 5 parts? (16 minutes)* (Note that some children may say 4 × 5 = 20 minutes. However, to make 5 parts, he only needs to make 4 cuts, so the answer is 16 minutes.)
- *Amira has the digits 1 to 7 written on cards. How many pairs of cards add up to 8? (three: 1 + 7, 2 + 6, 3 + 5) How many groups of three cards can you make that add up to 10? (four: 1 + 2 + 7, 1 + 3 + 6, 1 + 4 + 5, 2 + 3 + 5)*
- *Find a number between 1 and 10 that is not even and which gives a remainder of 1 when divided by 3. (7)*

Find the operation: Display a grid like this one on the board to reinforce and practise addition and subtraction.

25	45	19
10	15	23
50	21	15

Move a pointer across the grid and ask the children to work out the operation needed to get from one number to the next. You can jump vertically, horizontally or diagonally and move one or two places.
For example:
- From 45 one down to 15: answer minus 30
- From 45 across diagonally to 10: answer minus 35

You can use this as a game with the children working in pairs to show moves on a grid and give the operations. They can use a calculator to check the answers.

You can make this activity more difficult by increasing the size of the grid and extending the number range.

Calculation skills

Addition and subtraction: Use number grids like these to practise mental addition and subtraction.

The operations are given across and down. Children can work in any order to fill in the missing values. Choose values that suit the mental strategies you are teaching.

This addition table shows adding 10 and multiples of 10.

+ 30	+ 10 →				
↓	15	25			
	45	55			

This subtraction table uses 9 and 11 as this encourages subtracting 10 and then compensating by adding or subtracting 1.

− 9	− 11 →				
↓	59	48			
	50	39			

Magic squares: In a magic square the rows, columns and diagonals all have the same total. Here are three easy examples for the children to complete:

8		6
3	5	
4	9	

	18	8
14		6
		16

		6
9	7	
8		

These two magic squares are more open ended and will require a little more thought to make them work.

10		
6		
5	7	9

	10	
	17	
	12	

Target numbers: Display a number and then ask the children to write five or more number sentences which give that result.

Some children may work systematically and produce a number of similar number sentences. For example, to make 100, they may write 99 + 1, 98 + 2, 97 + 3 and so on. This shows quite a sophisticated understanding of bonds and complementary addition, so be sure to share ideas and methods as a class afterwards.

Operations: Display a set of 3- and 4-digit numbers and ask the children to perform an operation on each number. For example, add 99 to each number (focusing on near multiples of 10) or write the number that is 7 less than each number (subtract a small number including crossing hundreds).

Addition and subtraction to 100: Use a set of 1–99 number cards. Draw a card at random, for example, 36. Choose a child to say what number should be added to make 100. Remove that card and continue.

Alternatively, ask the whole class to jot down the answers as you draw cards.

For subtraction, say 100 minus (draw the number) and ask the children work out the answer.

Five-number targets: Display a number less than 100, for example, 52. Ask the children to show how to add four or five smaller numbers to get this total.

Making a litre: Display a measuring jug or scale with an amount marked in millilitres (using multiples of 50). Ask the children how much liquid you need to add to make 1 litre (1000 ml), or how much was poured out from a litre. You can adapt this by asking them to combine amounts to make 1 litre.

Target 1000: Display the multiples of 100 from 0 to 900. Ask the children to sit in pairs and each to write down a 3-digit number.

Select a multiple, for example, 200. Ask the children to add that to their number, compare answers with their partners and decide whose answer is closest to 1000. Repeat this, with pace, and allow children keep a tally of who 'wins' each round.

Multiplication tables: Display a grid of the numbers from 1 to 10 in random order, for example:

3	6	1	5	9
2	7	8	4	10

Ask the children draw their own blank grid and write the product of multiplying each number on your grid by 3. Repeat for other multipliers.

Multiplication and division grids: Prepare multiplication grids where the children have to multiply 2-digit numbers by 1-digit numbers. For example:

	12	15	13	24	31	42	50	57	100
× 4									
× 8									

Discuss strategies. In this grid, the strategy would be that once children have worked out 4 × each number, that they can find 8 × by doubling.

Use grids like this for division, by giving the solutions and asking the children to work out the divisors (the numbers in the top row).

Doubling:
- Display a set of 2-digit numbers and choose children to double them.
- Display two sets of mixed-up numbers where one set is the doubles of the other. Ask the children to pair the numbers by doubling and/or halving.

Fractions: Choose a fraction, for example $\frac{1}{4}$.

Display a matching grid like this, where the numbers in the right-hand column are all 4 × the numbers in the left-hand column, but in random order.

7		4
8		28
1	... is $\frac{1}{4}$ of ...	36
9		12
3		32

Ask the children to match the numbers in the two columns and say or write a number sentence, for example, *a quarter of 36 is 9; 6 is a quarter of 24.*

Repeat this for halves, thirds, fifths, eighths and tenths.

Ask questions where you give the fraction of the amount and ask the children what the original amount could be. For example, *$\frac{1}{5}$ of an amount is 20. What is the amount?*

Ordering fractions: Display a set of fractions with the same denominators, or the same numerators. Ask the children write them in ascending or descending order and ask them to explain how they decided.

Which is more?: Ask questions comparing fractions, such as: *Which is more: $\frac{3}{4}$ of 100 or $\frac{4}{5}$ of 100?*

Quizzes: Prepare a series of short quizzes (15 to 20 questions) with mixed operations and mental strategies to use as mental warm-ups. Read the questions aloud and display them one by one for the class. Allow 20 to 30 seconds for the children to answer before moving on.

Here are some examples, but make sure that you have taught the topics before using them.

Sample quiz 1
1. Write nine thousand and twenty-three in numerals.
2. What is the value of the 3 in 2038?
3. What is 456 rounded to the nearest 10?
4. Double 19.   5. $20 - 13 =$ 6. $4 - 9 =$
7. How many fours in 28?   8. $137 + 9 =$
9. $10 - 45 =$ 10. $450 \div 10 =$ 11. $4 + \boxed{} = 20$
12. What is half of 36?   13. What is $\frac{1}{3}$ of 15?
14. What number is 100 less than 876?
15. Round 1455 to the nearest 100.

Answers: 1 9023   **2** thirty   **3** 460   **4** 38   **5** 7   **6** −5
7 7   **8** 146   **9** −35   **10** 45   **11** 16   **12** 18   **13** 5
14 776   **15** 1500

Sample quiz 2
1. Write in words the number that is 10 more than 1232.
2. How many tens are there in 4000?
3. Round 1254 to the nearest 10.
4. What is $\frac{1}{3}$ of 39?   5. Write 0.5 as a fraction.
6. $6 - 9 =$ 7. $54 \div 6 =$ 8. $81 \div \boxed{} = 9$
9. $19 + 327 =$ 10. $400 + 321 =$
11. What is the time half an hour later than 3:40 a.m.?
12. How many weeks is 28 days?
13. Double 39.   14. Double 75.   15. $850 + \boxed{} = 1000$

Answers: 1 one thousand, two hundred and forty-two
2 400   **3** 1250   **4** 13   **5** $\frac{5}{10}$ or $\frac{1}{2}$ **6** −3   **7** 9   **8** 9
9 345   **10** 721   **11** 4:10 a.m.   **12** 4   **13** 78   **14** 150
15 150

Sample quiz 3
1. Write in order from smallest to greatest: $4\frac{1}{2}$, $3\frac{1}{4}$, $4\frac{1}{4}$
2. Write eight thousand, three hundred and four in numerals.
3. What is the difference between 750 and 1000?
4. How many jumps of 5 do you make to get from 0 to 80?
5. $65 + 66 =$ 6. $97 - 8 =$ 7. $1234 + 9 =$
8. $3267 - 8 =$ 9. $143 + \boxed{} = 200$ 10. $325 - 45 =$
11. $9 - 11 =$ 12. $34 + 19 =$ 13. $135 + 49 =$
14. Is 552 a multiple of 5?   15. $499 - 100 =$

Answers: 1 $3\frac{1}{4}$, $4\frac{1}{4}$, $4\frac{1}{2}$ **2** 8304   **3** 250   **4** 16
5 131   **6** 89   **7** 1243   **8** 3259   **9** 57   **10** 280
11 −2   **12** 53   **13** 184   **14** No   **15** 399

Calendars and time

Calendars
Calendar fractions: Ask questions involving fractions and periods of time. For example:
How many months are there in:
- *half a year* (6)
- *one-quarter of a year* (3)
- *one-third of a year* (4)
- *three-quarters of a year* (9)
- *two-thirds of a year?* (8)

You can extend this activity by asking the children to estimate how many weeks each is equal to (for example, multiplying by 4, or halving 52).

Calendar counting sequences: Use a calendar to develop counting sequences in sevens. For example:
Write the next three numbers in each sequence:
- *7, 14, 21, __, __, __* (28, 35, 42)
- *1, 8, 15, __, __, __* (22, 29, 36)
- *4, 11, 18, __, __, __* (25, 32, 39)

You can also ask: *What is the tenth number in each sequence?*

Calendar questions: Display a current calendar. Point out today's date and ask the children to write down:
- the dates of the next four Fridays
- the date on Saturday
- the dates of the next three Tuesdays
- the date a week ago today.

Calendar puzzles: Give children a blank calendar for a month with some dates filled in and let the children work on the missing ones (and only those). For example, fill in the 1st and then let the children work out the dates going down diagonally.

Time

Time calculations: Display a table like this one. The first row is completed as an example.

$\frac{1}{2}$ hour earlier	Time	Time on a digital clock	15 minutes later	$\frac{3}{4}$ hour later
2:30 a.m.	3 a.m.	3:00	3:15 a.m.	3:30 a.m.
	12:30 p.m.			
	3:45 p.m.			
	4:20 a.m.			
	Half past two			
	Quarter to ten			

Choose children to give the missing times.

Timetables: Display a simple timetable, for example, showing the times of television programmes or school activities:

```
TV programmes
4:15 Robot dolls
4:20 Let us talk
4:30 Learning about trees
4:45 Spelling quiz
5:10 Popstarz
5:20 Fashion buzz
5:40 Sports highlights
6:00 News and weather
```

Ask questions to practise reading and interpreting the timetable, for example:
- *Which starts earlier – Popstarz or Spelling quiz?* (*Spelling quiz*)
- *Which programme starts at quarter past four?* (*Robot dolls*)
- *Which programme ends at quarter to five?* (*Learning about trees*)
- *What programme can you watch at 10 past five?* (*Popstarz*)
- *How long is Sports highlights?* (*20 minutes*)
- *Which programme is the longest? How long is it on for?* (*Spelling quiz, 25 minutes*)
- *News and weather is 20 minutes long. When does it end?* (*6:20*)

Masses in grams: Display items with their mass in grams (multiples of 50). Ask questions that involve ordering, comparing, adding and subtracting the masses. For example:
- *Which items weigh less than half a kilogram?*
- *Which items are heavier than 600 grams?*
- *Which two items together weigh 1 kilogram?*
- *What is the total mass of item A and B?*
- *I need 1 kilogram of sugar. How much more will I need?*
- *What is half of this item's mass?*

Fractions of shapes: Display shapes with parts shaded, like this:

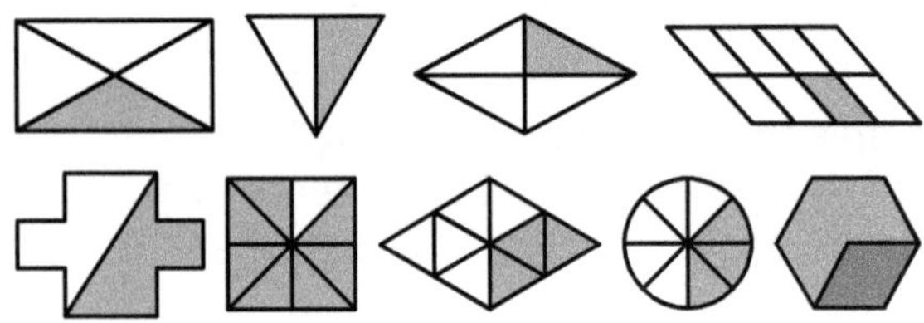

Ask the children to:
- name each shape based on its sides and/or angles
- say the fraction of the shape that is shaded
- order the fractions or find and name all the shapes with a particular fraction shaded.

Faces of solids: Display some solids and ask the children to draw all the faces of each one (for example, for a cube draw 6 squares).

Shapes around us: Display pictures of homes, public buildings and/or murals and decorative patterns from a range of sources. Ask children to identify and name the shapes, and any symmetry.

Describing position: Display a coordinate grid with objects, letters or shapes at coordinate points.
- Ask the children to give the coordinates of an object.
- Give the coordinates and ask the children to say what is found there.
- Give the children directions to get from one object to another, for example, 3 right and 2 down.

Section 5 Lesson notes
Think maths

Learning objectives

- Establish a classroom environment conducive to thinking and working mathematically
- Set up positive norms
- Establish key messages of growth mindset mathematics

Key words

cube   cuboid   cone   cylinder   prism   pyramid

Unit introduction

Teaching guidance

Unit 1 is different from all the other units in the course. It is a short introductory unit, which allows you to establish important norms for your class. This unit aims to prepare both the children and teachers for the work ahead. You should be able to complete this unit in one or two lessons.

As teachers of maths, it is important for us to examine and question our own ideas about who can do maths and about how mathematical ability and learning develop. From the earliest years, we need to find real, experiential ways for children to understand the following ideas:

- Asking questions is the best way to learn.
- Making mistakes is a key part of brain growth and learning.
- Everyone has the potential to learn maths. There is no such thing as a special 'maths brain'. If we simply tell children these things, they will not believe us. We need to show them evidence of the ways in which the brain grows and learns. We also need to show them that maths can include everyone.
- There is more information about the growth mindset on page 15.

What do you think?

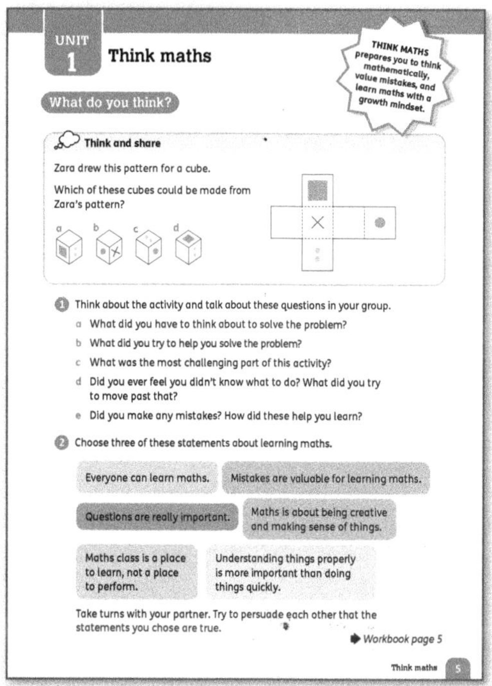

Materials

Thin (tracing) paper or squared paper; rulers; scissors.

Warm-up

- <u>Think and share</u>: Let the children work in pairs. Ask them to turn to **Pupil Book 4 page 5** and to consider the *cube* problem.
- Give them time to think. If necessary, allow them to trace or copy the *net*, then cut it out and fold it up to see which one is the correct cube.
- Take feedback to see which cube the children think is the correct one and let them say how they decided before agreeing that it is cube c.
- Explain that when we try to solve problems, we use our brains and the harder we think about the problem, the more connections we use and make in our brains. Explain that connections are links between different parts of our brains.

Focus

- For question 1, ask each pair to form a group with one or two other pairs. Explain that they are going to answer questions about the cube problem. You could let the children read the questions on their own and discuss them in their groups. Alternatively, you could read a question, then let the children talk about it and share their ideas before you move on to the next question. Whichever method you choose, take feedback from the discussion, focusing on how the children thought hard about the problem and stressing that mistakes are good and help us to learn.

- Ask the children to read through the statements in question 2 on their own. Ask them to each choose three statements and then give them time to think about how they could convince someone else that these statements are true. Have children talk in pairs about their ideas. Choose a few children to convince the class that a statement is true.
- Explain that scientific research has shown that our brains are active and that they grow when we have to try hard and make mistakes. The messages we tell ourselves can affect how we think about maths and can limit us if they are negative.

Follow-up

Use **Workbook 4 page 5** to address some limiting thoughts. Read through these with the class and let them choose more positive thoughts. They can select these from the examples on the page or make up their own positive messages. Encourage the children to discuss and find other limiting thoughts and to suggest more positive messages to replace them.

Interesting mistakes

Adults and children have implicit beliefs and stereotypes about mathematics. These assumptions may come from family or friends, from earlier experiences of the subject, from teachers and even from popular culture and the media. Here are a few myths and stereotypes:
- 'Maths is a difficult subject.'
- 'Only some people are truly good at maths.'
- 'If I get things wrong, it means I'm useless at maths.'
- 'Some people have a maths brain and others are better at creative/language subjects.'
- 'Maths is a boys' subject.'
- 'If I ask a question, everyone will know I don't understand and they might think I'm stupid.'
- 'Smart kids get everything right.'

None of these myths are true. Encourage the children to question them by focusing on the process of learning and the importance of asking questions and making mistakes.

Answers for Pupil Book 4 page 5

Think and share: cube c

1 a Individual answers. For example: When you fold the net into a cube, which faces will be next to each other?

   b Individual answers. For example: I tried to visualise the cube made from the net.

   c Individual answers. For example: trying to imagine the cube.

   d and e Individual answers

2 Individual answers

Answers for Workbook 4 page 5

1–**4** Individual answers

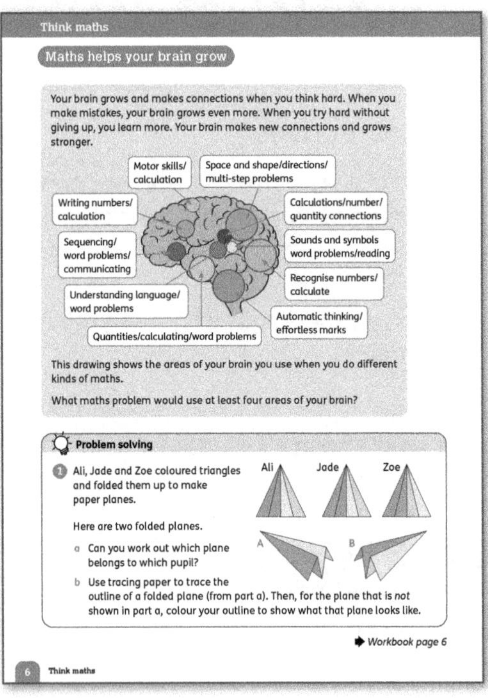

Materials

Plain paper and scissors; tracing paper; colouring pencils.

Warm-up

- Look at the diagram of the brain on **Pupil Book 4 page 6** and discuss how different kinds of mathematics thinking provide a 'workout' for different areas of the brain.
- Ask the class to suggest some kinds of mathematical activities that activate each part of the brain.

Focus

- Remind the class that they have learnt about the human brain and how it grows stronger by making connections when it is active. Explain that we can think of the brain as a muscle that gets bigger and stronger if we exercise it.
- Talk about each area of the brain and give examples from mathematics of the types of activity that would activate that part of the brain. For example, we use motor skills when we draw shapes, measure objects or fold up nets to make cubes.
- Let the class try to make up problems that would activate at least four areas of the brain. For example: *Sandra bought two items costing $2.50 and $4.90. How much change did she get if she paid with a $10 note?* The skills used include:
  - writing numbers
  - number/quantity connections
  - word problems/reading
  - understanding language/word problems
  - writing numbers
  - calculation
  - recognising numbers
  - working on a multi-step problem (and more).
- If the children suggest suitable problems, you could then ask the class to find the solutions.

- <u>Problem solving</u>: Let the children work in pairs to solve question 1 part a orally. Then let them make their own coloured diagram of the paper plane to answer part b.

Follow-up

Let the children complete **Workbook 4 page 6**. When they have finished, let them show their mind maps to a partner and talk through their ideas.

Answers for Pupil Book 4 page 6

1 a Jade (A) and Zoe (B)
  b Child's drawing of an airplane with colours to match those of Ali's plane

Answers for Workbook 4 page 6

1 and **2** Individual answers

Thinking about shapes

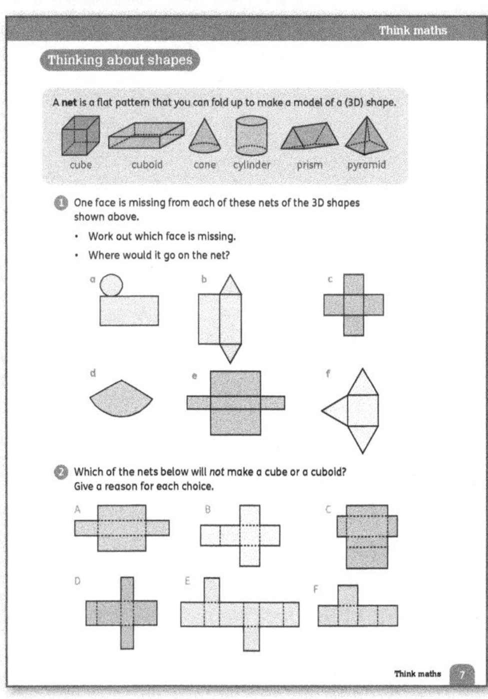

Materials

A box (unfolded and flattened to show the net of the box); sticky tape.

Warm-up

- Show the class the box you have opened out. Explain that this is called a *net*. Ask the children what shapes they can see on the net (for example, 2 squares and 4 rectangles).
- Explain that these shapes are the faces of the 3D shape that you can make with the net.
- Demonstrate how the net can be folded up to form a *cuboid* (or cube). Tape the edges together if necessary.
- Remind the children that the net is a flat pattern for making 3D shape. This follows on from the problem about the net of a cube on **Pupil Book 4 page 5**. The children will also have met the concept in earlier levels.

Focus

- Turn to **Pupil Book 4 page 7** and revise the names of the 3D shapes shown: cube, cuboid, *cone*, *cylinder*, *prism*, *pyramid*.
- Let the children work in small groups to do question 1 orally. There are different possible positions for the missing faces in some nets.
- Let the children work on their own to answer question 2 before asking them to share their ideas in groups.

Answers for Pupil Book 4 page 7

1 Possible answers:

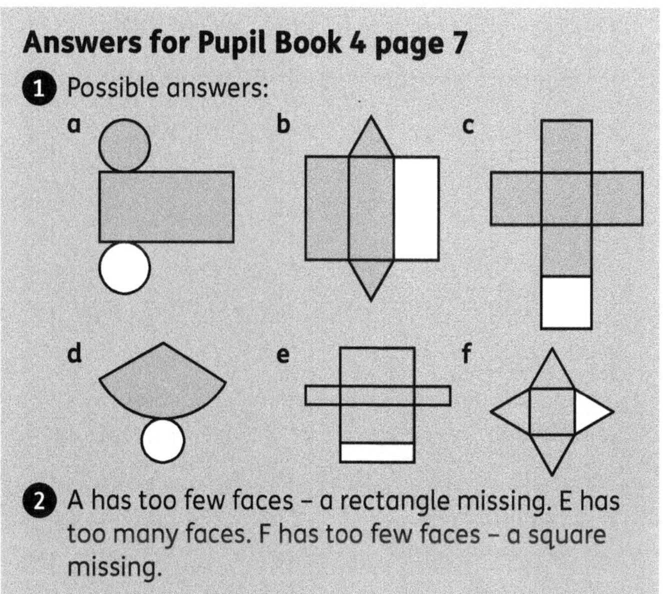

2 A has too few faces – a rectangle missing. E has too many faces. F has too few faces – a square missing.

End-of-unit check

Let the children prepare a poster containing thoughts and statements to help them to grow strong brains. They can work in groups to do this.

Ask questions related to some of the geometry concepts covered in this unit. For example:

- *What do we mean when we talk about the net of a 3D shape? (a flat/2D pattern that we can fold up to make a 3D shape)*
- *What do the 2D shapes on the net represent? (the faces of the 3D shape)*
- *A 3D shape has 6 faces. Is this enough information to make a net of this 3D shape? (No.) What other information do we need? (the shape and the side lengths of each face)*

Number and place value

Learning objectives

- Recognise the place value of each digit in a 4-digit number
- Order and compare numbers beyond 1000
- Identify, represent and estimate numbers
- Round numbers to the nearest 10, 100 and 1000
- Solve number and practical problems using place value
- Read and understand Roman numerals to 100

Key words

digit   number   place value   place   value   partitioning   expanded form   order   number line   estimate   round   rounded number   Roman numerals

Unit introduction

Materials

A blank 'Number of the day' chart (a poster or a whiteboard slide for display – like the one shown below, which can be used for any number up to 999); 0–9 digit cards; base-ten blocks (see page 21).

Teaching guidance

- Show the children the 'Number of the day' chart. Use *digit* cards to pick three digits and ask a child to make a *number*, for example, 437.
- Write this number in the oval on the chart and then work through the sections. Ask different children to complete the different sections of the chart like this:

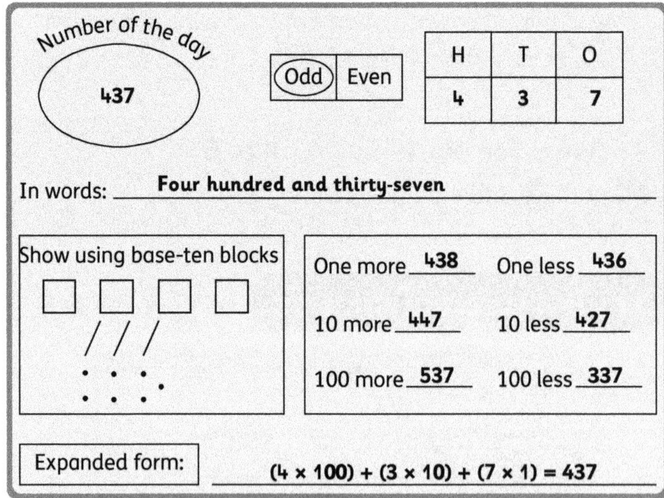

- During this activity, ask questions to revise the concepts and to introduce the necessary vocabulary. For example:
  - *Is the number odd or even? How do you know?* (*Look at the last digit. If it is 1, 3, 5, 7 or 9, the number is odd. If it is 0, 2, 4, 6 or 8, the number is even.*)
  - *What do we call this?* (*a place-value table*)
  - *How do you show the number in this table?* (*Write how many hundreds, tens and ones there are in the number.*)
  - *How do we say this number?* (*four hundred and thirty-seven*) *Can you write that in words?*
  - *How could we use base-ten blocks to show this number?* (*We could make the number with 100-flats, 10-rods and ones cubes.*) Discuss how to draw these. (See page 21 for information on base-ten blocks.)
  - *What is expanded form?* (*It is a way of writing the number as the sum of the values in each place.*)
  - *How do we write this number in expanded form?* ($4 \times 100 + (3 \times 10) + (7 \times 1) = 437$)
  - Discuss how you could find the number that is one more, one less and so on. (Example answers: *by counting on; by using the place-value table.*)

Revisit place value

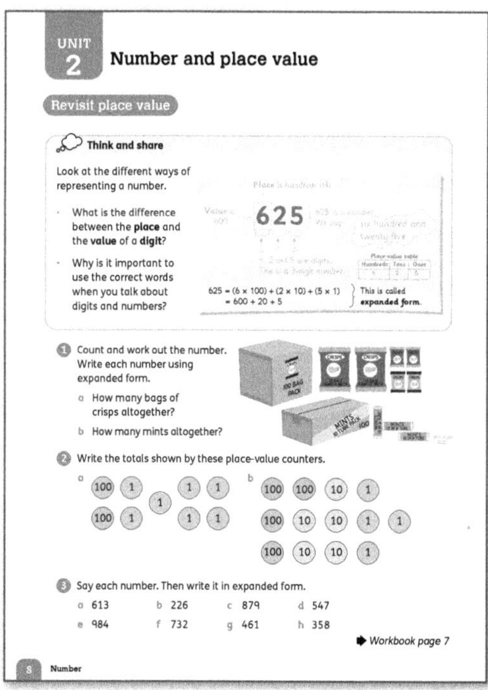

Materials

Place-value table with columns for thousands, hundreds, tens and ones (see page 22); place-value cards (thousands, hundreds, tens and ones) (see page 22); sets of 0–9 digit cards.

Warm-up

- As a mental warm-up, ask the children to select and write down any 3-digit number. Then ask them to write down the number that is:
  - one more
  - one less
  - ten more
  - ten less
  - one hundred more
  - one hundred less.
- You could also ask the children to make up three number sentences where the answer is the number they have chosen, for example: 100 + 234 = 334.

Focus

- <u>Think and share</u>: Use the activity at the top of **Pupil Book 4 page 8** to revise the concept of *place value* with the class. Give the children time to discuss the questions and share their ideas before taking feedback as a class.
- Make sure the children understand that *place* refers to the position of any digit in a number and that the *value* is how much a digit is worth in the number.
- Encourage the children to share their ideas about why it is important to use the correct words to refer to numbers and digits.
- Ask some additional questions to make sure that the children understand how the place-value table works and that they can partition numbers and write them in expanded form.
- Explain that *partitioning* simply means saying how many hundreds, tens and ones are in the number. For example:

352 = 3 hundreds, 5 tens, 2 ones.

- Explain that expanded form means the partitioned number expressed as multiples of hundreds, tens and ones. For example:
352 = (3 × 100) + (5 × 10) + (2 × 1)
- Round off the discussion by reminding the class that place value allows us to write any number using just the digits from 0 to 9.
- For question 1, let the children work in pairs to count the hundreds, tens and ones in each set. Next, they check and correct each other's answers.
- The children can work independently to write the numbers in question 2. Read out the number names and let them check their own work.
- Do the first part of question 3 orally (or let the children work in pairs to say the numbers to each other) and then let the children record the expanded form of each number.

Follow-up

- Use **Workbook 4 page 7** to consolidate understanding of place value to three digits. Let the children work in pairs to complete question 1 and question 2 and then they can work with another pair to check each other's work.
- <u>Problem solving</u>: Read through question 3 with the class and discuss how they can use diagrams (drawings of base-ten blocks to represent numbers) or models such as place-value tables, ten frames or other apparatus (see pages 21–22) to help them work out the numbers.
- Let the children complete the task and then share their answers, explaining how they worked to get the answer. This is a good opportunity to explore interesting mistakes and to discuss where the children got mixed up or if they disagree with each other's answers.

Challenge

- Let the children take part in a 'Place-value number hunt' by finding everyday numbers at home or at school that meet certain conditions.
- Let the class suggest conditions for numbers, such as: 'It has a 5 in the tens place.' 'It is greater than 400.' 'It has no tens or ones.'
- Ask the children to find examples of one or two numbers that meet each condition. They should list these and say where they found each number. For example: '255 is the number of the bus I catch.'

Support

Give the children a set of place-value cards or several sets of 0–9 digit cards. Say several numbers up to 999 and ask the children to lay out cards to show each number. Let them say the number when they have made it, to reinforce number names.

Interesting mistakes

Some children may simply say the digits in order instead of saying the correct number name. For example, they might say 'five four three' instead of 'five hundred and forty three'. If this happens, encourage them to identify the value of each digit and then say the numbers in expanded form.

Answers for Pupil Book 4 page 8

<u>Think and share:</u> Place (hundreds, tens, ones) is the position that a digit (from 0 to 9) occupies in a number. Value is what the digit is worth. It is important because the digit 7 is different from the number 7. The digit 7 value depends on its position. The number 7 has a value of 7 ones.

1   a $100 + 10 + 10 + 1 + 1 + 1 + 1 = 124$

     b $100 + 10 + 10 + 10 + 1 + 1 + 1 = 133$

2   a   207   b   454

3   a   six hundred and thirteen; $600 + 10 + 3$

     b   two hundred and twenty-six; $200 + 20 + 6$

     c   eight hundred and seventy-nine; $800 + 70 + 9$

     d   five-hundred and forty-seven; $500 + 40 + 7$

     e   nine hundred and eighty-four; $900 + 80 + 4$

     f   seven hundred and thirty-two; $700 + 30 + 2$

     g   four hundred and sixty-one; $400 + 60 + 1$

     h   three hundred and fifty-eight; $300 + 50 + 8$

Answers for Workbook 4 page 7

1   a   126   b   762

     c   Possible answers: 162, 172, 761, and so on

2   a   368   b   986

     c   Possible answers: 369, 689, 968, and so on

3   a   243   b   591   c   330   d   899

     e   520

Place value to thousands

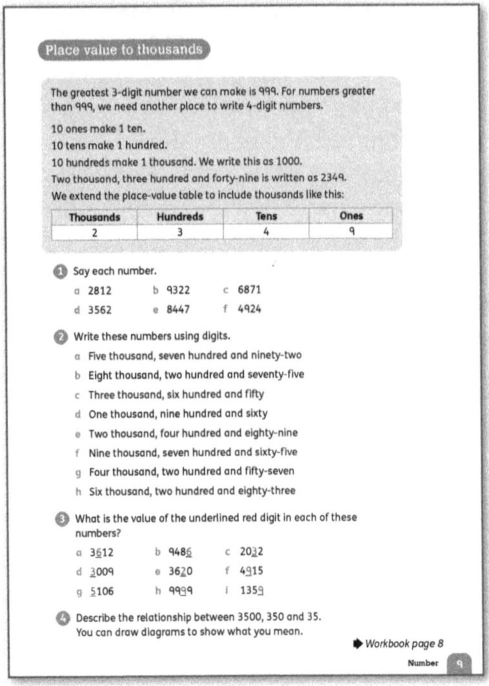

Place value to thousands

The greatest 3-digit number we can make is 999. For numbers greater than 999, we need another place to write 4-digit numbers.

10 ones make 1 ten.
10 tens make 1 hundred.
10 hundreds make 1 thousand. We write this as 1000.
Two thousand, three hundred and forty-nine is written as 2349.
We extend the place-value table to include thousands like this:

Thousands	Hundreds	Tens	Ones
2	3	4	9

1   Say each number.

   a   2812   b   9322   c   6871

   d   3562   e   8447   f   4924

2   Write these numbers using digits.

   a   Five thousand, seven hundred and ninety-two

   b   Eight thousand, two hundred and seventy-five

   c   Three thousand, six hundred and fifty

   d   One thousand, nine hundred and sixty

   e   Two thousand, four hundred and eighty-nine

   f   Nine thousand, seven hundred and sixty-five

   g   Four thousand, two hundred and fifty-seven

   h   Six thousand, two hundred and eighty-three

3   What is the value of the underlined red digit in each of these numbers?

   a   3612   b   9486   c   2032

   d   3009   e   3620   f   4915

   g   5106   h   9999   i   1359

4   Describe the relationship between 3500, 350 and 35. You can draw diagrams to show what you mean.

➧ *Workbook page 8*

Number 9

Materials

Place-value table (see page 22); large place-value chart (see 'Focus' below); four cards and adhesive tack (for covering numbers on the place-value chart); place-value cards (see page 22); sets of 0–9 digit cards.

Warm-up

- Spend some time counting on beyond 1000 before you show the children the place-value table with a Thousands column.
- Point out that 10 hundreds make 1000, and count on in thousands to teach the vocabulary.
- Make sure the children understand that 4-digit numbers have four places: thousands, hundreds, tens and ones. Clearly demonstrate the value of each digit using place-value cards and/or other concrete apparatus (see pages 20–22).

Focus

- Display a place-value chart like the one below.

1000	2000	3000	4000	5000	6000	7000	8000	9000
100	200	300	400	500	600	700	800	900
10	20	30	40	50	60	70	80	90
1	2	3	4	5	6	7	8	9

- Use cards to cover numbers on the chart and ask the children to say the numbers aloud. For example, cover 2000, 300, 20 and 3.
- Start with 3-digit numbers and extend to include 4-digit numbers. It is important to say the number, for example, *two thousand, three hundred and twenty-three*, so the children link place value with the counting system.
- After covering a few numbers for the children to say, swap roles so you say a number and ask some children to come and stick cards onto the place-value chart to represent that number.
- Read through the information on **Pupil Book 4 page 9** with the children to consolidate the ideas before asking them to complete the questions.
- The children work in pairs to say each number in question 1.
- The children work independently to write the numbers in question 2. Let them check and correct each other's work.
- Do question 3 orally with the class.
- Let the children discuss question 4 in groups before presenting their explanations to the class.

Follow-up

Use **Workbook 4 page 8** to check that the children can work with numbers in the thousands. They can write the answers to question 3 as homework.

Challenge

Draw four empty boxes in a row on the board. Ask the children to copy the boxes on to paper. Pick a card from a set of digit cards and then return it to the pack and shuffle. Continue until you have picked four cards. Each time, the children must select a box to put the number in. Once placed, the number cannot be moved.

The aim is to create the largest number when all four boxes have been filled. Discuss the strategies that the children used to make decisions about where to place digits. Repeat, this time trying to make the smallest number.

Support
Use place-value cards to make an 'incomplete' number. (See 'Place value with place-value cards' on page 24 for some ideas on how to use place-value cards.) For example, display 1000, 40 and 3, then ask the children: *What card is missing for the number one thousand three hundred and forty-three?* Repeat this for several numbers that have the same digit in two or more places, to reinforce the idea of place value in 4-digit numbers.

Interesting mistakes
The children may make mistakes when writing and/ or saying numbers with zero as a place holder, such as 4056 or 1304. Ask the children to enter numbers into a place-value table and then say them in expanded form, emphasising where there is a zero in a column and what that means.

Answers for Pupil Book 4 page 9
1
a two thousand, eight hundred and twelve
b nine thousand, three hundred and twenty-two
c six thousand, eight hundred and seventy-one
d three thousand, five hundred and sixty-two
e eight thousand, four hundred and forty-seven
f four thousand, nine hundred and twenty-four

2 a 5792   b 8275   c 3650   d 1960
e 2489   f 9765   g 4257   h 6283

3 a six hundred   b six   c thirty
d three thousand   e twenty   f nine hundred
g five thousand   h ninety   i nine

4 Possible answer: 3500 is ten times greater than 350, and 350 is ten times greater than 35. 35 is ten times less than 350 and 100 times less than 3500. As you go from 35 to 3500, the '3' and '5' digits each move one place to the left in the place-value table. The children may draw a place-value table to support their answer.

Answers for Workbook 4 page 8
1 a 200 (provided as an example)   b 0
c 1000   d 3000   e 50   f 7000
g 6000   h 100   i 60

2 a 2843 (provided as an example)   b 6065
c 8015   d 7220   e 4004

3 Individual answers

Expanded form

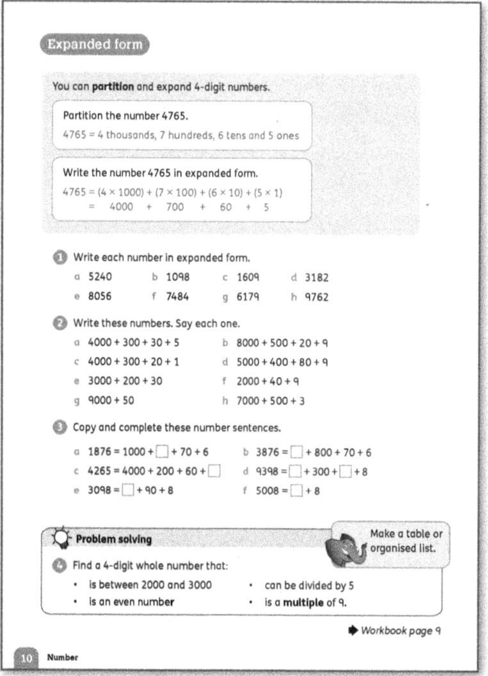

Materials
Place-value cards (see page 22); calculators.

Warm-up
Revise expanded form. Use place-value cards to demonstrate partitioning numbers into thousands, hundreds, tens and ones. Choose one or more of the 'Place value with place-value cards' activities on page 24.

Focus
- Let the children work independently to read through the worked examples on **Pupil Book 4 page 10** before completing the questions independently.
- For question 1, be clear about whether you want the children to use the longer form (with multiplications and brackets) or the shorter form (the sum of values in each place). Discuss how to deal with 0 in a particular place: they can either write (0 × 1) or just leave it out, as any number multiplied by 0 will result in an answer of 0.
- Once the children have written the numbers in question 2, they can say them to themselves or to a partner.
- The children can complete question 3 by writing the number sentences in their books.
- Problem solving: Remind the children to work systematically to complete question 4. Encourage them to read through the entire problem before they do any working and allow them to ask questions if they don't understand any of the conditions. Encourage them to use a calculator if they are going to count up in nines.

Follow-up
Use **Workbook 4 page 9** to consolidate and reinforce understanding of expanded form.

Answers for Pupil Book 4 page 10

1 a 5000 + 200 + 40   b 1000 + 90 + 8
   c 1000 + 600 + 9   d 3000 + 100 + 80 + 2
   e 8000 + 50 + 6   f 7000 + 400 + 80 + 4
   g 6000 + 100 + 70 + 9   h 9000 + 700 + 60 + 2

2 a 4335; four thousand, three hundred and thirty-five
   b 8529; eight thousand, five hundred and twenty-nine
   c 4321; four thousand, three hundred and twenty-one
   d 5489; five thousand, four hundred and eighty-nine
   e 3230; three thousand, two hundred and thirty
   f 2049; two thousand and forty-nine
   g 9050; nine thousand and fifty
   h 7503; seven thousand, five hundred and three

3 a 1876 = 1000 + 800 + 70 + 6
   b 3876 = 3000 + 800 + 70 + 6
   c 4265 = 4000 + 200 + 60 + 5
   d 9398 = 9000 + 300 + 90 + 8
   e 3098 = 3000 + 90 + 8
   f 5008 = 5000 + 8

4 Possible answers: 2070, 2160, 2250, 2340, 2430, 2520, 2610, 2700, 2790, 2880, 2970

Answers for Workbook 4 page 9

1 a 5792 = 5000 + 700 + 90 + 2 (provided as an example)
   b 3650 = 3000 + 600 + 50
   c 8275 = 8000 + 200 + 70 + 5
   d 1960 = 1000 + 900 + 60
   e 2009 = 2000 + 9   f 6090 = 6000 + 90

2 a 1543 = 1000 + 500 + 40 + 3 (provided as an example)
   b 3412 = 3000 + 400 + 10 + 2
   c 4088 = 4000 + 80 + 8
   d 8799 = 8000 + 700 + 90 + 9
   e 9200 = 9000 + 200
   f 7999 = 7000 + 900 + 90 + 9

3 a 4308   b 2054   c 8400   d 4200
   e 5545   f 1919

Challenge

Challenge the children to find all the possible answers to question 4 on **Pupil Book 4 page 10** and to explain how they know they have found them all. There are 11 possible answers. The children can use patterning to find these if they work systematically.

Compare and order numbers

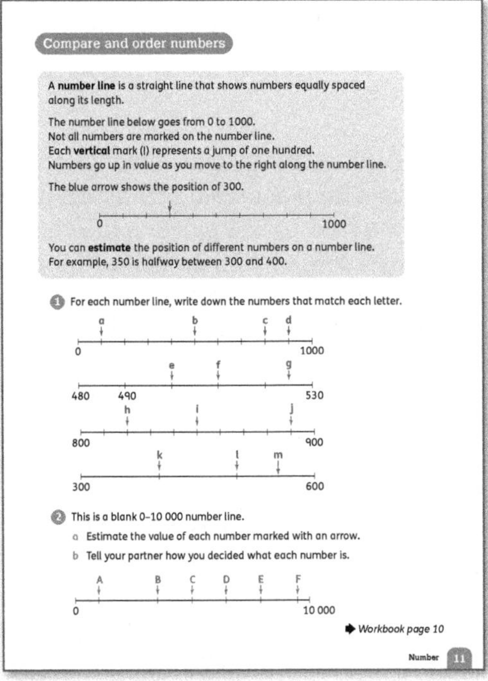

Materials

Set of 1–9 digit cards; large number line with ten divisions; a set of cards each showing a different 3-digit number; string or ribbon; 0–10 number cards and pegs.

Warm-up

- Play the following game, in which the children have to make the largest possible 4-digit numbers using digits that you display.
- Ask the children to draw a set of six rows of four boxes. Each row shoud look like this:

- Shuffle the digit cards and display one card. The children choose one of the boxes in the first row and write the number in it. After four digits, let the children compare numbers with the others in their group and decide who has the greatest number.
- Repeat this six times, using a different row of boxes each time.
- Next, ask the children to share the strategies they used for placing the digits.

Focus

- Ask a group of children to stand up. Give each child in the group a card showing a number less than 1000. Ask the rest of the class to instruct the group so they stand showing the numbers in *order* from smallest to greatest or greatest to smallest.
- Give another child a different number card and ask them to position themselves in the line. Ask the class to say a number between two of the numbers in the line.
- Revise the concept of a *number line* with the class. Show the class a blank number line with ten divisions.

- Write a *multiple* of 100 at each end of the number line. For example, write 500 at the start and 600 at the end. Explain that the number line goes from 500 to 600. Ask the class questions to make sure they understand that not every number between 500 and 600 is shown on the line. For example:
  - *How many spaces are on the number line? (10)*
  - *Are there only ten numbers between 500 and 600? (No.)*
  - *So, what does each space represent? (a jump of 10)*
  - *What number does this mark represent? (Point to 550, for example.) (550)*
  - *How do you know? (It's halfway between 500 and 600. / You can count up in tens.)*
  - *How would you count from 500 to 600 on this number line? (510, 520 . . . 590, 600)*
  - *Where would you place 515? (halfway between 510 and 520)*
- Show the class how to work with a blank number line marked from 0 to 10 000. Discuss where to put 5000 on the number line, and then 9000 and 9999. Give the children time to explain how they work out or *estimate* where a number will go.
- Place the number 3000 on the line. Ask the children how this can help them work out where to place 1500 and so on.
- Before the children start work on **Pupil Book 4 page 11**, remind them to work out what the intervals shown on unmarked number lines represent before they decide what the numbers are. They will use this skill in measuring as well.
- The children can discuss question 1 in pairs before writing the answers on their own.
- For question 2, let the children write their estimates before they discuss with their partners how they decided what the numbers were.

Follow-up

Use **Workbook 4 page 10** to check that the children can estimate values from a number line, and position numbers correctly on a number line.

Challenge

Give the children a set of four cards, each with a different digit on it (choose any four digits). Ask them to use the cards to make and list as many 4-digit numbers as possible.

Once they have listed the numbers they have made, give them a 0–10 000 number line marked in thousands and ask them to place the numbers they have made in the correct positions on the number line.

Support

- Make a string or ribbon number line for display in the classroom. Place a set of 0–10 number cards close to the number line.
- Ask the children where they would place 0. (Make sure that they realise it should go at the left-hand end of the line.) Let one child peg the card in the position they have chosen.

- Next, ask them where they would put 10. (It can go anywhere to the right of 0, but the children may place it on the other end of the line.)
- After they have placed 0 and 10, discuss where they would place the other numbers and why. It is useful to start by placing 5, as that makes equal parts to place the other numbers.
- Once the children have placed the cards on the line, move on to a different number set to check their understanding. For example, use multiples of 10, 100 or 1000.
- You can extend this to pencil and paper activities, giving the children progressively more challenging number lines. Start with simple questions, such as: *Where would you place 2, 7 and 9 on this number line?*

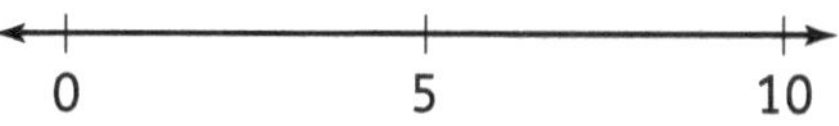

- Repeat, using different numbers, and then move on to higher ranges, using number lines like these:

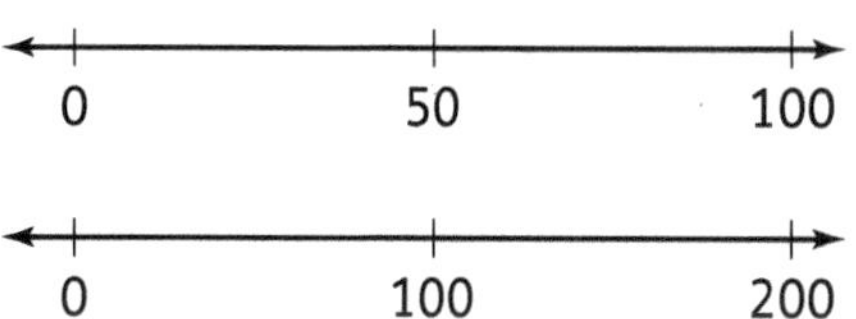

Interesting mistakes

Some children may have difficulty positioning numbers on a number line. Help them by pointing out the start and end points and then working systematically through the smaller divisions. Provide several different number lines for additional practice.

Answers for Pupil Book 4 page 11

1 a = 100, b = 500, c = 800, d = 900; e = 500, f = 510, g = 525; h = 820, i = 850, j = 890; k = 400, l = 500, m = 550

2 a and b Possible answers: a = 1000 (about one tenth of the way between 0 and 10 000), b = 3500, c = 5000 (about half-way between 0 and 10 000), d = 6500, e = 8000 (equally spaced about c), f = 9500 (close to 10 000)

Answers for Workbook 4 page 10

1 a 1000 (provided as an example)
 b 3000
 c 5000
 d 5500
 e 8000
 f 9500

2 Second number line
 10 000 — 9999
 — 9000
 — 8000
 7800
 — 6500
 — 3000
 — 2150
 0

Use < and > to compare and order numbers

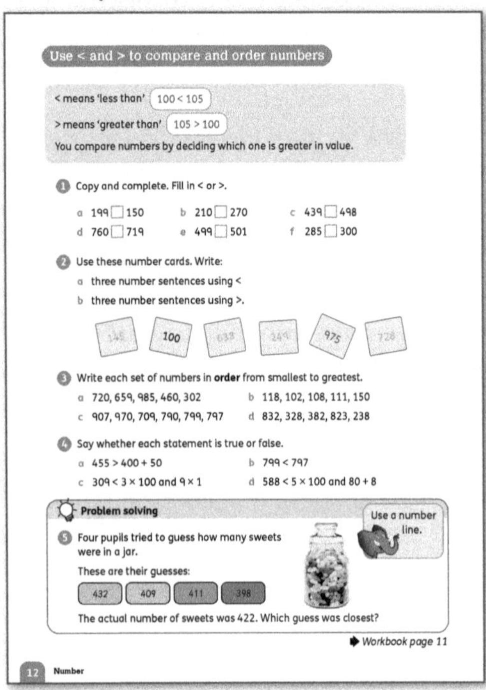

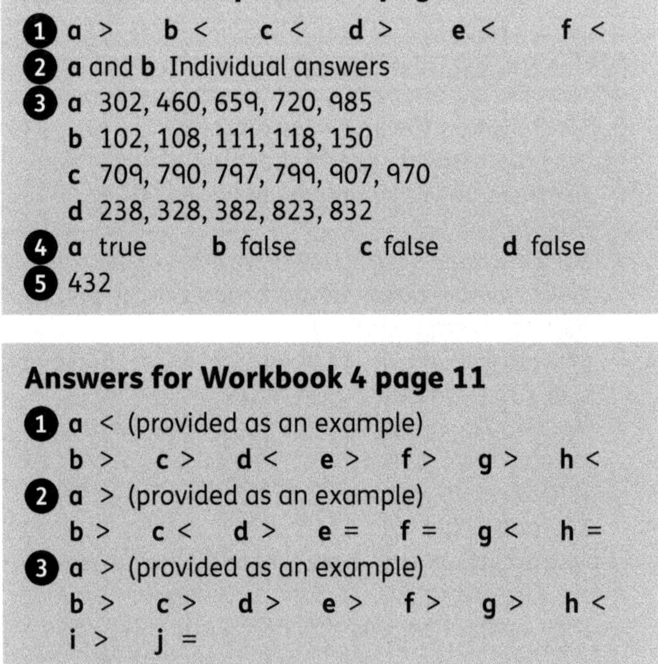

Answers for Pupil Book 4 page 12

1. a >   b <   c <   d >   e <   f <
2. a and b Individual answers
3. a 302, 460, 659, 720, 985
   b 102, 108, 111, 118, 150
   c 709, 790, 797, 799, 907, 970
   d 238, 328, 382, 823, 832
4. a true   b false   c false   d false
5. 432

Answers for Workbook 4 page 11

1. a < (provided as an example)
   b >   c >   d <   e >   f >   g >   h <
2. a > (provided as an example)
   b >   c <   d >   e =   f =   g <   h =
3. a > (provided as an example)
   b >   c >   d >   e >   f >   g >   h <
   i >   j =

Compare and order larger numbers

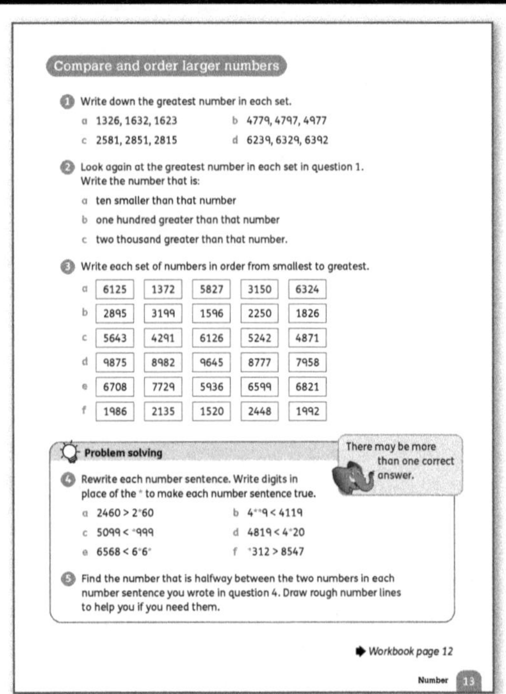

Materials
0–9 digit cards (one set for each pair of children).

Warm-up
- As a mental warm-up, ask the children to shuffle the digit cards and place them face down. They then take three cards each. They should look at their cards and make a 3-digit number.
- The children compare their number with their partner's. The child who made the greater number scores a point.
- Repeat this five times – the children shuffle the cards and take three cards each time.

Focus
- Revise the use of the < and > symbols as necessary. Let the children work independently to complete questions 1–4 on **Pupil Book 4 page 12**. Check the answers as a class and discuss any interesting mistakes.
- <u>Problem solving</u>: Encourage the children to use a number line to help them work out question 5.

Follow-up
Use **Workbook 4 page 11** to provide additional practice and consolidation.

Interesting mistakes
Some children may not use the < and > symbols correctly. Discuss how to remember which is which (for example, use crocodile mouths or write on the board: L means less and < is less).

Once you are satisfied that the children can order and compare 3-digit numbers, move on to 4-digit numbers.

Materials
Atlases (for the Workbook activity); number lines or place-value tables (see page 22); place-value cards (see page 22).

Warm-up
- As a mental warm-up, draw five balloons on the board and write a set of 4-digit numbers that look similar in the balloons. For example, you could write 4404, 4440, 4044, 4040 and 4400.

- Tell the children that they are going to play a game in which they pop three balloons and get the points displayed on the balloons. The winner is the person who gets the highest score.
- Ask: *Which three balloons do we need to pop to get the highest score?* (4440, 4404 and 4400)
- List the three numbers in a row with spaces between them. Ask: *What symbols can we write in the spaces to compare these numbers?* (4440 > 4404 > 4400)

Focus

- Give the children time to discuss the questions on **Pupil Book 4 page 13** and to think about what strategies they can use to try to solve them. Discuss their strategies with them. Their approaches to question 3 in particular will give you an indication of how well they have understood and mastered using place value to compare and order numbers.
- Let the children work on their own to find the largest number in each set in question 1. Check the answers as a class.
- The children can use a number line or place-value table to help them find the numbers in question 2.
- Use question 3 to check that the children can compare and order 4-digit numbers.
- Problem solving: Let the children find their own answers to question 4 and then compare with others in their group. Talk about the strategies they used to replace the digits.
- Show the children how to find the halfway point by working through one or two examples with the class before asking them to do question 5 on their own.

Follow-up

- Use **Workbook 4 page 12** to reinforce how we use and order numbers in daily life. Encourage the children to find the countries and mountains in an atlas before they complete question 1.
- Problem solving: Read through the problems in questions 2 and 3 with the class and let them discuss how they can solve these before letting them find their own solutions.

Support

It is useful for the children to have a set of place-value cards (see page 22). They can use these to make 4-digit numbers and separate the cards to show the expanded notation. This allows them to compare the numbers place by place. These cards can also be used to help the children make numbers that are 10, 100 or 1000 more or less than the given number. Place-value cards really help children who do not yet understand place value and related number concepts.

Interesting mistakes

Some children may make mistakes when reading 4-digit numbers with a zero in one or more places. Help them by building numbers using place-value cards (as described in the 'Support' section above) and asking the children to say the numbers they have made.

Round numbers to the nearest 10

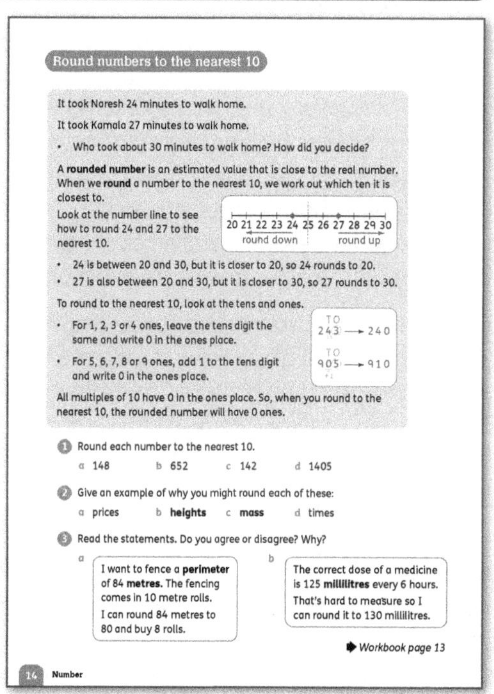

Materials

Sticky notes or index cards and adhesive tack; reusable laminated number line or interactive online number line.

Warm-up

Use one of the 'Place value and number sense' activities on pages 23–27 as a mental warm-up.

Focus

- Use a number talk (see pages 17–18) to revise and discuss the concept of *rounding* numbers. Give the class the following problem to consider:

> Maria and Zara pick three numbers from a box without looking at them.
> Maria picks 8, 12 and 17.
> Zara picks 9, 14 and 11.
>
> If the number is closer to 10, they get 5 points.
> If the number is closer to 20, they get 10 points.
>
> What does each child score?
> How did you decide?
> What is the score for picking 15? Why?

- Encourage the children to discuss the problem with a partner and let them indicate when they have an answer to share.
- Spend some time talking about how they decided and what the score is for 15.
- Display a number line on the board. Using sticky notes or index cards, write a number on each note or card and put it on the number line.
- Ask: *How should we round the number to the nearest 10? Do we round up or down?*
- Draw a vertical dotted line at the halfway point between the tens (as shown at 25 on the number line on **Pupil Book 4 page 14**) .
- Ask: *What does the dotted line mean?* (*It means that this number is halfway between the start and end numbers.*) Explain that in maths, numbers that are halfway or further along the number line round up to the next 10.
- Demonstrate with a few examples how to round to the nearest 10. Then change the start and end numbers on the line and let several children come to the front of the class and complete some more examples.

Work through the question at the top of **Pupil Book 4 page 14**. The children's answers will indicate whether they have a good understanding of rounding up and down. They should realise that Kamala took about 30 minutes because 27 is closer to 30 than it is to 20.

- Work through the examples and explanations with the class to consolidate the discussion and practical demonstration of rounding before doing the questions.
- Let the children complete question 1 independently. Check their answers.
- Have a class discussion about the quantities and measurements in question 2 and when and how these might be rounded in real life.
- Let the children work in groups to discuss the statements in question 3. Take feedback from the class. Make sure that the children understand that the context is important when you decide to round a number or not.

Follow-up

- The children can fill in the first column of question 1 on **Workbook 4 page 13** or you may wish to wait until after the next lesson on rounding to the nearest 100.

- Problem solving: The children could use a number line to help them decide which numbers round to 1500 to the nearest 10 in question 3b. The answers to **Workbook 4 page 13** are on page 43.

Challenge

Give the children prepared measuring task sheets like the one shown below. If you have time, the children can measure objects themselves to reinforce measuring skills, but if not, give the children the lengths and ask them to round these measurements.

Object to measure	Length in cm	Number line showing which two 10s this measurement lies between	Measurement rounded to the nearest 10 cm
Length of an arm			
Height of a desk			
Length of the board			
Width of the door			

Support

Allow the children to use reusable laminated number lines to position numbers and to decide how to round them. You could also use an interactive online number line.

Interesting mistakes

- The children may not understand that a *rounded number* has about the same value as the number you started with. A rounded number is just a less exact value. Using number lines to show the position of the rounded number and the starting number can help to address this.
- The children may also find it confusing when they have to round a number that has a 9 in the rounding place, for example, when rounding 199 to the nearest 10. Using a number line allows the children to view the divisions on the number lines as tens and this makes it easier to round to the next 10, which is 200 (20 tens).

Answers for Pupil Book 4 page 14

1   a   150   b   650   c   140   d   1410

2   Possible answers:
- a   To make sure you have the enough money to pay for something
- b   To work out if you are tall enough to go on a theme park ride
- c   When cooking
- d   When estimating how long something will take or when you will arrive somewhere

3   a   Disagree: 80 metres is less than 84 metres so the fence wouldn't be complete.
- b   Disagree: Medicines can be dangerous if you take too much.

Round numbers to the nearest 100

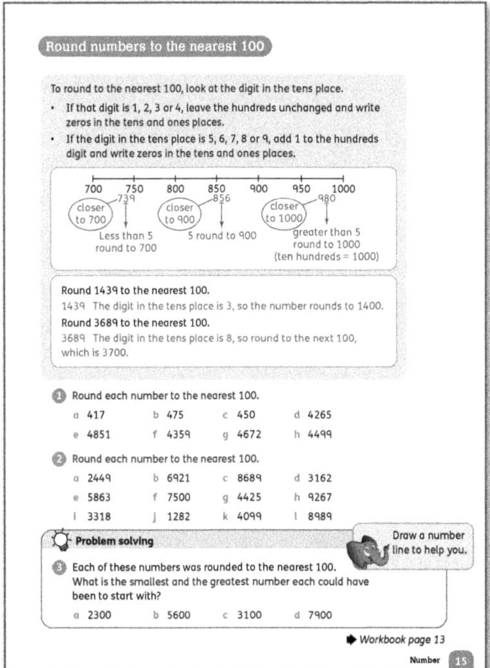

Materials

A large number line with ten divisions (for display); cards with numbers (for demonstration); reusable laminated number line or interactive online number line.

Warm-up

Select a 'Rounding' activity from page 26 as a mental warm-up.

Focus

- Work through the examples on **Pupil Book 4 page 15** to teach the children how to round to the nearest 100. Remind them that this works in the same way as rounding to the nearest 10. Place some 3- and 4-digit numbers on a number line and ask the class to explain how to round each number up or down to the nearest 100.
- Encourage the children to use number lines to help with question 1 if they need to.
- Let the children work on their own to round the numbers in question 2, then compare with a partner and check each other's answers.
- <u>Problem solving:</u> Spend some time discussing the answers to question 3 and encourage the children to share how they worked out their solutions. If they used number lines, let them show these to the class to reinforce the concepts from this lesson.

Follow-up

Use **Workbook 4 page 13** as an informal assessment task. Let the children work independently to complete the questions and then work in pairs to check their answers.

Challenge

Adapt the measuring activity from the previous lesson to include rounding to the nearest 100, by using longer lengths or grams or millilitres.

Support

As with rounding to the nearest 10, allow the children to use erasable/reusable number lines to position numbers and to decide how to round them. You could also use an interactive online number line.

Interesting mistakes

When the children have to work out the smallest and greatest value that will round to a given multiple of 100, they may not realise that the smallest value is less than the rounded number (in other words, it rounds up). Encourage them to use a number line marked with the hundreds. Show them how to find the middle value and use it to work out which numbers round up and which numbers round down.

Answers for Pupil Book 4 page 15

1
a 400	b 500	c 500	d 4300
e 4900	f 4400	g 4700	h 4500

2
a 2400	b 6900	c 8700	d 3200
e 5900	f 7500	g 4400	h 9300
i 3300	j 1300	k 4100	l 9000

3
- a smallest 2250, greatest 2349
- b smallest 5550, greatest 5649
- c smallest 3050, greatest 3149
- d smallest 7850, greatest 7949

Answers for Workbook 4 page 13

1

Number	To the nearest 10	To the nearest 100	
317	320	300	(provided as an example)
233	230	200	
671	670	700	
894	890	900	
2185	2190	2200	
3347	3350	3300	

2

1456 <u>1399</u> 1045 <u>1599</u> (1302) 1350
1500 1000 1600

1428 1528 1889 1010 (1309) 1480
1900 1000 1500

1375 1357 (1328) 1488 1444 (1333)

1555 1455 (1299) 1250 1245 1899
1600 1500 1300 1200 1900

3 a 1450   b 1504

Round numbers to the nearest 1000

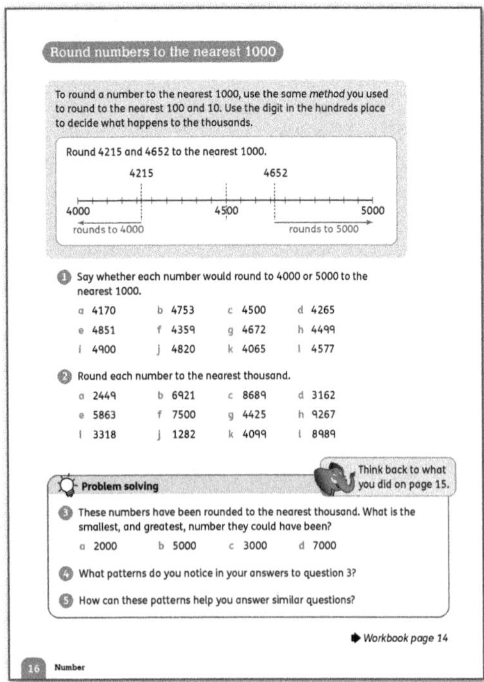

Materials
Number lines.

Warm-up
- Revise rounding to the nearest 10 and 100 as a mental warm-up.
- Write a multiple of 10 on the board. Tell the children that you are thinking of a number that when rounded to the nearest 10 is the number on the board.
- Invite the children to tell you what the number could be.
- Repeat with multiples of 100.

Focus
- Use the example at the top of **Pupil Book 4 page 16** to revise the fundamental rules of rounding.

> - Make sure that the children understand that the value of the digit to the right of the one you are rounding to determines what you do to round the number.
> - Once the children understand the rules for rounding, they can round to any place, including rounding decimals to the nearest whole number.

- Let the children work in pairs to do question 1 orally before they write the answers.
- Let the children work on their own or in pairs to round the numbers in question 2. Check the answers orally with the class before moving on.
- <u>Problem solving</u>: Question 3 is similar to the problem-solving activity in the previous lesson. Make sure that the children realise this and let them recap the strategies they found useful. Then ask them to work in pairs to find the solutions.

- For questions 4 and 5, have a class discussion about the patterns in question 3. Then share ideas about how the patterns the children have identified can help them to solve similar problems.

Follow-up
Use **Workbook 4 page 14** as a consolidation activity. The children can check each other's work.

Support
- Continue to use number lines to support children who need help to apply the rules and who need a visual reminder of which multiple of 1000 their number is closer to.
- You may also find it useful to show the children how to cover the tens and ones (with a ruler or a piece of paper) so that they only focus on the digits in the thousands and hundreds places.

Interesting mistakes
Often, if children see 'high digits' such as 8 or 9 in the tens and ones places, they are tempted to round up. It is important to keep asking them which place they are rounding to and then making sure that they look only at the digit in the place to the right of it.

Answers for Pupil Book 4 page 16

1
a 4000	b 5000	c 5000	d 4000
e 5000	f 4000	g 5000	h 4000
i 5000	j 5000	k 4000	l 5000

2
a 2000	b 7000	c 9000	d 3000
e 6000	f 8000	g 4000	h 9000
i 3000	j 1000	k 4000	l 9000

3
a smallest 1500, greatest 2499
b smallest 4500, greatest 5499
c smallest 2500, greatest 3499
d smallest 6500, greatest 7499

4 The smallest number is 500 less, the greatest number is 499 more

5 Individual answers, for example: The smallest numbers end in 500. The 'thousands' digit is always 1 less than the rounded number. Largest numbers end in 499. The 'thousands' digit is always the same as the rounded number.

Answers for Workbook 4 page 14

1 800 and 1250 linked to 1000
2263 and 1790 linked to 2000
2863 and 3125 linked to 3000
3850 and 4275 linked to 4000
4982 and 5459 linked to 5000
6129 and 5823 linked to 6000
6944 and 7095 linked to 7000
8350 (provided as an example) and 7500 linked to 8000
8978 and 9160 linked to 9000

Work with rounded numbers

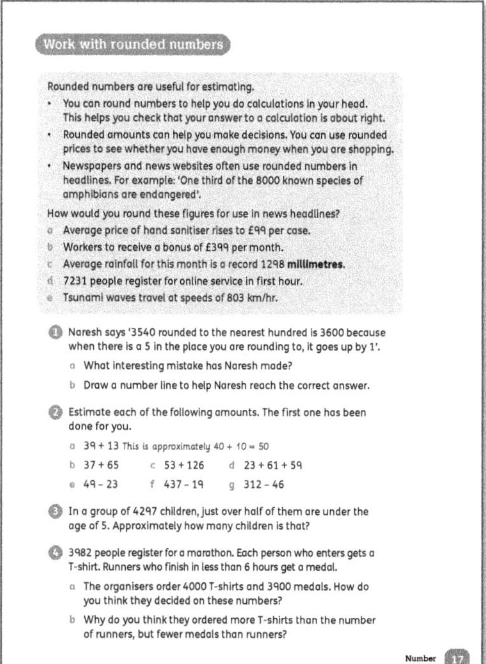

Work with rounded numbers

Rounded numbers are useful for estimating.
- You can round numbers to help you do calculations in your head. This helps you check that your answer to a calculation is about right.
- Rounded amounts can help you make decisions. You can use rounded prices to see whether you have enough money when you are shopping.
- Newspapers and news websites often use rounded numbers in headlines. For example: 'One third of the 8000 known species of amphibians are endangered'.

How would you round these figures for use in news headlines?
a   Average price of hand sanitiser rises to £99 per case.
b   Workers to receive a bonus of £399 per month.
c   Average rainfall for this month is a record 1298 **millimetres**.
d   7231 people register for online service in first hour.
e   Tsunami waves travel at speeds of 803 km/hr.

1. Naresh says '3540 rounded to the nearest hundred is 3600 because when there is a 5 in the place you are rounding to, it goes up by 1'.
   a   What interesting mistake has Naresh made?
   b   Draw a number line to help Naresh reach the correct answer.

2. Estimate each of the following amounts. The first one has been done for you.
   a $39 + 13$ This is approximately $40 + 10 = 50$
   b $37 + 65$ c $53 + 126$ d $23 + 61 + 59$
   e $49 - 23$ f $437 - 19$ g $312 - 46$

3. In a group of 4297 children, just over half of them are under the age of 5. Approximately how many children is that?

4. 3982 people register for a marathon. Each person who enters gets a T-shirt. Runners who finish in less than 6 hours get a medal.
   a   The organisers order 4000 T-shirts and 3900 medals. How do you think they decided on these numbers?
   b   Why do you think they ordered more T-shirts than the number of runners, but fewer medals than runners?

Number   17

Materials

A selection of headlines that use rounded numbers (such as cut-outs from newspapers or magazines, or examples written on slides for display).

Warm-up

Revise estimating as a mental warm-up, for example, using the 'Estimating answers' activity on page 27.

Focus

This lesson deals with how we use rounded numbers in daily life, including for estimation.
- Ask the children to turn to **Pupil Book 4 page 17**. Discuss the different ways we use rounded numbers that are described at the top of the page.
- Let the children give examples to support each use of rounding. For example, to check your answer to 22×9, you could estimate by rounding the 9 to 10 ($22 \times 10 = 220$).
- Explain to the class that often the media use rounded numbers because they are thought to be easier to picture and read so, for example, a newspaper report might say 'about 6000 people were left stranded' when a cruise was cancelled rather than give the exact number of people.
- Encourage the children to think about how they would round the numbers, including which place they would round to. Let them explain their decisions. (Suggested answers: **a** £100, **b** £400, **c** 1300 millimetres, **d** 7200 **e** 800 km/hr)
- Have a class discussion about Naresh's statement in question 1 and let the children think about their answers before they share their ideas.

- For question 2, let the children complete the estimates on their own. They do not need to do the actual calculations, but you can let them do these and compare them with their estimates if you have time.
- Question 3 and question 4 can be done as small-group discussions. The children can share their ideas, reach a shared answer and then feed back to the class.

Interesting mistakes

When children have to work with estimated numbers, they sometimes find it difficult to decide what place they are rounding to and whether or not they have to round all of the numbers in the calculation.

For example, in a calculation with 5827, this number can be rounded to the nearest ten (5830), the nearest hundred (5800), and the nearest thousand (6000). There is a significant difference between each rounded value and this will affect the estimated answer. To avoid confusion when you ask the children to estimate, try to tell them the place they are rounding to and which numbers to round. Explain that in real life we round to the value that is easiest for us to work with but that is also as close to the exact number as possible.

Answers for Pupil Book 4 page 17

1. a   She has used the digit in the hundreds column to round instead of the digit to its right in the tens column.
   b   rounds down to 3500

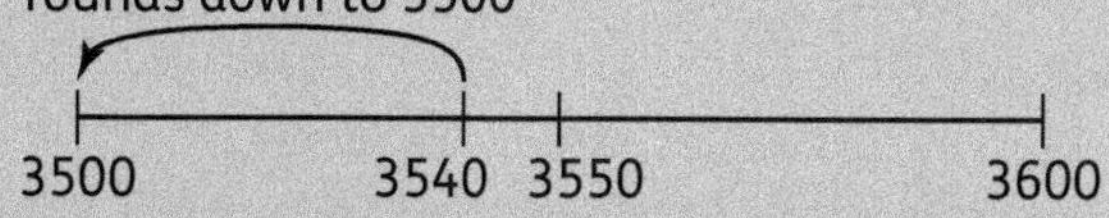

   3500   3540 3550   3600

2. a $40 + 10 = 50$ (provided as an example)
   b $40 + 70 = 110$ c $50 + 130 = 180$
   d $20 + 60 + 60 = 140$ e $50 - 20 = 30$
   f $440 - 20 = 420$ g $310 - 50 = 260$
3. 2150
4. a   Possible answer: They rounded the number of people who registered up to the nearest 1000 or 100. They rounded the number of people who registered down to the nearest 100.
   b   Possible answer: Because everyone who enters gets a T-shirt but not everyone will finish in less than 6 hours.

Roman numerals

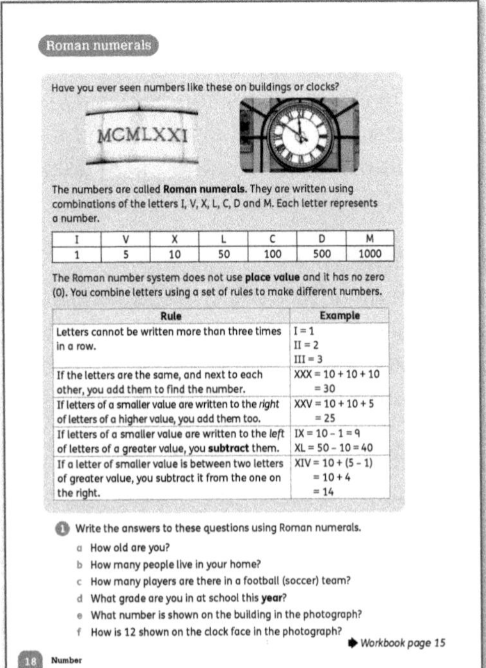

Materials

A real clock or picture of a clock face with Roman numerals; sets of flashcards with the Roman numbers I, V, X, L, C, D and M.

Warm-up

- Show the class a clock face with *Roman numerals*. Let them use their knowledge of clocks and telling the time to work out what the Roman numerals represent.

Focus

- Tell the class that Roman numerals are the system of numbers that the ancient Romans used from about 3000 years ago.
- Remind the class that our number system is called the Hindu–Arabic system and that it uses the digits 0–9 and place value to write numbers. Roman numerals are made by combining seven letters in different ways to write numbers. The Roman numeral system had no zero and it did not use place value.
- Display some numbers from 1 to 12 and let some children come to the board and write the Roman numeral equivalent. Do the same with some Roman numerals and let the children say what number each represents.
- Use the flashcards to show the children how to build numbers using Roman numerals. Start with I, V and X. Explain that they will sometimes need to add and sometimes need to subtract to work out what the number is.
- Give the groups sets of flashcards and let them investigate how to build Roman numerals by applying the rules given on **Pupil Book 4 page 18**.
- Then let them complete question 1. Check that they are able to write each number. As an additional challenge, ask the children to order the numbers from smallest to greatest.

Follow-up

- **Workbook 4 page 15** provides more practice with Roman numerals. You may prefer to leave this until after the next lesson 'More Roman numerals'. The answers to **Workbook 4 page 15** are on page 47.

Answers for Pupil Book 4 page 18

1 **a** and **b** Individual answers

   c XI   **d** Possible answer: IV

   e 1971   **f** XII

More Roman numerals

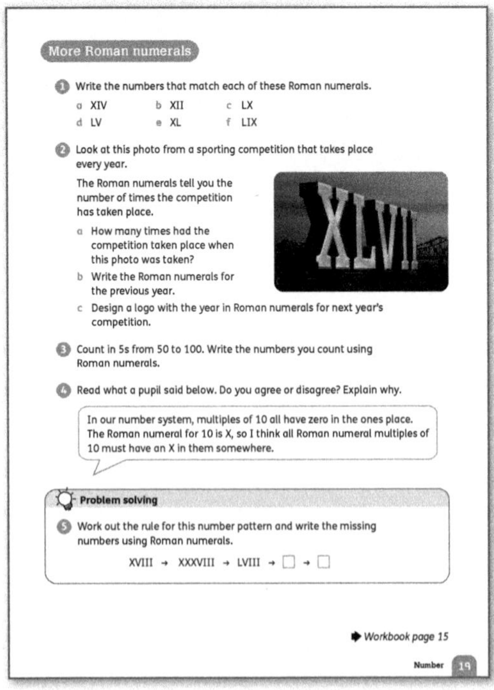

Materials

Drawing equipment and paper (for designing logos); interactive online games to practise Roman numerals.

Warm-up

Draw a clock face on the board with the numbers missing. Ask the children to help you fill in the missing numbers as Roman numerals.

Focus

- Ask the children to turn to **Pupil Book 4 page 19**. Here, the children can apply what they have already learnt to complete more challenging tasks using Roman numerals. No specific teaching is required.
- Let the children write their answers to question 1 and then ask them to find a 1–100 chart of Roman numbers online and use it to check their own answers.
- The Superbowl is an American sporting event that uses Roman numbers to indicate the number of times it has taken place. Let the children work in pairs to answer question 2.
- The children can do question 3 orally in groups before writing the numbers.
- The children can do question 4 orally in groups or together as a class.

- <u>Problem solving</u>: For question 5, remind the children that a number pattern follows a rule, so they have to work out what the rule is before they can work out and write down the missing terms.

Follow-up

Let the children work through **Workbook 4 page 15** as an informal assessment task.

Challenge

The children can use the Internet or other resources to do their own research into how to write large Roman numerals, such as years.

Support

You can find a number of interactive games and printable sheets to consolidate and reinforce work on Roman numerals by searching online.

Interesting mistakes

The children may list the values of each letter separately or they may treat the letters as individual values rather than as a number. For example, they may treat XIX as 10, 1 and 10 rather than as the symbol for 19.

Remind them that they are dealing with letters as digits that make up a number. (Romans only had seven digits, not the ten that we have.) The rules for making the numbers are different because our number system uses base ten place value. Continue to provide support and allow the children to check the rules for making numbers as they work. You could display these in the classroom.

Answers for Pupil Book 4 page 19

1 a 14   b 12   c 60   d 55
    e 40   f 59
2 a 47   b XLVI   c Individual answers
3 L, LV, LX, LXV, LXX, LXXV, LXXX, LXXXV, XC, XCV, C
4 Disagree. Possible answers: multiples of 10 that are also multiples of 50 or 100 will only have L or C in them. L, C, D, M are all multiples of 10 with no X.
5 LXXVIII, XCVIII; The pattern is 'add 20 to get the next number'.

Answers for Workbook 4 page 15

1 II = 2; VII = 5; VIII = 8; X = 10: L = 50; LXX = 70; LXXXIX = 89; XCIX = 99; C = 100
2 IV (4), XI (6), IX (9), XXIV (24), LV (55), XC (90), CX (110)
3 70 = LXX, 75 = LXXV, 80 = LXXX, 85 = LXXXV, 90 = XC, 95 = XCV
4 a IX   b VII   c II

End-of-unit check

Use some or all of these questions to assess how well the children have understood the concepts in this unit.

- *What is the value of each digit in the number 1372? (one thousand, three hundred, seventy, two)*
- *What is the value of the 4 in 1408? (four hundred)*
- *Write 5387 in expanded form. ((5 × 1000) + (3 × 100) + (8 × 10) + (7 × 1) = 5387)*
- *Write these numbers in order from smallest to greatest: 954, 949, 994, 998, 940. (940, 949, 954, 994, 998)*
- *What is 75 rounded to the nearest 10? (80)*
- *What is 444 rounded to the nearest 100? (400)*
- *(Give a number.) Which digit is in the ones/tens/ hundreds place? Does that mean you round it up or down to get to the nearest 10/100/1000?*
- *Write these numbers using Roman numerals: 7, 12, 16, 20, 19, 45, 62, 90, 100. (VII, XII, XVI, XX, XIX, XLV, LXII, XC, C)*
- *(Show some Roman numbers between 1 and 100.) What number does each Roman numeral represent?*
- *Find some Roman numerals that are the same from front to back and back to front. (II, III, XX, XXX)*
- *Naresh says that LXI and XLI are the same number because Roman numerals don't use place value. What mistake has he made? (He has not realised that the order of the letters is important. A smaller value written before a larger value means that the smaller value is subtracted from the larger value. LXI = 61 and XLI = 41)*

Learning objectives

- Compare and classify 2D shapes based on their properties
- Identify and name regular and irregular polygons
- Recognise and name different types of quadrilaterals
- Use Carroll diagrams to sort and classify quadrilaterals
- Investigate properties of squares and rectangles
- Solve problems involving 2D shapes and their properties

Note that symmetry is covered in Unit 8 and angles and triangles are covered in more detail in Unit 11.

Key words

square rectangle pentagon hexagon
octagon triangle circle sides regular
irregular polygon heptagon properties
length angle right angle vertex/vertices
quadrilateral rhombus trapezium
parallelogram kite Carroll diagram

Unit introduction

Materials

Photocopies of a tangram puzzle; scissors.

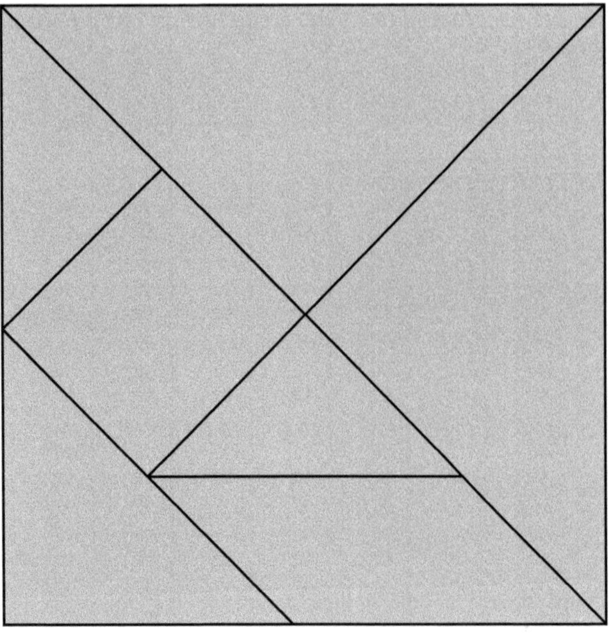

Teaching guidance

- Hand out copies of the tangram puzzle and let the children cut out the shapes accurately. Help the children to do this if necessary.
- Give the children time to play with the tangram.
- Challenge them to try to make the following shapes in more than one way: *square*, *rectangle*, *pentagon*, *hexagon*.
- Allow them to watch each other and to discuss what they are doing.

Revisit 2D shapes

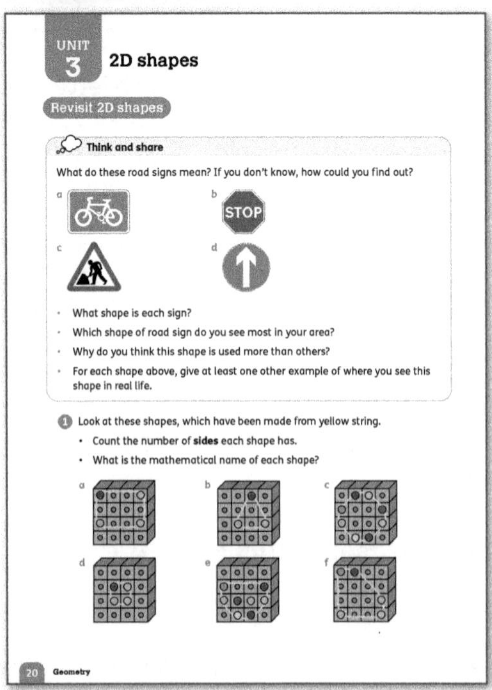

Materials

Slides or pictures of shapes in the environment (including traffic signs); flashcards with the names of shapes (triangle, quadrilateral, pentagon, hexagon, heptagon, octagon, circle); a dictionary or access to online resources to find the names of shapes.

Warm-up

- <u>Think and share</u>: Turn to **Pupil Book 4 page 20** and ask the children to say what each road sign means. If they are not sure, let them suggest ways of finding out (for example, asking adults who drive or finding the information online).
- Bear in mind that different countries may have slightly different road signs.

Focus

- <u>Think and share</u> (continued): Display the flashcards with the names of the shapes and read through these with the class. Ask the children to make a small sketch of each shape if they know what it is, but don't go into detail at this stage.

- Ask different children to say what shape the signs are: rectangle or *quadrilateral, octagon, triangle, circle*. They should remember most of these from last year. If not, revise the shape names.
- Let the children talk in groups about which shapes are most common for traffic signs and why.
- Take feedback and encourage discussion if groups don't agree. In most places, the most common shapes are circles, triangles and rectangles. Stop signs are a different shape on purpose so that drivers can recognise them easily and stop immediately.
- Ask different children to give examples of where they might find these four shapes used in other areas of real life. Examples might include warning symbols on cleaning products or film classification signs.
- Before the children complete question 1 on **Pupil Book 4 page 20**, remind them that geoboards (sometimes called pinboards) have pegs and we can use elastic bands to make different shapes. Let the children work independently to count the *sides* and name the shapes. If they aren't sure of the names, let them look these up (on **Pupil Book 4 page 21**, in a dictionary or online).

Answers for Pupil Book 4 page 20

<u>Think and share:</u>

a a sign to show a cycling lane **b** a stop sign
c a sign to show there are road works ahead
d a sign to show drivers they can drive straight ahead only

- Shapes of the road signs:
  a rectangle **b** octagon **c** triangle **d** circle
- Individual answers to the remaining questions

1 **a** 4 sides; rectangle **b** 3 sides; triangle
  c 8 sides; octagon **d** 4 sides; square
  e 6 sides; hexagon **f** 5 sides; pentagon

Shapes and their properties

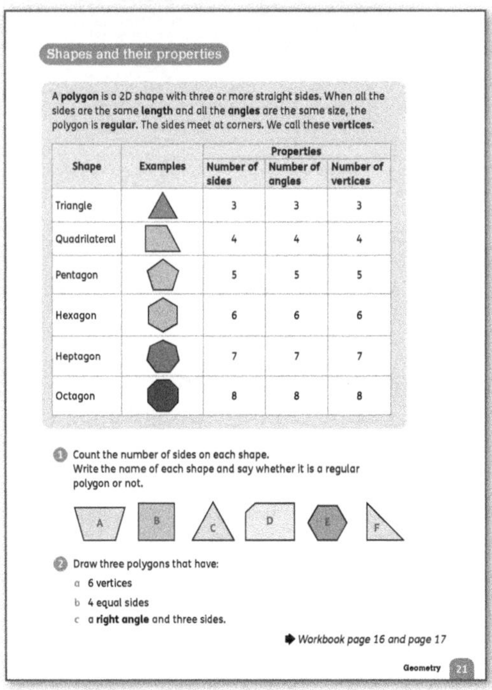

Shape	Examples	Properties		
		Number of sides	Number of angles	Number of vertices
Triangle		3	3	3
Quadrilateral		4	4	4
Pentagon		5	5	5
Hexagon		6	6	6
Heptagon		7	7	7
Octagon		8	8	8

Materials
A selection of card or plastic 2D shapes (including regular and irregular polygons and some circles); set of multiplication and division fact questions related to shapes (optional); a shape wordsearch puzzle (see 'Support').

Warm-up
Choose any suitable 'Calculation skills' activity from pages 27–29. Alternatively, prepare a set of multiplication and division fact questions related to shapes. Ask, for example: *How many sides do 9 triangles have?* (27) *If there are 18 sides, how many triangles are there?* (6) *How many sides are there in 7 octagons?* (56) *There are 55 sides, how many pentagons is this?* (11)

Focus
- Give groups a variety of *regular* and *irregular* 2D shapes. Tell them that they need to decide on different ways of sorting and re-sorting these shapes. Structure this task by providing guideline questions such as:
  - *Can you put the shapes into two groups? What properties did you use to decide where the shapes went?*
  - *Now make three groups. How did you sort the shapes now?*
  - *How else can you sort the shapes?*
  - *I've made two groups: circles and all other shapes. How are the shapes in these two groups different?*
- Allow time for discussion and encourage the children to share their thinking about the shapes and how they have grouped them.
- Use **Pupil Book 4 page 21** to revise the names of *polygons* the children already know and to teach those that they don't know or cannot remember, such as *heptagon*.
- Review the properties we use to identify polygons: the number of sides, angles and *vertices*.
- Ask the children to draw three or four different versions of each shape as you discuss it to consolidate the properties.
- Make sure the children understand that regular polygons have sides of the same *length* and *angles* of the same size. Bear in mind that the square is the only regular quadrilateral. (If a *rhombus* has equal angles, it is a square.)
- Let the children work on their own to complete question 1. Tell them that they can refer to the table if they are not sure of the names or the spelling of the different polygons.
- You may need to remind the children that a *right angle* is a quarter turn (square corner). Tell them to use rulers and pencils to draw the three shapes in question 2. Let them compare and check each other's drawings.

Follow-up
Ask the children to complete **Workbook 4 page 16** to consolidate the names of polygons and to make sure that they can classify regular and irregular polygons.

Set **Workbook 4 page 17** as a homework task or let the children work in pairs to complete the table summarising the properties of shapes. Check they are able to do this before moving on to the next lesson.

Challenge

The children can play a 'Guess the shape' game in pairs. Give each child a plastic or card polygon. They should not show their shape to their partner. Ask them to try to guess which shape their partner has by asking questions about it. The partner may only answer 'Yes' or 'No' to questions. So, they cannot ask: 'How many sides does it have?' but they can ask: 'Does it have more than 3 sides?'

Support

Prepare a shape wordsearch puzzle like the one below with visual clues for the shape words that the children need to find. When they find the name of a shape, they can either write it next to the matching shape or tick the shape.

Interesting mistakes

Some children may be confused by the names of shapes. Make links between the names and the properties of the shapes to help the children remember them. Say, for example, *Tri– means three. A tricycle has three wheels. A triangle has three sides and three angles.* Make sure that you model the correct terms and label large display shapes. Gently correct and encourage the children to use accurate terms to talk about shapes.

Answers for Pupil Book 4 page 21

1 A: 4 sides, quadrilateral, not regular; B: 4 sides, equilateral, regular; C: 3 side, triangle, regular; D: 5 sides, pentagon, not regular; E: 6 sides, hexagon, regular; F: 3 sides, triangle, not regular

2 Individual answers:
   a (any regular or irregular hexagon)
   b (any rhombus or square)
   c (any right-angled triangle)

Answers for Workbook 4 page 16

1 **a** quadrilateral (provided as an example)
   b triangle   **c** quadrilateral   **d** pentagon
   e hexagon   **f** quadrilateral   **g** quadrilateral
   h hexagon   **i** quadrilateral   **j** hexagon
   k triangle   **l** quadrilateral
2 regular polygons: b, d, g, h, i; irregular polygons: a, c, e, f, j, k, l

Answers for Workbook 4 page 17

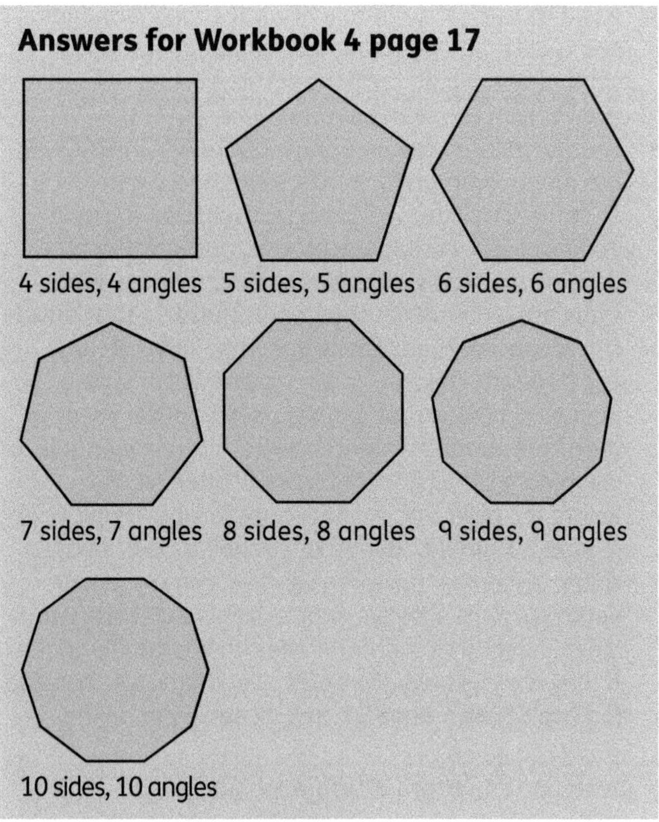

4 sides, 4 angles   5 sides, 5 angles   6 sides, 6 angles

7 sides, 7 angles   8 sides, 8 angles   9 sides, 9 angles

10 sides, 10 angles

Shapes and angles

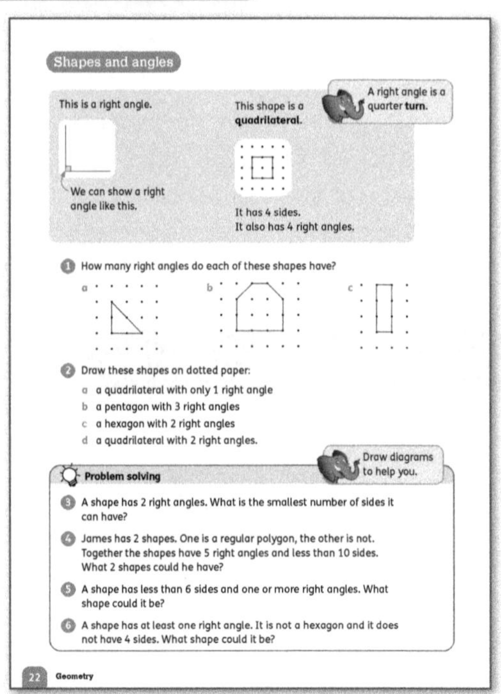

Materials

Dotted paper (see **Pupil Book 4 page 22** question 1); geoboards and elastic bands (if available).

Warm-up

You can repeat the multiplication and division facts activity from the previous lesson (using shapes and numbers of sides). Alternatively, choose any of the 'Place value and number sense' activities on pages 23–27.

Focus

Revise the concept of an angle as a measure of turn.

- Ask a few children to come up and to demonstrate a full, half and quarter turn on the board. Explain that a quarter turn makes a square corner. The term for this is a *right angle*.
- Show the children how to fold a piece of paper twice to make a right angle.
- Let them put four folded pieces together to show that the four quarter turns make a full turn.
- Then ask them to use their folded paper to find examples of right angles in the classroom.
- Turn to **Pupil Book 4 page 22**. Explain that there are angles inside shapes (where the sides meet). Encourage the children to check that the angles are right angles by comparing them with the paper angle they made.
- The children should be able to see the right angles in question 1 clearly, but they can use their angle measure to check.
- For question 2, make sure that everyone has dotted paper, a ruler and a pencil. Let the children work on their own to draw the shapes. Next, let them compare their shapes with a partner or in groups. This is an important step as it allows them to see that:
  - there are several different possible shapes that fit each description
  - the right angles can be in different positions in the shape
  - there can be right angles in shapes of different sizes and with different side lengths.

  Check the shapes that the children have drawn.
- <u>Problem solving</u>: Read through the problems in questions 3-6 with the class before the children work through the questions in pairs.
- The children can use their dotted paper to investigate question 3.
- For question 4, remind the children that they can turn back to **Pupil Book 4 page 21** and look at the table of shapes if necessary.
- Let the children explore question 5 and remind them that some problems have more than one solution.
- The children should realise quite quickly that there are several possible answers to question 6.

Answers for Pupil Book 4 page 22

1 a 1   b 2   c 4

2 Individual answers (drawings of shapes to fit the specified criteria)

3 4 sides

4 Possible answers: a square and a right-angled triangle; an (equilateral) triangle and hexagon with five right angles; a square and a quadrilateral with one right angle; a square and a pentagon with one right angle

5 a right-angled triangle, a quadrilateral with at least one right angle or a pentagon with at least one right angle

6 a right-angled triangle or a pentagon, heptagon or octagon with at least one right angle

Quadrilaterals

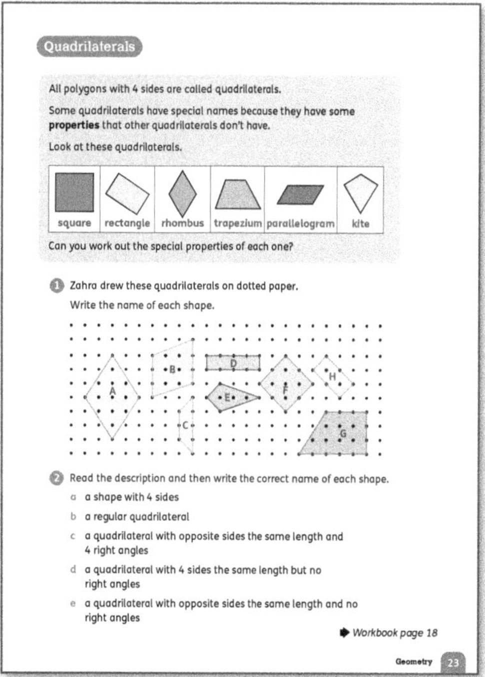

Materials

A selection of card or plastic 2D shapes (including regular and irregular polygons and some circles); angle measure (the right angle that the children made for the questions on **Pupil Book 4 page 22**); geoboards and elastic bands (if available); a set of mixed quadrilaterals and coloured marker pens.

Warm-up

Choose suitable activities from the 'Activity Bank' on pages 23–30 as a warm-up. As there are no calculations in this unit, select activities that reinforce number facts and mental calculation skills.

Focus

- Turn to **Pupil Book 4 page 23**. Explain that *quad-* means four and that the word *quadrilateral* is used to describe all shapes with four sides.
- Refer the children to the pictures of the shapes and read the names aloud, focusing on the new names: rhombus, *trapezium*, *parallelogram* and *kite*. Use the diagrams, but also show them cut-out or plastic shapes in different orientations. The children must be able to recognise shapes in any position or orientation.
- Give the children time to work out what special properties each type of quadrilateral has. Take feedback and correct any misconceptions.
- Go through the names of the shapes again and point out the special properties, for example:
  A square has four equal sides and four right angles.
  A rectangle has opposite sides that are the same length and four right angles.
  A rhombus has four equal sides, but no right angles.
  A trapezium has two parallel sides, but they are not the same length.
  A parallelogram has opposite sides the same length, but no right angles.
  A kite has two pairs of sides that are the same length, these sides are adjacent (next to each other).

- For question 1, the children write the letters A–H and identify the special properties of each shape to name it. Encourage them to measure side lengths if they are not sure whether they are equal and to use their angle measure to check for right angles.
- Let the children work in pairs to read and discuss the descriptions in question 2 before they write down the name of each shape.

Follow-up
Let the children work in pairs to draw 16 different quadrilaterals on **Workbook 4 page 18**. Allow them to use geoboards to model their answers before they draw them if necessary.

Challenge
Give the children a sheet of dotted or squared paper with four identical right-angled triangles drawn, as shown below, or prepare sets of four cardboard triangles for each pair. Note the lengths of the sides next to the right angle if you draw these yourself and make them a different size – the shorter side is half the length of the longer one.

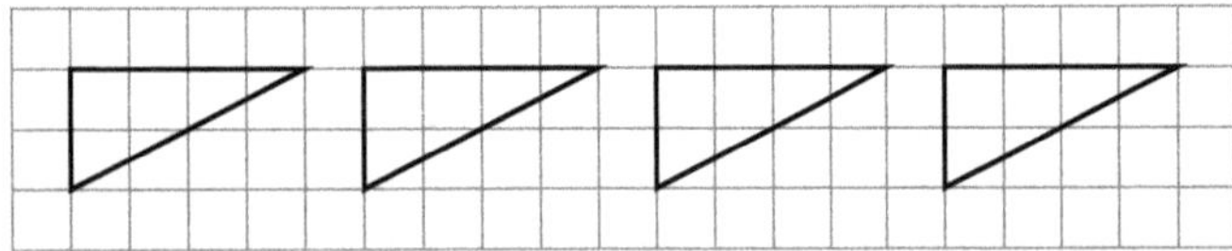

Ask the children to combine the four triangles to make as many different types of quadrilaterals as they can. They should sketch these on the rest of the paper. Note that some triangles will need to be flipped (reflected) to make some of the shapes.

Support
Give the children a set of mixed quadrilaterals either cut-outs or printed on a sheet of paper. Let them use coloured marker pens to mark the equal sides, and let them draw square angle markings to show right angles. They should use their angle measure to test for right angles, and a ruler to check that sides are equal. If they are not able to measure accurately, they can check equal sides by folding or by marking the length against a paper strip and using it to compare the lengths of the other sides.

Interesting mistakes
Children sometimes have difficulty identifying shapes in different orientations. For example, they will have no trouble identifying the first square below, but they may say that the square on the right is a kite or a rhombus (or, incorrectly, a diamond).

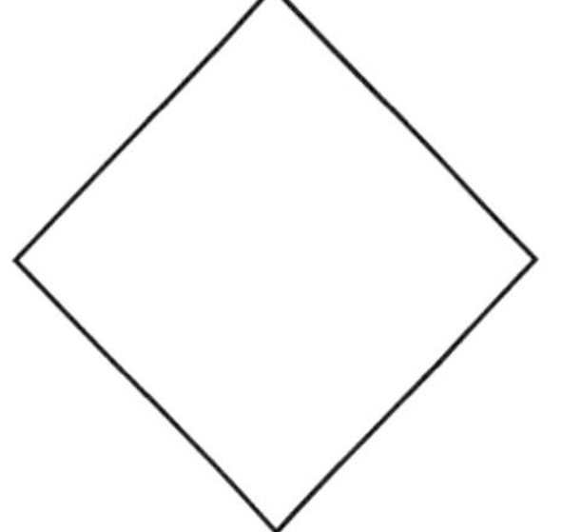

It is important to work with shapes in different orientations. Lots of practice with a geoboard and cut-out shapes will help with this.

Answers for Pupil Book 4 page 23
1 **A** rhombus   **B** parallelogram   **C** trapezium
  D rectangle   **E** kite   **F** square
  G trapezium   **H** rectangle
2 **a** quadrilateral   **b** square   **c** square/rectangle
  d rhombus   **e** rhombus/parallelogram

Answers for Workbook 4 page 18
1–**3** Possible answers:

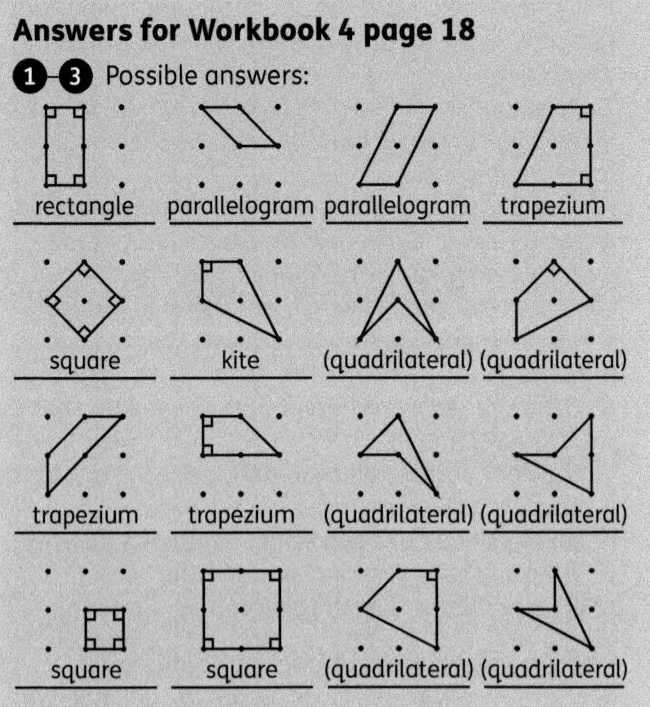

More about quadrilaterals

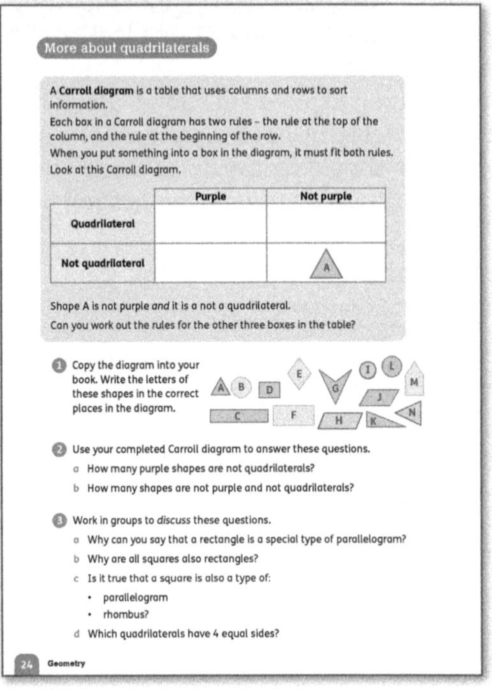

Materials
A large Carroll diagram for display (see 'Warm-up').

Warm-up

As a mental warm-up and to revise *Carroll diagrams* with the class, display a large Carroll diagram like the one below. You can select different criteria to suit the children's abilities and to reinforce different tables.

	Rounds to 20	Does not round to 20
Is a multiple of 3		
Is not a multiple of 3		

Point to different boxes in the Carroll diagram and ask the children to suggest a number that will fit into the box. For example, 18 fits into the top left box because it is a multiple of 3 and it rounds to 20.

Focus

- Read through the information on **Pupil Book 4 page 24** and remind the children that we can only put an item in a box if it fits both rules (the rule at the top of the column and the rule at the beginning of the row).
- Give different children a chance to say what the rules are for the other boxes in the Carroll diagram at the top of the page. These are:
  top left: is purple *and* is a quadrilateral
  top right: is not purple *and* is a quadrilaterial
  bottom left: is purple *and* is not a quadrilateral.
- For question 1, the children copy the Carroll diagram and use it to sort shapes A–N. Let them compare and check each other's diagrams before moving on.
- For question 2, the children have to work out which box of their Carroll diagram each question is referring to.
- Place the children in groups for question 3 and give them time to discuss the questions. Some of the relationships between the shapes might not be clear to all children and you may need to offer extra support.

Answers for Pupil Book 4 page 24

1

	Purple	Not purple
Quadrilateral	H	C, D, E, F, G, J
Not quadrilateral	I, N	A, B, K, L, M

2 a 2   b 5

3 a A parallelogram is a 4-sided shape with opposite sides parallel and equal. A rectangle fits this description, so is a special type of parallelogram that also has 4 right angles.

   b Squares are also rectangles because they are 4-sided shapes with two pairs of parallel opposite sides and four right angles.

   c Both are true. A parallelogram is a 4-sided shape with opposite sides parallel and equal. A square is a parallelogram that also has four right angles and side lengths equal. A rhombus is a quadrilateral with four equal sides. A square is a rhombus that also has four right angles.

   d square and rhombus

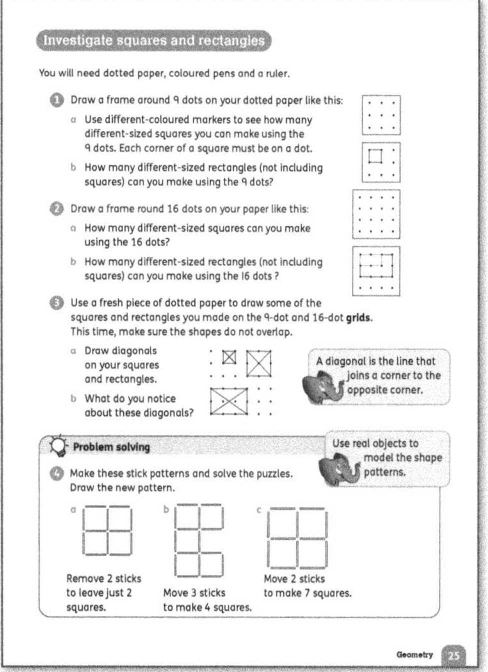

Materials

Dotted paper (see **Pupil Book 4 page 25** question 1); coloured marker pens; toothpicks or other small items of the same length; geoboards and elastic bands (if available).

Warm-up

As a mental warm-up, select one of the 'Rounding' activities on page 26 to reinforce rounding to the nearest 10, 100 or 1000.

Focus

Children often have difficulty with the idea that a square is also a rectangle (a rectangle with equal sides).

- **Pupil Book 4 page 25** provides opportunities for the children to investigate squares and rectangles using dotted grids. It is important to allow time to discuss what the children find out.
- It may be useful to work through question 1 as a class so that the children know what they have to do.
- For question 2, let the children work in pairs to investigate squares and possible rectangles on a 4 × 4 frame.
- Before the children start question 3, remind them or tell them that a diagonal is a line from one vertex to another in a shape. Give them time to draw in some diagonals and then discuss what they notice.
- <u>Problem solving</u>: Provide toothpicks or other items of the same length for question 4 so the children can model these stick patterns.

Support

- Use geoboards and elastic bands as follows to help the children investigate rectangles and squares before they start to draw them.

- Let the children construct rectangles on a geoboard. Ask how they know that their shapes are rectangles. Revise the properties of rectangles: four sides (quadrilateral), opposite sides equal, angles are right angles (square corners).
- Ask the children to construct squares on a geoboard. Discuss the properties of a square: four equal sides, four right angles and so on. Explain that a square is a special type of rectangle.

Challenge

Give the children squared paper and ask them to draw four sets of 4 × 4 square frames:

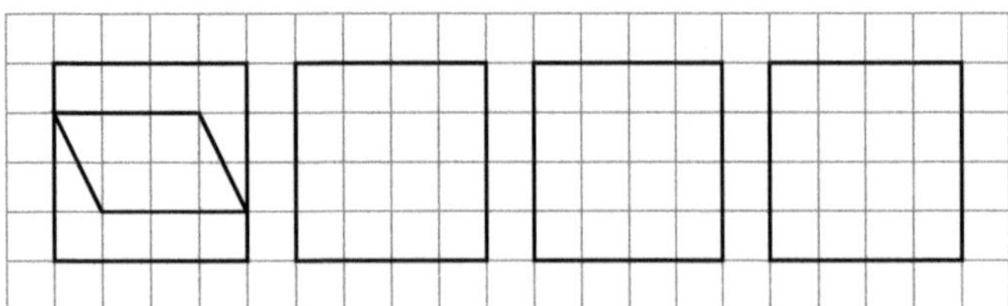

Ask the children to draw a specified quadrilateral in the first 4 × 4 square frame, by saying, for example: *Draw a parallelogram that touches at least two sides of the square frame.* Then ask them to turn the shape (rotate it) by a quarter turn and to draw the shape in its rotated position in the second frame. They repeat this, rotating the shape and drawing it. This helps them to develop ideas about rotational symmetry and to view and draw shapes in different orientations.

Answers for Pupil Book 4 page 25

1 a
  b 1; 1 square by 2 squares

2 a 4
  b 22

3 a Individual answers
  b Possible answer: The diagonals of a square cross at right angles; also, they cross in the middle, making two equal pieces. Diagonals of a rectangle cut each other into equal parts, but they do not meet at right angles. Diagonals of both shapes are the same length. The diagonal of a square is a line of symmetry. This is not the case in rectangles (if the children fold the rectangle, the two parts won't fit onto each other).

4 Possible answers (in any orientation):
  a   b   c

Use some or all of these questions to assess how well the children have understood the concepts in this unit.

- (Say the name of a shape: hexagon, quadrilateral, pentagon, and so on.) *How many sides does this shape have? How many vertices?* (hexagon: 6 sides, 6 vertices; quadrilateral: 4 sides, 4 vertices; pentagon: 5 sides, 5 vertices)
- (Show a set of different polygons.) *How could you group these shapes?* (for example: groups of shapes with the same number of sides; groups of regular and irregular shapes)
- (Show a 2D shape.) *How can you describe this shape?* (for example: the number of sides; the number of angles; the number of equal sides/angles; regular/ irregular)
- (Show some different quadrilaterals.) *What is the correct name for this quadrilateral? What are its properties?*
- Give the children a set of different shapes to sort into a Carroll diagram.

4 Time

Learning objectives

- Read and record time accurately using both analogue and digital clocks
- Write times in different formats, including 24-hour notation
- Understand that units of time are not decimal and know the relationship between them
- Convert hours to minutes, minutes to seconds, years to months and weeks to days
- Solve problems involving times and conversion of time units

Key words

second minute hour analogue digital a.m. p.m. before noon after noon 24-hour notation timetable day week month year leap year

Unit introduction

Materials

Stopwatch (for example, a smartphone app or online stopwatch).

Teaching guidance

- In these activities, the children investigate time in *seconds*. Start by asking the children to find a stopwatch. They can use the stopwatch function on a smartphone or an online stopwatch.
- Discuss how a stopwatch works. (*It counts time in seconds. When you press stop, it tells you how many seconds have passed since you started the stopwatch.*)
- Discuss what the display shows. Different stopwatches may have different displays but most smartphone stopwatch apps show seconds and *minutes* and count in hundredths of seconds.
- Let the children list a few activities that take a short time. If they cannot think of many activities, here are some examples: tying your shoelaces, counting in fives to 100, washing your hands properly (for at least 20 seconds), taking out your maths equipment, walking to the classroom door.
- Once they have listed some activities, ask them to estimate how many seconds they think each activity will take. They should write down their estimates.
- Let the children work in pairs to carry out some of the activities they have listed (as long as it is safe and practical to do so). They can take turns to do each activity while their partner times it in seconds and records the result next to their estimate.

- Point out that to estimate seconds, we can say: *One thousand and one, one thousand and two . . .* with each count taking a second to say. Let the pairs test this, with one child saying the numbers and the other timing for 5 or 10 seconds to see how accurate this counting method is for measuring seconds.

Analogue clocks

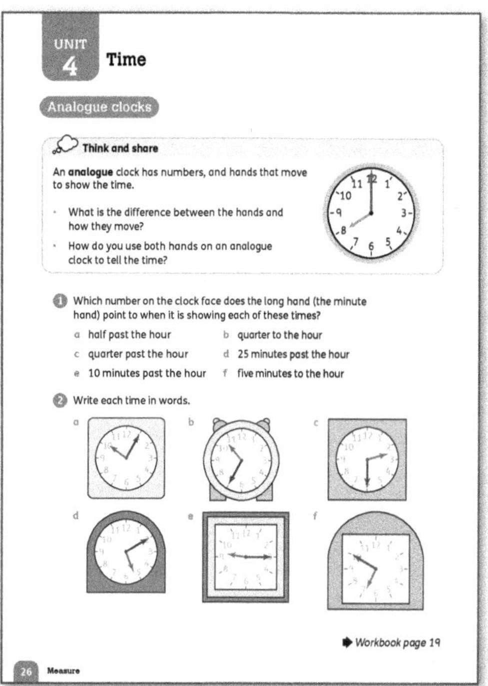

Materials

A large analogue clock face with moveable hands; a real analogue clock that can be adjusted manually; a range of digital clocks (clock displays on smartphones, as well as actual clocks and watches).

Warm-up

- <u>Think and share</u>: Turn to **Pupil Book 4 page 26**. Show the children the picture of a clock. Also show them a real *analogue* clock.
- Give the children time to discuss the questions in groups or pairs and then have a class discussion about how analogue clocks work and what the children use to tell the time.

Focus

- Invite pairs of children to the front of the class. Ask one child to say a time and the other to demonstrate the time on a large clock face. Alternatively, ask one child to set the clock and the other to tell the time.
- Say events of the *day* to the children, such as *lunch time* or *the end of the school day*, and ask them to set the clock face to the time when the event takes place.
- Include events that may be different for different children or on different days, and ask the children to show possible times and explain the decisions they make.

- Let the children answer the questions in question 1 orally, checking on the analogue clock face if they need to.
- Work orally through question 2 and make sure that the children know how to read and say the times before asking them to work independently to write each time in words.

Follow-up

- The children could draw the hands on the analogue clocks on **Workbook 4 page 19**, but you may prefer to leave this until the next lesson, when they can also complete the digital clock times. The answers to **Workbook 4 page 19** are on page 57.

Support

- If necessary, before the children do question 1, revise where the minute hand is on an analogue clock face at different times past and to the *hour*.
- Show the class the clock face at 'five past' and then ask different children to come and move the display to show different positions for the minute hand.
- Focus on times to the hour, such as '20 to' and '10 to', because these are sometimes confusing, especially as most children will be more familiar with *digital* than analogue clocks.

Answers for Pupil Book 4 page 26

<u>Think and share:</u> Possible answer: The short hand (blue) shows the hours and the long hand (red) shows the minutes. The hour hand moves slowly between two hours when the minutes hand make a full turn (beginning at 12 and finishing at 12).
First, I check the hour hand to know the hour. If the hour hand is on a number or after it but before the next number, that is the hour. Then I check the minute hand to check the minutes. For the minutes hand, 12 is 0 minutes, 1 is 5 minutes, 2 is 10 minutes, 3 is 15 minutes, 4 is 20 minutes, 5 is 25 minutes, 6 is 30 minutes, 7 is 35 minutes, 8 is 40 minutes, 9 is 45 minutes, 10 is 50 minutes, 11 is 55 minutes and 12 is 60 minutes.

1 a 6   b 9
   c 3   d 5
   e 2   f 11

2 a five past ten   b twenty-five to eleven
   c half past two   d ten past five
   e quarter past nine   f ten to seven

Digital clocks

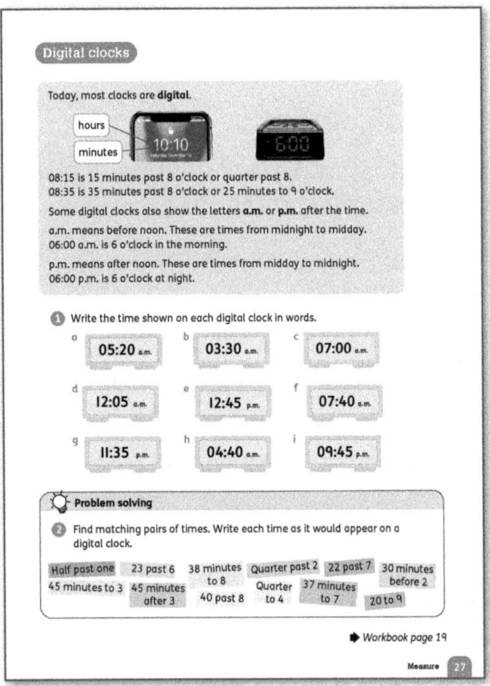

Materials

A range of digital clocks (clock displays on smartphones, as well as actual clocks and watches); blank pieces of card; paper plates and split pins.

Warm-up

Ask the children to mentally work out durations in the day by asking questions such as: *What time is it now? How much longer is it until lunch time?* They can give the answers in minutes or mixed units such as 1 hour 15 minutes, or one and a quarter hours.

Focus

The children should be familiar with digital clock faces, as they will see and use digital clocks in many different ways.

- Read through the information on **Pupil Book 4 page 27** with the class. Focus on the different way we say and read the time on a digital clock. It is important that children understand that, when we read a time such as 8:35 on a digital clock, we say the hour and the minutes past the hour. Using an analogue clock, we would say 25 to 9.
- Make sure that the children can move between the two ways of saying the same time.
- Remind the class that in 12-hour notation we use the terms *a.m.* and *p.m.* to distinguish between times before and after noon. The abbreviations are short for the Latin terms for *ante meridiem (before noon)* and *post meridiem (after noon)*. It is generally acceptable to write the abbreviations with or without full stops.
- Let the children do question 1 on their own. Check that everyone is able to read and write the times.

- <u>Problem solving</u>: For question 2, the children can write the times on pieces of card and physically move them around to sort them. Once they have matched the pairs of times, they should write the equivalent digital time for each pair in their books.

Follow-up
Use **Workbook 4 page 19** to assess whether the children can work with both analogue and digital clocks to show and write times.

Challenge
As a fun challenge, ask the children to design a clock face where the numbers 1 to 12 are shown as mathematical operations. For example, 1 could be 5 – 4. Challenge them to make each calculation as difficult as they can. Let them compare and check each other's calculations, using a calculator if they need to.

Support
Let the children make a paper plate clock face with cardboard hands attached with a split pin that they can use to check and model times on an analogue clock. If it helps, they can write the times past and to the hour next to the numbers, for example, '5 past' next to 1, 'quarter to' next to 9, '10 to' next to 10.

Interesting mistakes
Many children find it very difficult to tell the time because the numbers on a clock face have different 'values', depending on whether we are looking at the hour hand or minute hand. The children need to experiment with what happens as the hands move around a clock face. The children need to tell the time regularly, as well as work out how long it is since something happened or how long until something is going to happen.

Answers for Pupil Book 4 page 27
1
  a twenty minutes past five in the morning
  b half past three in the morning
  c seven o'clock in the morning
  d five minutes past twelve in the morning
  e quarter to one in the afternoon
  f twenty minutes to eight in the morning
  g twenty-five minutes to twelve at night
  h twenty minutes to five in the morning
  i quarter to ten at night
2 half past one, 30 minutes before 2 01:30
    quarter past 2, 45 minutes to 3 02:15
    quarter to 4, 45 minutes after 3 03:45
    38 minutes to 8, 22 past 7 07:22
    40 past 8, 20 to 9 08:40
    23 past 6, 37 minutes to 7 6:23

Answers for Workbook 4 page 19
1 hands drawn to show three o'clock; 03:00
2 hands drawn to show half past one; half past one
3 04:45; quarter to five
4 hands drawn to show twenty minutes past eleven; 11:20
5 hands drawn to show to show quarter past three; quarter past three
6 twenty to seven; 06:40
7 hands drawn to show ten minutes past eight; 08:10
8 hands drawn to show twelve o'clock; 12:00

24-hour notation

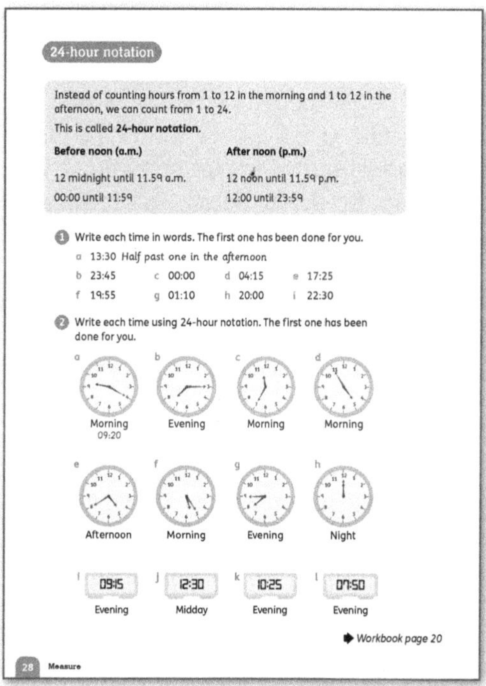

Materials
Several different local timetables (such as bus, metro, train timetables, TV schedules), including timetables that use 24-hour notation; paper plates.

Warm-up
- Ask the children how many hours there are from midnight to the next midnight. Introduce the idea of a 24-hour clock, with the day beginning and ending at 00:00 (24:00).
- Discuss with the class where they are likely to see times in *24-hour notation*. Examples include timetables, displays at bus stations, train stations and airports and digital displays on appliances. Give the children time to look at the timetables that you have brought to the lesson and see what they notice about the times.

Focus

- Turn to **Pupil Book 4 page 28** and explain that the 24-hour clock is only different from the 12-hour clock for times after 12 noon. It may help to draw a timeline showing hours marked in 12-hour format from 1 a.m. to 12 p.m. and then write the 24-hour format times below these in the matching positions.
- Talk about how we read and say times in the 24-hour format. Encourage the children to read a time such as 19:00 as 'nineteen oh oh' or 'nineteen zero zero' rather than '19 hundred hours' to prevent them from getting the misleading idea that time is decimal and that there are 100 minutes in an hour.
- Explain that all 24-hour times are written as four digits. The children should put a zero to the left of times between midnight and 9 a.m., for example, 00:00 and 09:52.
- Discuss how you can convert times by adding or subtracting 12. Start with an example such as 17:00. Explain that this is 5 p.m. Ask the children to work out the difference between the times (*12*).
- Repeat this with a few different times and ask the class to suggest a method for changing a 24-hour time to a p.m. time. Make sure they realise that the times from 00:00 to 12:00 do not need converting. 11:30 a.m. in 24-hour notation is 11:30.
- Repeat this for addition, explaining that 4:00 p.m. is 16:00 and, again, look at the difference (*12*). The children should be able to say that you can change between p.m. time and the 24-hour equivalents by adding 12.
- The children read the times in question 1 and then write the times in words. They can do this independently or in pairs.
- Make sure that the children read the time of day below each clock face in question 2 to work out whether it is before or after noon.

Follow-up

Use **Workbook 4 page 20** to consolidate converting between 12-hour and 24-hour notation, and reading and saying time in words.

Support

As a practical activity, give the children a paper plate and let them design a clock-face showing 24-hour times. They can use this to help them convert between times.

Interesting mistakes

A common error when using 24-hour times is to mistake the second digit for the hour time. For example, some children may think 14:00 is 4 o'clock and that 15:00 is 5 o'clock. Emphasise counting on from 12. For children who have grasped this quickly, you could explain that our time system is a base-12 rather than a base-10 system. This is why we count hours from 0 to 12, then 13 represents 12 + 1, 14 represents 12 + 2 and so on.

Answers for Pupil Book 4 page 28

1 a half past one in the afternoon (provided as an example)
   b quarter to twelve at night
   c midnight
   d quarter past four in the morning
   e twenty-five minutes past five in the afternoon
   f five minutes to eight in the evening
   g ten minutes past one in the morning
   h eight o'clock in the evening
   i half past ten at night

2 a 09:20 b 19:15 c 10:35 d 03:55
   e 16:40 f 05:25 g 19:45 h 00:00
   i 21:15 j 12:30 k 22:25 l 19:50

Answers for Workbook 4 page 20

1 hands drawn to show half past eight, 08:30 (provided as an example); hands drawn to show three o'clock; hands drawn to show quarter to eight, 19:45; hands drawn to show seven o'clock, 07:00; hands drawn to show half past ten, 22:30; hands drawn to show quarter past four; 16:15; hands drawn to show quarter to one, 12:45; hands drawn to show twenty-five to five, 16:35; hands drawn to five minutes to eleven, 22:55

Timetables

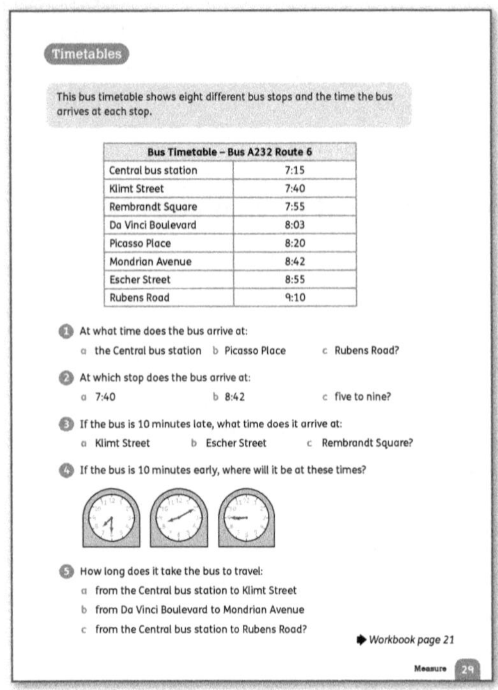

Materials

Timeline (see 'Support').

A timetable is a table with one or more lists of times with some details of what happens at each time. The children have probably already worked with class timetables and television schedules so the concept should not be unfamiliar. You will need to teach them that they can use the times on the timetables to work out time intervals (how much time passes from one point to another) and the duration of events (how long something takes). It is always useful to relate this work to everyday timetables that the children may use, such as bus or train timetables, tide tables and television schedules.

Warm-up

Let the children develop a *timetable* of their own to show what they do on a typical day at the weekend. Before they start, explain that a timetable shows the time at which an event starts. It does not give the duration of the event, but you can work out how long something takes by looking at when the next event starts.

Focus

- Use the bus timetable on **Pupil Book 4 page 29** to make sure that the children understand how to read a simple timetable.
- Read through the questions with the class, discussing how they would find the answers. Then let the children complete the questions in their books.
- For question 1, the children read the times from the timetable and record them.
- For question 2, make sure that the children realise they have to find the times (and they may look different) and then link each to the correct stop.
- The children need to add 10 minutes to each time to work out question 3.
- For question 4, the children need to read the analogue clock time and then work out where the bus might be (given that it is 10 minutes early).
- The children can draw number lines and draw jumps on them to answer the questions in question 5.

Follow-up

Let the children work in pairs or small groups to complete **Workbook 4 page 21**.

Challenge

Ask the children to do some research into the bus stop names on **Pupil Book 4 page 29**. (The streets are named after well-known artists.) Encourage them to choose one or two of the artists and to find out more about them and how they used mathematics in their art. Mondrian and Escher are particularly good examples.

Support

Provide a timeline like the one below on which the children can model the movement of a bus or train to show how the timetable works. This will also help them to see where the bus or train is at any time.

Central Bus Station

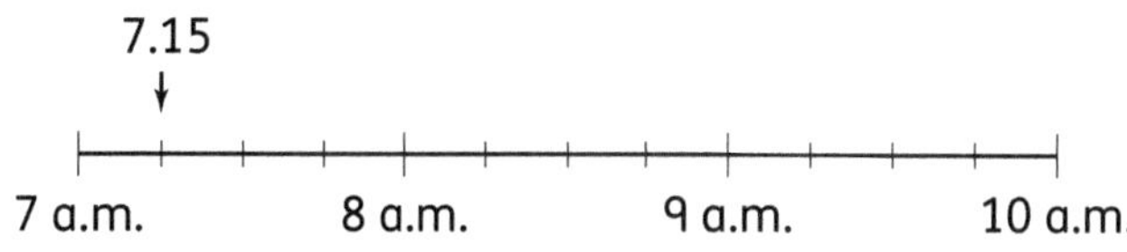

Answers for Pupil Book 4 page 29

1 **a** 7:15   **b** 8:20   **c** 9:10
2 **a** Klimt Street   **b** Mondrian Avenue
  c Escher Street
3 **a** 7:50   **b** 9:05   **c** 8:05
4 Klimt Street, Picasso Place, Escher Street
5 **a** 25 minutes   **b** 39 minutes
  c 1 hour and 55 minutes

Answers for Workbook 4 page 21

1 **a** Maluri   **b** Kajang
  c Muzium Negara   **d** Pasar Seni
2 (Sungai Buoh) 07:40 (provided as an example); (Surian) 9:25; (Muzium Negara) 11:00; (Pasar Seni) 12:15; (Merdeka) 12:30; (Bukit Bintang) 13:00; (Maluri) 14:00; (Kajang) 14:45
3 7 hours and 5 minutes

Reading timetables

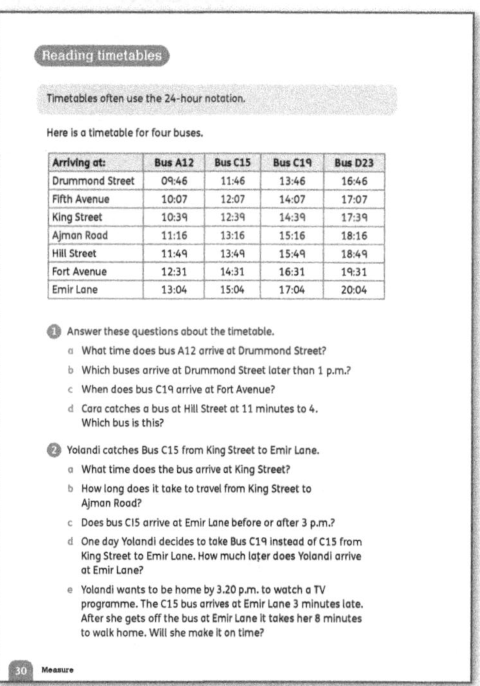

Arriving at:	Bus A12	Bus C15	Bus C19	Bus D23
Drummond Street	09:46	11:46	13:46	16:46
Fifth Avenue	10:07	12:07	14:07	17:07
King Street	10:39	12:39	14:39	17:39
Ajman Road	11:16	13:16	15:16	18:16
Hill Street	11:49	13:49	15:49	18:49
Fort Avenue	12:31	14:31	16:31	19:31
Emir Lane	13:04	15:04	17:04	20:04

Materials
Several different local timetables (such as bus, metro, train timetables, TV schedules), including timetables that use 24-hour notation; number line.

Warm-up
Use the 'Timetables' activity on page 30 as a warm-up, using the timetable given or a local timetable.

Focus
- Discuss the bus timetable on **Pupil Book 4 page 30** with the class and display this timeline on the board:

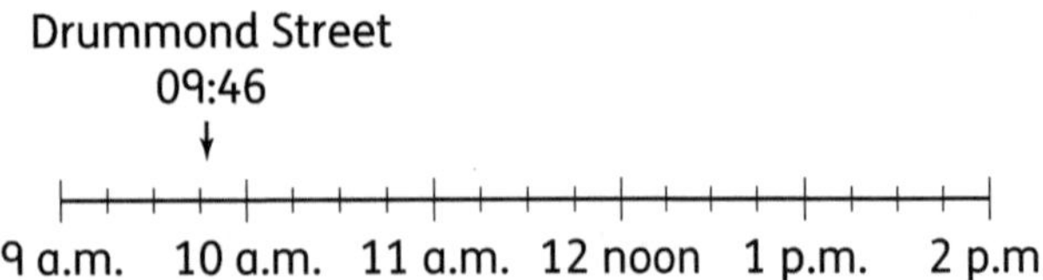

Explain that the timeline represents the journey of bus A12. Discuss where to place the other stops on the line.
- Draw a similar line from 11 a.m. to 4 p.m. and ask the children to place the stops of bus C15 to make sure that they can make sense of the 24-hour notation on the timetable.
- The children can work in pairs to answer question 1 orally. Check their answers as a class before moving on.
- Let the children work independently to complete question 2.

Support
Some children find questions such as 'Which buses arrive at Drummond Street later than 1 p.m.?' difficult because the time in the question is in a different format to the times in the timetable. Support children to convert the 12-hour clock times to 24-hour clock times before attempting such questions.

Using a number line and saying the 24-hour clock times times aloud will help children to see which times are earlier and which are later than the given time.

Answers for Pupil Book 4 page 30
1 a 09:46   b C19 and D23   c 16:31
   d C19
2 a 12:39   b 37 minutes   c after   d 2 hours
   e Yes, she arrives home at 15:15 or 3:15 p.m.

Converting units of time

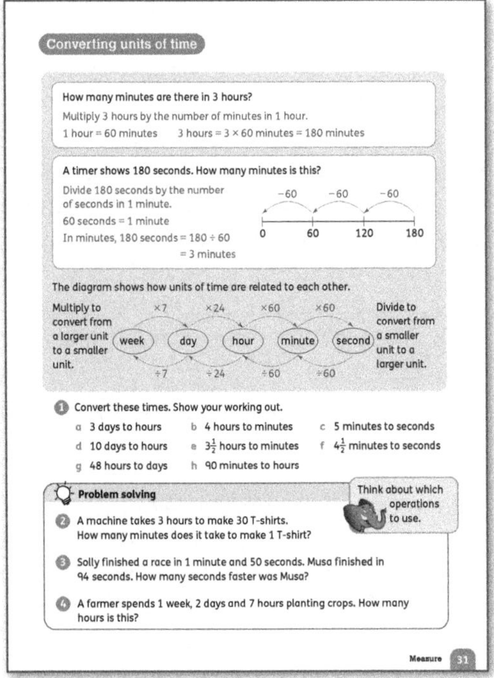

Materials
A set of flashcards with pairs of equivalent times in different units for the children to match up; calculators.

You could use the times shown below on the flashcards, or adapt the times as necessary to make the level of difficulty appropriate for your class.

1 hour	90 seconds	90 minutes	2 days	1½ hours	240 minutes
4 hours	120 seconds	48 hours	1½ minutes	60 minutes	2 minutes

Warm-up
As a mental warm-up, select any of the 'Calculation skills' activities on pages 28–29.

Focus
- Start by asking the class: *What are units of time?* List the children's ideas on the board.
- Explain that you are going to focus on some units we use every day, and learn how to convert from one unit to another.
- Place the children in pairs to read through the examples on **Pupil Book 4 page 31**. Let them talk about how to convert between units of time and ask a few children to explain their thinking using an example they have made up.
- Ask questions to check that the children know the relationship between the units of time. For example: *How many days are there in a week? How many hours are there in a day? How many seconds are there in a minute? How many minutes are there in an hour? How many minutes are there in half an hour?*

- Display the prepared flashcards that show different units. Let the children take turns to make equivalent pairs (or sets).
- Turn to the diagram on **Pupil Book 4 page 31**, which shows how units of time are related. It is important that the children understand the relationships and that they don't just try to memorise rules for converting units. Stress that units of time are not decimal like other metric units (that is, we do not convert between them by multiplying or dividing by 10, 100, 1000).
- Let the children work in pairs to complete question 1. Encourage them to check their answers using a calculator. If the children are not able to do the multiplication or division needed, they can find the answers using a number line and repeated addition or subtraction. Check the answers before moving on to the problem-solving questions.
- Problem solving: Use question 2 to show the children that drawing a bar model (see page 22) can help them to work out what operation they need to solve the problem. For example:

3 h	60 min	60 min	60 min
30 T-shirts	10 Ts	10 Ts	10 Ts

$60 \div 10 = 6$

6 min per T-shirt

- Make sure the children understand that if they want an answer in minutes they can convert hours to minutes before they start.
- The children can solve question 3 on their own. Remind them to draw a bar model if they need to.
- For question 4, remind the children that different units require different conversions.

Challenge

Give the children sets of three different times and ask them to order these from shortest to longest. Mix units to provide the challenge. For example:

a 15 minutes and 205 seconds, 13 minutes and 300 seconds, 18 minutes and 40 seconds
b 705 seconds, 11 minutes, 9 minutes and 145 seconds

Support

If the children find the conversions in the Pupil Book difficult, allow them to use a calculator to do the calculations. The focus here is on how to convert the units, rather than calculation skills.

Answers for Pupil Book 4 page 31

1 a 72 hours b 240 minutes c 300 seconds
  d 240 hours e 210 minutes f 270 seconds
  g 2 days h 1.5 hours
2 6 minutes
3 16 seconds
4 223 hours

More time conversions

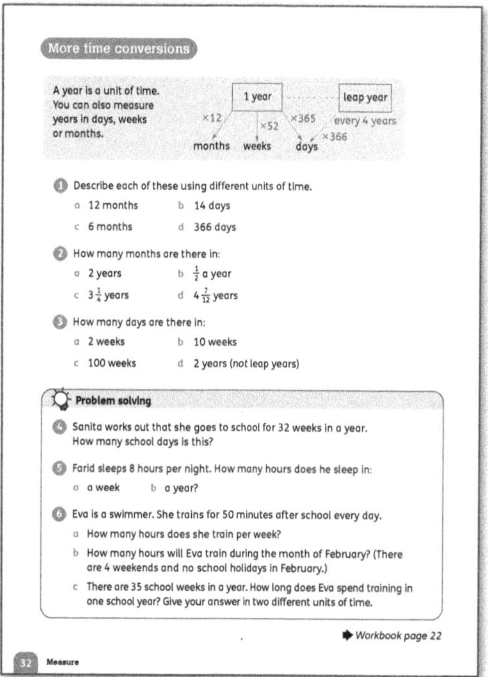

Materials

Calculators; calendar.

Warm-up

Start the lesson with a mental warm-up selected from the 'Place value and number sense' activities on pages 23–27.

Focus

- Turn to **Pupil Book 4 page 32** and ask the children to consider the diagram and what they can learn from it about the different units. Ask different children to share some of their ideas.
- Introduce the concept of a *leap year* if the children don't know what this is. Every fourth *year* an extra day is added to February and that year has 366 days. This is because the Earth actually takes $365\frac{1}{4}$ days to go around the Sun and the four quarters are added up to make the extra whole day.
- Discuss whether this year is a leap year and, if not, when the next one will be.
- The children can do question 1 orally in groups. Let them share their ideas with the class. The children might give correct answers that you don't expect. For example, 12 *months* could be 1 year, 52 weeks or 365 days.
- Encourage the children to draw number lines or bar models (see page 22) if they need help with the conversions from years to months in question 2.
- The children can use calculators to complete question 3 if they need to.
- Check the children's work and make sure that they are able to relate units and convert between them before asking them to complete the problem-solving questions in pairs or small groups.

- Problem solving: Have a calendar available to help the children with the conversion in question 4. You may wish to show the children how to draw a bar model to help them see how many school days there are in a week and to work out that they need to multiply that number by 32 to solve the problem.
- The children should show their working for questions 5 and 6.

Follow-up
Use **Workbook 4 page 22** to consolidate understanding of the relationships between units of time and to check that the children can convert between units.

Challenge
Provide the children with a number of problems involving days, weeks and years as well as hours, minutes and seconds. For example:

- *School started at 07:30. The school day was 270 minutes long. What time did school finish? (12:00)*
- *I started training for a race 3 days and 18 hours ago. When was that?*
- *What was the day and time 50 hours ago?*
- *I played computer games for 7860 seconds. I started playing at 18:08. What time did I finish? (20:19)*
- *A whale was first tracked on radar at 06:00 on 3rd September. The next time it was tracked it was 180 hours later. What were the date and time? (10th September, 18:00)*

Support
As in the previous lesson, if the children find the calculations difficult, allow them to use a calculator to do the conversions. The focus here is on how to convert the units, rather than calculation skills.

Interesting mistakes
Some children may treat time like a decimal quantity and may think that there are 100 minutes in an hour (especially if they learn 24-hour times as '14 hundred hours' and so on). Explain that time units were developed a long time before metric measurements and display charts to remind the children what the relationships are.

Answers for Pupil Book 4 page 32

1   **a** one year   **b** two weeks
    c 26 weeks, half a year
    d leap year, a year and a day

2   **a** 24   **b** 6   **c** 39   **d** 55

3   **a** 14   **b** 70   **c** 700   **d** 730

4   160 days

5   **a** 56 hours   **b** 2920

6   **a** 250   **b** 1000
    c 8750 minutes or 145 hours and 50 minutes

Answers for Workbook 4 page 22

1 day = a period of 24 hours (provided as an example); year = 12 months ; second = $\frac{1}{60}$ of a minute; leap year = 365 days; week = 7 days; 1 minute = 60 seconds

2 48 hours = 2 days; 3.5 days = 84 hours; 730 days = 24 months; 3 weeks = 21 days; 36 months = 3 years; 300 minutes = 5 hours; 6 weeks = 42 days; 120 seconds = 2 minutes; 49 days = 7 weeks

3   **a** Possible answer: $60 \times 24 = 1440$ minutes
    b Possible answer: $60 \times 60 = 3600$ seconds
    c Possible answer: $60 \times 60 \times 20 \times 4 = 86\,400$ seconds
    d Possible answer: $420 \div 60 = 7$ minutes

End-of-unit check

Use all or some of these questions and activities to assess how well the children have understood the concepts in this unit.

- Show times on analogue and digital clocks and ask: *What time does this clock show?*
- Give the child a clock with moveable hands and ask them to show some different times. For example: *Show 10:53 on this clock.*
- *What is quarter to 3 in digital time? (2:45)*
- *A plane arrives at the airport at 05:45. Is that a.m. or p.m.? (a.m.)*
- *A nurse finishes work at 19:30. What time is that on an analogue clock? (half past 7)*
- Show a TV guide. *What time does the news start? What is on TV at 17:30?*
- Ask the children to make up a time story. (*For example: Ayeisha left her house at 12:34 p.m. She reached the bus stop at 1.00 p.m., got off the bus at 1:23 p.m. and arrived at Sally's house at 1:56 p.m. It took her 1 hour 22 minutes to reach Sally's house.*)
- Use the bus or metro timetables from the Pupil Book or Workbook. Show a time on an analogue clock face and ask the children to say where the bus or train is at that time.
- As a more challenging question, show a time that is between two of the times on the timetable.
- *Which is longer: 150 seconds or $3\frac{1}{2}$ minutes? ($3\frac{1}{2}$ minutes)*
- *What is a leap year? (a year with 366 days)*
- *What do people mean when they say 'just over three weeks'? (one or two days more than three weeks, so 22 or 23 days)*
- *How many days could '3 to 4 weeks' be? (any number of days between 21 and 28)*
- *How many years is 36 months? (3 years)*

UNIT 5 Decimals

Learning objectives

- Use decimal notation for tenths and hundredths
- Extend place-value tables to include tenths and hundredths and recognise the value of digits in decimal fractions
- Locate decimals on a number line
- Count forward and backwards in tenths and hundredths
- Compare and order decimals
- Round decimals to the nearest whole number (ones)
- Solve a range of problems involving decimal amounts, including money and measurements

Key words

fraction   tenths   hundredths   decimal
decimal point   counting sequence   round
whole number   kilometre   approximately

Unit introduction

Materials

Food and drink containers labelled with different masses and capacities; tape measures; coins and notes.

Teaching guidance

This is the first time the children are formally introduced to decimal notation and the concept of extending the place-value table to include *fractions* (*tenths* and *hundredths*). This foundational learning is very important as the children will work with *decimals* to increasing degrees as they move up the school levels. At this stage, the concepts are taught and then applied in the context of money and measurements so that the children get a good sense of how and why decimal fractions are used. Children usually have a fairly well-developed concept of money and how it works (no matter where they live or what currency they use) but they may not understand that a money amount such as £1.50 means the same as 150 pence or one-and-a-half pounds.

- Use the 'Making a litre' activity on page 28 as a warm-up.
- Have a number talk (see pages 17–18) using decimal amounts of mass, capacity, length and money.
- Show the children some food and drinks containers of different masses and capacities. Discuss what it means if a bottle says it holds 1.5 litres, and how much 0.5 kg is.
- Hand out tape measures and let the children measure their own heights in centimetres.
- Point out that 1 metre is divided into 100 centimetres, so each centimetre is $\frac{1}{100}$ of a metre. Discuss how to write heights as decimal numbers in metres. Give the children time to talk about this and to share their suggestions.
- Finally, talk about money. Choose a price or amount such as £5.99. Demonstrate that this is 5 pounds (ones), 9 ten-pence coins and 9 pence. Use different amounts and ask the class to say how many whole pounds are involved and what fraction of a pound is given after the decimal place (in hundredths).
- Look at amounts such as £1.50. Stress that this is 1 pound and $\frac{50}{100}$ of a pound. When we work with money, it is a 'rule' (convention) to write the 0 on the end if there are no pence in the hundredths place. This is mainly so that amounts of money can be aligned neatly in columns, which makes adding or subtracting them much easier.
- Ask the class how to write one pound and five pence. Stress that 5p is $\frac{5}{100}$ of a pound. They should begin to realise that we write 0 as a place holder in the tenths column to show that there are no tenths because the pence amounts are written as 2-digit figures.

Decimals are fractions

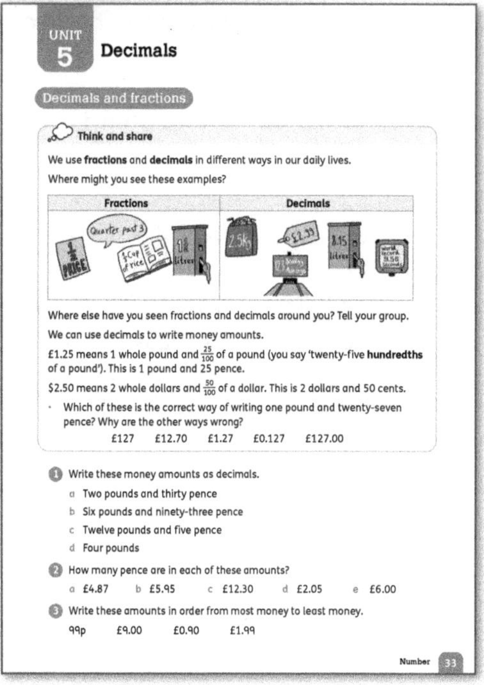

Materials

Food packaging with examples of decimal quantities; photographs showing decimals in the local environment (for example, masses on vehicles, money, road distances); flashcards showing different decimal amounts; cut-out cardboard coins (pounds and pence).

Warm-up

- <u>Think and share</u>: Look at the examples of fractions and *decimals* in the table on **Pupil Book 4 page 33**. Invite the children to say where they might see these fractions and decimals. Ask them to think of other places where they might see fractions and decimals. Let them share their ideas in groups and write a list.
- Take feedback. Then show the class the examples of decimals on food packaging and in the photographs you have brought. Ask the children to add these examples to their lists.

Focus

- Remind the children that they have talked about money as a decimal and let them consider the question in the 'Think and share' section. Once you have established that £1.27 is correct, ask different children to say what is wrong with the other ways of writing the amount.
- Let the children discuss question 1 if they need to before writing the amounts.
- If the children find question 2 difficult, encourage them to circle the pence in each amount or to draw diagrams to represent the pounds and pence. For example:

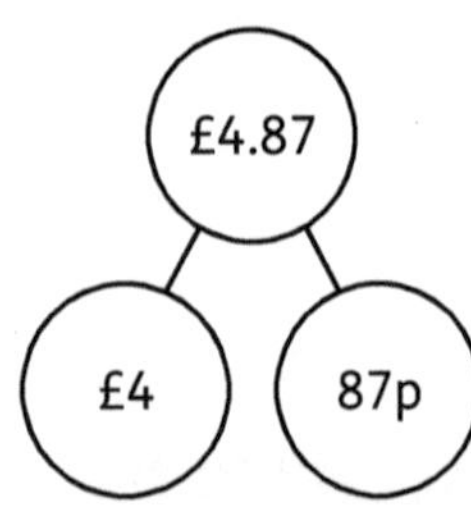

- If the children cannot order the written amounts of money in question 3, allow them to use coins to model the amounts.

Answers for Pupil Book 4 page 33

<u>Think and share</u>: Individual answers
£1.27. The others are wrong because the decimal point is in the wrong position or there is no decimal point to show one pound and twenty-even pence.

1   a £2.30   b £6.93   c £12.05   d £4.50

2   a 487   b 595   c 1230   d 205
   e 600

3 £0.90, 99p, £1.99, £9.00

Decimal place value

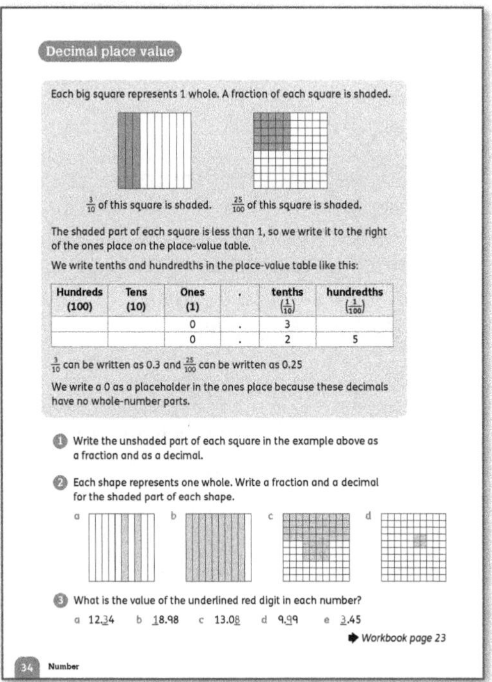

Decimal place value

Each big square represents 1 whole. A fraction of each square is shaded.

$\frac{3}{10}$ of this square is shaded. $\frac{25}{100}$ of this square is shaded.

The shaded part of each square is less than 1, so we write it to the right of the ones place on the place-value table.

We write tenths and hundredths in the place-value table like this:

Hundreds (100)	Tens (10)	Ones (1)	.	tenths ($\frac{1}{10}$)	hundredths ($\frac{1}{100}$)
		0	.	3	
		0	.	2	5

$\frac{3}{10}$ can be written as 0.3 and $\frac{25}{100}$ can be written as 0.25

We write a 0 as a placeholder in the ones place because these decimals have no whole-number parts.

1 Write the unshaded part of each square in the example above as a fraction and as a decimal.

2 Each shape represents one whole. Write a fraction and a decimal for the shaded part of each shape.
   a   b   c   d

3 What is the value of the underlined red digit in each number?
   a 12.<u>3</u>4   b <u>1</u>8.98   c 13.0<u>8</u>   d 9.<u>9</u>9   e <u>3</u>.45

➡ *Workbook page 23*

Materials

Interlocking cubes or base-ten blocks (see pages 20–21); large place-value table that includes tenths and hundredths (see page 22); place-value cards that include tenths and hundredths (see page 22).

Warm-up

Use the 'Counting in given steps' activity on page 23 as a mental warm-up. Give the children a range of different starting numbers and ask them to count forwards and backwards in tens and hundreds.

Focus

- Give the children some interlocking cubes. They should join the cubes together in tens of the same colour. Each stick of ten cubes represents a 'one'.
- Alternatively, use base-ten blocks. Establish that each length of ten cubes is a 'one' and the loose cubes are 'tenths'. Ask: *Into how many equal pieces is each stick divided?* Talk about tenths, for example, ask: *Show me one and two-tenths of a stick.*
- Turn to **Pupil Book 4 page 34**. Work through the explanations with the class. Focus on the extended place-value table. Ask the class why it is extended to the right of the ones place (the fractions are less than 1, and places to the left are ten times greater than the ones next to them).
- Use the 'one stick' the children have made and let them work in pairs to show numbers with a decimal part (tenths) and to say what each decimal is.
- Use place-value cards to show some decimals under 1. Start with tenths and let the children model the decimals you show them using their 'one sticks'.
- Move on to showing some hundredths and ask the children to read these and to say how many hundredths you are showing. Discuss how they know that 0.99 is $\frac{99}{100}$ ths and that 99 is 9 tens and 9 ones. It is important that they understand place value before moving on.

- Do question 1 orally with the class before asking them to write the fractions.
- Ask the children to work on their own to complete question 2.
- Make sure the children realise that the numbers in question 3 are greater than 1 whole with a decimal part.

Follow-up

Let the children work in pairs, using their 'one sticks' and the place-value table if necessary to complete **Workbook 4 page 23**. Spend some time, after they have done this, checking their answers and discussing and resolving any misconceptions.

Interesting mistakes

There are many common misconceptions about decimals. Some children see the numbers after the *decimal point* as a mirror of those before, thus thinking of 34.56 as 'thirty-four point fifty-six'. This leads to incorrect comparison of numbers. For example, the children might think that 34.56 is greater than 34.7 because 'fifty-six is greater than seven'.

To avoid this mistake, it is important to use place-value tables or cards to emphasise the value of digits in a number, and regularly practise how to say numbers correctly.

Answers for Pupil Book 4 page 34

1. $\frac{7}{10}$, 0.7; $\frac{75}{100}$, 0.75
2. a $\frac{2}{10}$, 0.2   b $\frac{9}{10}$, 0.9   c $\frac{42}{100}$, 0.42   d $\frac{4}{100}$, 0.04
3. a three tenths   b ten   c eight hundredths
   d nine tenths   e three

Answers for Workbook 4 page 23

1. 0.1 (provided as an example)   0.2   0.3   0.4
   0.5   0.6   0.7   0.8   0.9   1.0
2. a

   | 0   0.1   0.2   0.3   0.4   0.5   0.6   0.7   0.8   0.9   1 |

   b

   | 2.6   2.7   2.8   2.9   3   3.1   3.2   3.3 |

   c

   | 7.9   8.0   8.1   8.2   8.3   8.4 |

   d

   | 10   10.1   10.2   10.3   10.4   10.5   10.6 |

   e

   | 27.4 27.5 27.6 27.7 27.8 27.9 28.0 28.1 28.2 |
3. a any 5 parts shaded   b any 7 parts shaded
   c any 1 part shaded
   d one whole circle shaded and any 5 parts of other circle shaded

Modelling decimals

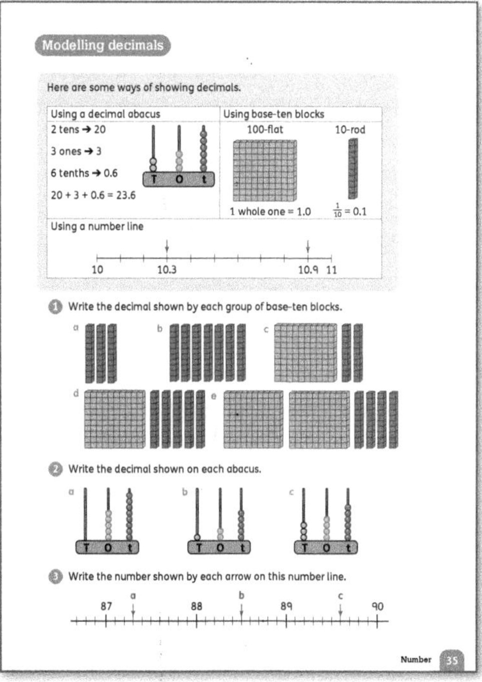

Materials

100 charts (a large one for display and smaller ones for the children to use); a number line with ten divisions.

Warm-up

As a mental warm-up, show a large 100 chart with some numbers circled or highlighted. Ask different children to say one of the circled numbers and then to give the number 10 more or 10 less. This is good practice for working with hundredths in the next lesson.

Focus

- Turn to **Pupil Book 4 page 35**. Ask the children to work in groups to look at the different ways of showing decimals. Next, discuss which model they think is clearest, and let them say why. Let the children suggest any other ways in which decimals could be modelled. Share the children's ideas as a class, allowing for discussion and suggestions for improving models.
- For question 1 remind the children that the 100-flat represents 1 whole and that the 10-rod represents $\frac{1}{10}$ of a whole.
- For question 2, explain that the abacus is similar to a place-value table with counters.
- Before the children do question 3, make sure they realise that the number line goes up in ones and ask them to say what the divisions represent (tenths).

Support

- Provide any children who need support with a blank 100 chart, which they can use to model and count hundredths.
- If any children need additional support, ask them to colour different fractions in different colours to help them understand the relationship between the blocks and the decimals, for example: colour 1 hundredth blue, colour 25 hundredths red and so on.

- To reinforce the relationship between fractions and decimals and to position numbers on a number line, give the children a number line with ten divisions (see below).
- Give them two consecutive *whole numbers*, such as 6 and 7, and ask them to write these numbers at the ends of the number line.
  Let them choose any five points along the number line and write that number as a fraction (for example, $\frac{61}{10}$) above the line and as a decimal below it. 6.1.

Challenge

Let the children work in pairs using a 0–1 number line divided into hundredths (like the one below) and a set of 2-digit decimal cards with a selection of decimals less than 1.

0   1

The children take turns to select a card and indicate using dots or arrows where it should be positioned on the number line.

Once they have done this, ask them to work out what the difference is between the two decimals they have placed. They can work this out by counting or they can subtract the decimals using a calculator.

Answers for Pupil Book 4 page 35

1 a 0.3   b 0.7   c 1.2   d 1.5
    e 2.4
2 a 5.8   b 12.5   c 34.6
3 87.3, 88.5, 89.6

Count in tenths and hundredths

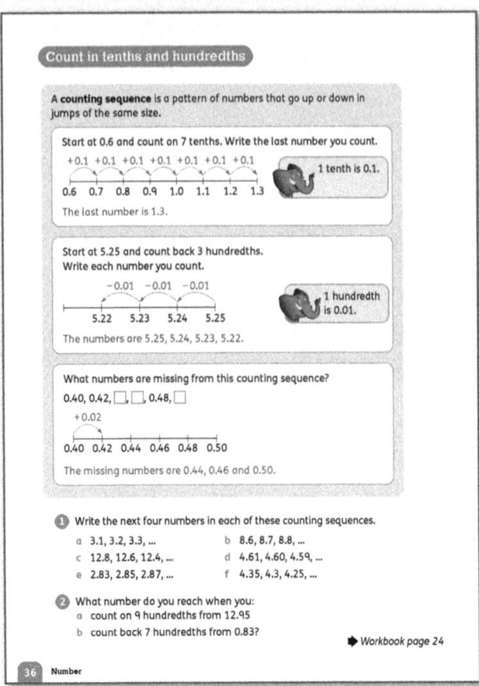

Materials

Charts showing decimals in tenths, from 0.1 to 10.0 (see 'Warm-up'); blank number lines (some divided into tenths and some into hundredths); calculators.

Warm-up

As a mental warm-up, display a chart with some decimals in tenths marked on it. For example:

0.1	0.2							0.9	1.0
		1.3	1.4						
2.1					2.6				
3.1									
				4.5					
				5.5					
						6.7	6.8		
								7.9	8.0
					9.6			9.9	
9.1									10.0

Ask the children to count down or across from different starting numbers. Alternatively, ask the children to complete sections of the chart on the whiteboard.

Focus

- Turn to **Pupil Book 4 page 36**. Remind the children that a *counting sequence* is a pattern that increases or decreases in steps of the same size (as in 5, 10, 15 and so on).
- Explain that we can use a number line to count on or back in decimals (of any size). Work through the examples with the class to reinforce how to do this. Pay attention to the third example with the missing values and make sure the children can see that they have to work out what the steps are before they can work out the values.
- Ask the children to work on their own to complete the counting sequences in question 1. They can use number charts or number lines if they need to.
- Let the children use a calculator to check their answers to question 2.

Follow-up

Use **Workbook 4 page 24** to informally assess how well the children have grasped the concepts in this lesson. Let them work independently to complete the questions and then they can compare and check their answers in pairs or small groups.

<u>Problem solving:</u> The children could make or use a chart showing decimals in hundredths, from 0.01 to 1.00, to help them solve question 3.

Answers for Pupil Book 4 page 36

1 a 3.4, 3.5, 3.6, 3.7   b 8.9, 9.0, 9.1, 9.2
    c 12.2, 12.0, 11.8, 11.6   d 4.58, 4.57, 4.56, 4.55
    e 2.89, 2.91, 2.93, 2.95   f 4.2, 4.15, 4.1, 4.05
2 a 13.04   b   0.76

Answers for Workbook 4 page 24

1 a 7.5 b 8.1 c 9.2

2 3.1 3.6 4.2 4.9

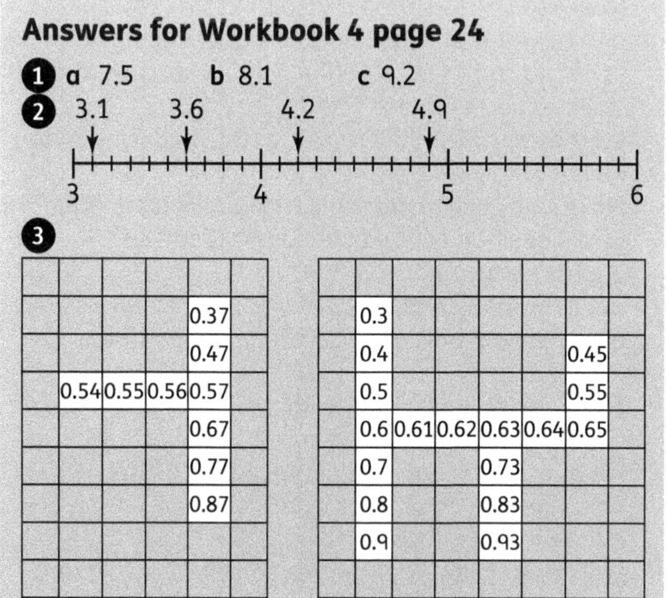

3

				0.37	
				0.47	
	0.54	0.55	0.56	0.57	
				0.67	
				0.77	
				0.87	

	0.3						
	0.4					0.45	
	0.5					0.55	
	0.6	0.61	0.62	0.63	0.64	0.65	
	0.7		0.73				
	0.8		0.83				
	0.9		0.93				

More tenths and hundredths

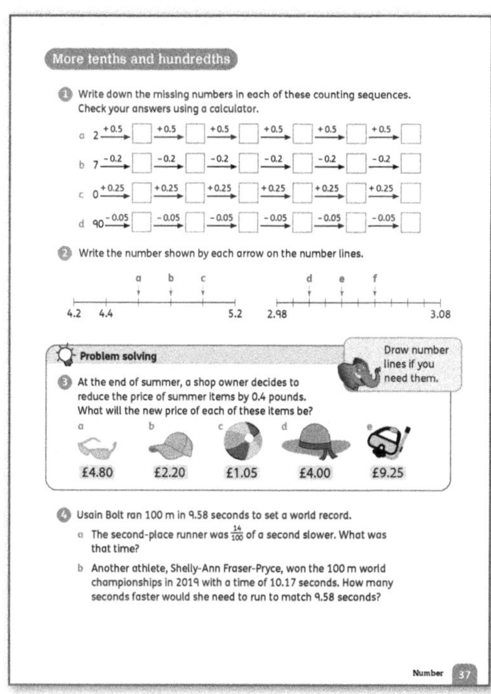

Materials

Charts showing decimals from 0.01 to 1.00; calculators.

Warm-up

As a mental warm-up, repeat the counting on and back activity from the previous lesson but using a hundredths chart from 0.01 to 1.00. Extend this by pointing to a number and asking questions such as:

* *What number is 3 hundredths more/less than this one?*
* *If I start here and count 12 hundredths, which number will I end on?*

Focus

* Turn to **Pupil Book 4 page 37**. No new concepts are introduced in this lesson. The children will work with more challenging sequences and use what they have learnt to solve problems involving counting on or back in decimal amounts.

* Make sure that the children try to work out the missing numbers in question 1 on their own before using a calculator. Remind them that + 0.5 means a jump of 0.5, particularly if they think they cannot add or subtract decimals.
* Point out that the number lines in question 2 are marked in different ways and so the children need to work out what the divisions mean before they start.
* <u>Problem solving</u>: Allow the children to use a calculator to check their answers to question 3.
* Let the children work in pairs to discuss and then solve the problems in question 4. Drawing bar models (see page 22) may help them see what they need to do. Make sure they realise that a slower runner will take more time than a faster runner.

Challenge

Provide some information cards with decimal quantities on them. Let the children work in pairs to make up their own problems involving decimals. They can exchange these and try to solve each other's.

Some useful contexts are: Olympic records; unit prices (price per gram or kilogram); land speed records. You could also use regulation heights or masses of different sporting equipment. For example, the mass of a table tennis ball is 2.7 g, squash balls can be between 23 and 25 g, cricket balls must be between 155.9 and 163.0 g, official basketballs are 623.7 g. The children could research this on their own and use what they find out to develop their problems.

Interesting mistakes

A common error occurs in sequences of decimals when children reach 1 whole. For example, the children might continue this sequence of tenths, 0.6, 0.7, 0.8, 0.9, . . ., with 0.10. The reason is that they count 'six tenths, seven tenths, eight tenths, nine tenths, ten tenths'.

Ask the children what 10 tenths means (1 whole) and use diagrams to help them visualise this. You can also show the children the blocks or sticks they made earlier and ask them to count up in tenths: . . . '08, 0.9, 1 whole stick'.

Diagrams and numbers lines can help the children to see that 1.0 is the same as 1 whole. Remind the children that the most you can put in any column in a place-value table is 9. Say that this also applies to the places for tenths and hundredths.

Answers for Pupil Book 4 page 37

1 a 2.5, 3, 3.5, 4, 4.5, 5

b 6.8, 6.6, 6.4, 6.2, 6, 5.8

c 0.25, 0.5, 0.75, 1, 1.25, 1.5

d 89.95, 89.9, 89.85, 89.8, 89.75, 89.7

2 a 4.6 b 4.8 c 5.0 d 3.00

e 3.02 f 3.04

3 a £4.40 b £1.80 c £0.65 d £3.60

e £8.85

4 a 9.72 seconds b 0.59 seconds

Compare and order decimals

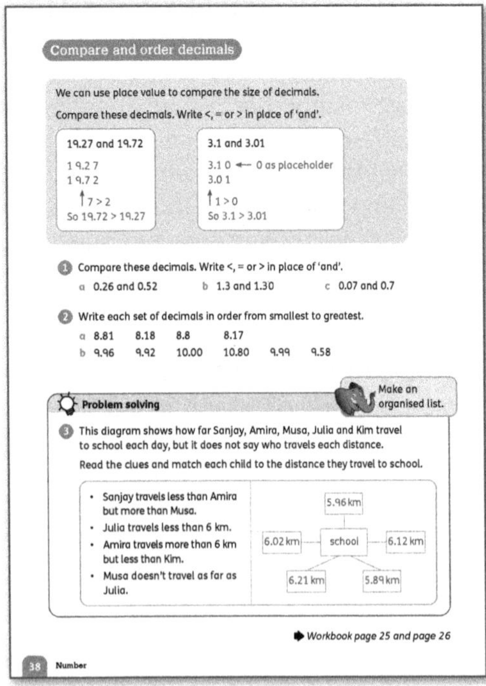

Materials

Decimal place-value cards (see page 22); flashcards with <, > and = signs; the number lines used in previous lessons; sets of digit cards in two colours.

Warm-up

Use any of the 'Comparing numbers' activities on page 25 to start this lesson.

Focus

- Start with a number talk (see pages 17–18). Ask the class to think about these problems:
  - *Selma says 0.8 is less than 0.80 because 8 is less than 80. Is she correct?*
  - *The paint on a door is 0.1 mm thick. The side of the paint can is 1.0 mm thick. Which is thicker? How did you decide?*
- Give the children some time to think and talk, and then take feedback. When a child answers, ask the others whether they agree or not and whether they have any different ways of thinking.
- Remind the class that we compare numbers to decide which is greater or less and that we use the < and > signs to show what we decide.
- Use the place-value cards to demonstrate that 0.8 = 0.80 and that 0.1 mm < 1.0 mm. We can also order numbers by size. Remind the children that decimals are numbers, if necessary.
- Turn to **Pupil Book 4 page 38**. Give the children time to read the information and work through the examples in groups.
- Ask: *How is comparing decimals similar to comparing whole numbers? (We use place value and compare digit by digit to find the first place with a different digit. Then we decide which is greater or less).*
- Next, ask whether there are any differences.

- Make sure the children understand that you can add 0 to the right of a decimal without changing its value so, for example, 0.7 = 0.70 = 0.700 and so on. If the children don't understand this, show them using 10-rods and 100-flats that 7 tenths is equivalent to 70 hundredths.
- Make sure they understand that you cannot add 0 as a place holder to whole numbers because 1 is not equivalent to 10.
- Use **Workbook 4 page 25** at this point to consolidate ordering decimals and positioning them on a number line.
- Turn to **Pupil Book 4 page 38**. Let the children use a number line to help them complete question 1.
- For question 2, again, encourage the children to use a number line to place the decimals – this will make it easier to sort them.
- Problem solving: The children can discuss question 3 in pairs and use a number line to organise the information.

Follow-up

Use **Workbook 4 page 26** to informally assess how well the children are able to compare and order decimals, including money amounts and measurements.

Challenge

You can make the comparing decimals activity more challenging by asking the children to each draw a blank frame like this one:

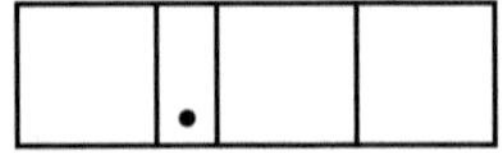

Give each child one set of 0–9 digit cards (a colour each) and ask them to shuffle the cards. Then they take turns to draw one card at a time and place it in any place on their frame. They continue taking one card at a time in turns until they have filled all three places. The aim is to make a decimal greater or less than their partner's. Tell the children in advance which they are aiming for. Again, you can award points to turn it into a game.

Support

Place the children in pairs and give each pair two sets of digit cards, with each set a different colour. Ask them to each draw a large frame like this one for making decimals:

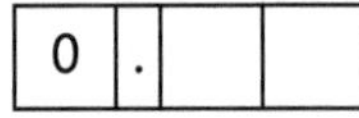

Tell the children to keep the cards of each colour separate. They should shuffle each colour set and then take turns to choose one card from each pile and use them to make a two-place decimal. The children then compare their decimals and decide which is greater/less. You can make this into a game by awarding points for greatest/smallest.

Answers for Pupil Book 4 page 38

1. **a** 0.26 < 0.52   **b** 1.3 = 1.30   **c** 0.07 < 0.7
2. **a** 8.17, 8.18, 8.8, 8.81
   b 9.58, 9.92, 9.96, 9.99, 10.00, 10.80
3. Musa 5.89 km, Julia 5.96 km, Sanjay 6.02 km, Amira 6.12 km, Kim 6.21 km

Answers for Workbook 4 page 25

1. **a** C = 5.3, D = 6.5, E = 7.4
   b F = 1.2, G = 8.7, H = 9.3
   c I = 15.7, J = 20.5, K = 26.1
   d L = 10.5, M = 11.2, N = 25.3
   e O = 1.3, P = 2.5, Q = 4.9
   f R = 8.3, S = 16.8, T = 25.6
   g U = 12.3, V = 22.5, W = 73.1
   h X = 1.7, Y = 62.5, Z = 73.4

Answers for Workbook 4 page 26

1. **a** circle 6.5 ℓ, underline 9.6 ℓ
   b circle 1.2 m, underline 2.5 m
   c circle $4.58, underline $5.84
   d circle $1.24, underline $123.40
   e circle 0.04 kg, underline 14.4 kg
   f circle 1.25 m, underline 123.5 m
   g circle 3.75 kg, underline 7.35 kg
   h circle 2.07 kg, underline 2.73 kg
2. **a** 0.59, 0.94, 0.95   **b** 2.33, 3.13, 3.3
   c 2.05, 2.55, 5.55
3. **a** >   **b** <   **c** <   **d** <
   e <   **f** =

Round decimals to the nearest whole number

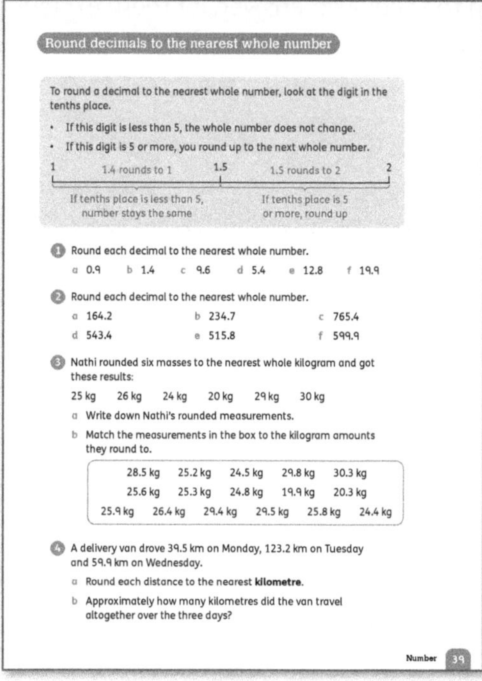

Materials

Blank number lines divided into tenths; sticky notes (optional).

Warm-up

As a mental warm-up, choose one of the 'Rounding' activities on page 26. Only work with whole numbers in the warm-up.

Focus

- Ask the children to think about how they might *round* decimals to the nearest whole number. Draw a number line from 0 to 5 with the tenths marked between each whole number on the board. Remind the children about the idea of rounding to approximate numbers, and how they have rounded whole numbers to the nearest 10 and 100. Explain that you are now going to look at rounding decimals in a similar way.
- Mark 3.7 on the line. Ask the children which *whole number* this is nearest to.
- Tell them that this means that 3.7 rounded to the nearest whole number is 4.
- Repeat for other examples.
- Mark 2.5 on the line and ask the children what they think they should do.
- Remind the children that when we round whole numbers, if they fall halfway between two numbers, we round up.
- Explain that we do the same with decimals, so 2.5 rounded to the nearest whole number would be 3.
- Read through the explanation and additional examples on **Pupil Book 4 page 39** to formalise the rules you have demonstrated to the class.
- For question 1, encourage the children to use a blank number line if they need to.
- For question 2, you may need to explain that it doesn't matter how big or small the whole number part is; we only need to think about the ones digit. Unless the ones digit is 9 and they need to round up, they don't need to worry about digits in the tens and hundreds places.
- For question 3, the children could write the measurements on sticky notes and physically move these around to sort the measurements before writing their answers. They could draw a number line from 20 to 30 to help them.
- Question 4 involves numbers rounded to the nearest *kilometre* to estimate and answer. Make sure the children understand that *approximately* means they should find an answer that is 'more or less' correct.

Challenge

Place the children in pairs. Ask them to each draw a 3 × 3 grid and to write 9 decimals on the grid. Both their grids should be the same. The children take turns to choose (and circle) a decimal on the grid without letting their partner see what they have chosen. The partner can ask three questions (or more, depending on the children) that can only be answered by 'Yes' or 'No' to try to guess which decimal their partner has chosen. For example,

they might ask: 'Would the decimal number round up?',
'Is the whole number less than 6'? If they guess correctly,
the partner ticks that circle. The winner is the person who
gets the most correct out of the three questions.

Support
- Print a blank number line like the one below for each
  child or ask them to draw their own.

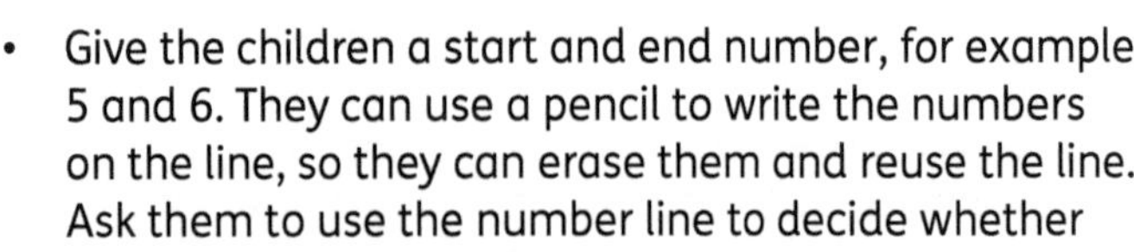

- Give the children a start and end number, for example
  5 and 6. They can use a pencil to write the numbers
  on the line, so they can erase them and reuse the line.
  Ask them to use the number line to decide whether
  these decimal numbers round to 5 or to 6:

  5.9   5.1   5.4   5.6   5.5   5.7

- Give them time to work out how to position the
  numbers and decide which number they round to.
- Explain that they can use this number line for any
  two whole numbers.

Interesting mistakes
The children may think that rounding down means they
have to subtract from the whole number. It helps if you
talk about rounding to the nearest whole number, rather
than rounding up or rounding down.

Answers for Pupil Book 4 page 39
1. a 1 b 1 c 10 d 5
   e 13 f 20
2. a 164 b 235 c 765 d 543
   e 516 f 600
3. 25 kg: 25.2 kg, 24.5 kg, 25.3 kg, 24.8 kg
   26 kg: 25.6 kg, 25.9 kg, 26.4 kg, 25.8 kg
   24 kg: 24.4 kg
   20 kg: 19.9 kg, 20.3 kg
   29 kg: 28.5 kg, 29.4 kg
   30 kg: 29.8 kg, 30.3 kg, 29.5 kg
4. a Monday, 40 km; Tuesday, 123 km; Wednesday,
     60 km
   b 223 km

Solve problems involving decimals

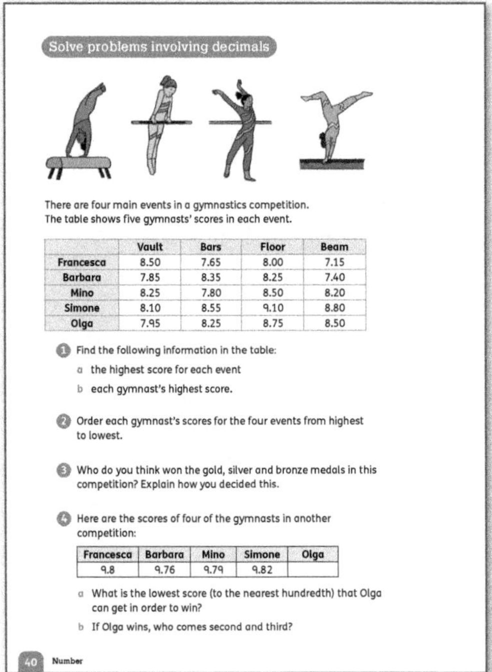

Solve problems involving decimals

There are four main events in a gymnastics competition.
The table shows five gymnasts' scores in each event.

	Vault	Bars	Floor	Beam
Francesca	8.50	7.65	8.00	7.15
Barbara	7.85	8.35	8.25	7.40
Mino	8.25	7.80	8.50	8.20
Simone	8.10	8.55	9.10	8.80
Olga	7.95	8.25	8.75	8.50

1. Find the following information in the table:
   a the highest score for each event
   b each gymnast's highest score.

2. Order each gymnast's scores for the four events from highest
   to lowest.

3. Who do you think won the gold, silver and bronze medals in this
   competition? Explain how you decided this.

4. Here are the scores of four of the gymnasts in another
   competition:

Francesca	Barbara	Mino	Simone	Olga
9.8	9.76	9.79	9.82	

   a What is the lowest score (to the nearest hundredth) that Olga
     can get in order to win?
   b If Olga wins, who comes second and third?

40 Number

Warm-up
As a mental warm-up, choose any of the 'Mental maths'
activities on pages 27–29.

Focus
This lesson gives the children a chance to apply their
skills in a problem-solving context (decimal scores).

- Turn to **Pupil Book 4 page 40.** You may need to work
  through some examples and explain how scores
  are given in contests such as the Olympic Games or
  talent shows on television.
- For question 1, the children need to read the table
  and write down the required information.
- Before the children complete question 2, remind
  them how to order decimals.
- Let the children answer question 3 orally in groups.
  When they have an answer, let them explain how
  they decided.
- Let the children work out question 4 on their own.

Answers for Pupil Book 4 page 40
1. a Vault 8.50; Bars 8.55; Floor 9.10; Beam 8.80
   b Francesca 8.50; Barbara 8.35; Mino 8.50;
     Simone 9.10; Olga 8.75
2. Francesca 8.50, 8.00, 7.65, 7.15; Barbara 8.35,
   8.25, 7.85, 7.40; Mino 8.50, 8.25, 8.20, 7.80
   Simone 9.10, 8.80, 8.55, 8.10; Olga 8.75, 8.50,
   8.25, 7.95
3. gold: Simone (highest total score 34.55 and won
   3 out of 4 events), silver: Olga (second highest
   score of 33.45); bronze: Mino (third highest score
   of 32.75)
4. a 9.83
   b Simone is second, Francesca is third

Use some or all of these questions and activities to assess how well the children have understood the concepts in this unit.

- Point to a digit in a decimal. *What is the value of this digit?*
- Provide a set of decimals and ask the children to put them in order from smallest to greatest.
- *Describe how you ordered the decimals. What did you do first? How did you compare them? (For example: I wrote the decimals in a column with the decimal points lined up. I compared the digits, starting from the left. I stopped at the first digit that was different and compared them.)*
- *Which numbers do you find hardest to order? Why?*
- *Tell me a number that is between 3.2 and 3.3. (any of 3.21, 3.22, 3.23, 3.24, 3.25, 3.26, 3.27, 3.28, 3.29) How many answers do you think there are to this question? (Using hundredths, there are 9 possible answers.) (There is actually an infinite number of answers, as we can keep making more numbers by going to the next decimal place, but the children are not expected to know this.)*
- Provide a number line and a set of decimals for the children to place on the line.
- Provide a set of decimals for the children to show on an abacus.
- Give pairs of decimals, for example: 0.22 and 0.3; 0.48 and 0.51; 0.19 and 0.91. Ask the children to say which decimal in each pair is greater. Allow them to use 100 charts or diagrams to do this. (*0.3; 0.51; 0.91*)
- *How could you give instructions for how to round decimal fractions to the nearest whole number? (Look at the digit in the tenths place: if it is less than 5, round to the previous whole number; if it is 5 or more, round to the next whole number.)*
- Present a problem that involves reasoning and explaining. For example: *A forest manager gives the height of two trees as 4 m to the nearest whole number. Is it possible that one tree is 0.7 m taller than the other? Explain why or why not. (Yes. For example, the trees could be 3.6 m and 4.3 m.)*

UNIT 6 — Measures and money

- Estimate, compare and calculate lengths, mass and capacity
- Use knowledge of fractions to read and interpret measuring scales
- Use decimal notation for measurements and money amounts
- Convert between different units of measure
- Solve problems involving units of measurement
- Estimate, compare and calculate amounts of money

Key words

length units millimetre (mm) centimetre (cm) metre (m) kilometre (km) height capacity volume millilitre (ml) litre (ℓ) mass gram (g) kilogram (kg) estimate measure scale notes coins currency bar model remainder

Materials

Books; tape measures; masking tape, chalk or adhesive tack; metre sticks.

Teaching guidance

- Draw three stick figures inside different shapes on the board – a square, a tall rectangle and a wide rectangle, as shown here:

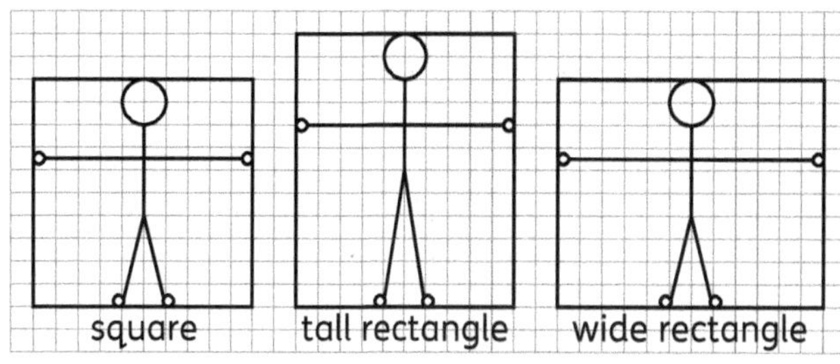

- Explain to the class that people generally fit into one of these three shapes based on their *height* and their arm span. Spend some time talking about what the differences are between the shapes and the people in them.

- Make sure the children realise that the square means the height and arm span are more or less the same. The tall rectangle means the height is greater than the arm span and the wide rectangle means the arm span is greater than the height.
- Ask the children to guess what shape they would fit into if they stood with their arms outstretched. Give them time to think about this and let them stand up and stretch out their arms to help them decide.
- Ask what they would need to *measure* to confirm (*their height and the width of their arm span*) and then discuss how they could measure these two *lengths* in *centimetres*.

> These ways of measuring work well:
>
> To measure height, one child stands against a wall or door and holds a book flat on the top of their head. Their partner then measures the height from the floor to the book. Ensure that the children respect each other's space.
>
> To measure arm span, one child stands against a wall with their arms outstretched. Their partner marks the wall at the end of their fingers (using masking tape, chalk or adhesive tack to avoid marking the wall) and measures the distance between them.
>
> If you need to adapt the activity for any reason, ask the children to find things around the classroom that are different shapes.

- Place the children in small groups and give each group a metre stick or tape measure with centimetres marked on it. Remind them that they measured in centimetres in Year 3.
- Let the children measure and record the two measurements.
- Once they have done this, let them decide which shape they would fit into. Allow for some discussion about whether a person who fits into the square would need to have exactly the same height and arm span, or whether these two measures just need to be very close. Reach an agreement as a class.
- If you have time, tally the shapes and draw a block graph to show the number of children that are each shape in your class.

Metric units of length

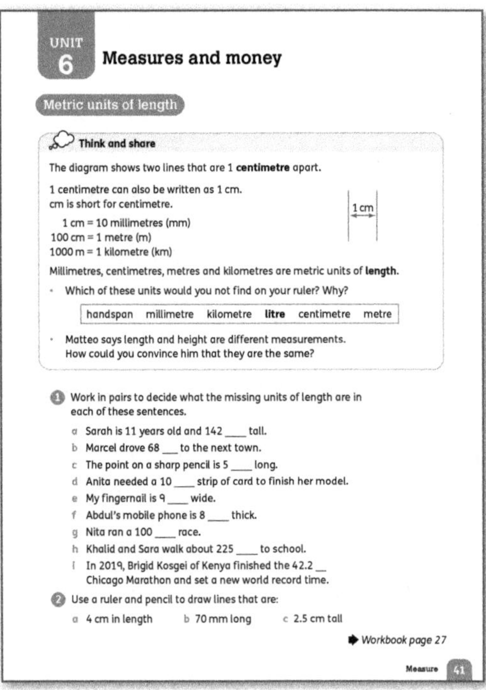

Materials

Metre sticks, rulers, tape measures, chalk, string and any other instruments for measuring length (such as a trundle wheel, callipers).

Warm-up

- As a mental warm-up, revise multiplication by 10 and 100, as the children will use these skills to convert *units* of measurement.
- Display ten starter numbers. Point to any number (for example, 6). Say a child's name and ask: *What is 10 × 6?*
- Repeat for 100 × 6 and carry on choosing numbers and children until you have covered them all.

Focus

- Ask the children what units they know for measuring length. They should remember *kilometres, metres,* centimetres and *millimetres*.
- Write these terms on the board. Ask the children to suggest things that we would need to use metres, centimetres or millimetres to measure.
- Write some lengths and distances on the board, for example, the length of a pencil, the distance between these two towns. Ask the children to suggest a sensible unit to use for measuring each.
- Show the class some different instruments that we use to measure length, such as tape measures, rulers, metre rules.
- If you can get hold of a trundle wheel and a pair of callipers, show the class how we use these to measure different kinds of distances. A trundle wheel has a circumference of 1 m, and clicks after each rotation. A pair of callipers is useful for very precise measurement of tiny items.

- Have a class discussion about which instruments we use to measure particular kinds of lengths and distances.
- Use **Workbook 4 page 27** to summarise and consolidate the work on units and measuring instruments.
- <u>Think and share:</u> Read through the information on **Pupil Book 4 page 41** with the class and then give them time to answer the questions in groups. Share the answers as a class before moving on.
- Place the children in pairs for question 1 and let them decide what the missing units are. Ask different pairs to give their answers. Allow for discussion if the children have different ideas.
- Do some practical measuring and drawing lengths before moving on.
- Show the class a ruler. Point out how each centimetre is divided into millimetres. Write on the board: 10 mm = 1 cm. Ask a child to come to the board and draw a line of 5 cm, using a ruler to measure it out. Write: 5 cm = ___ mm. Then have another child come up and measure it in millimetres, and fill in the answer. Repeat with some other lengths.
- Write a length on the board, for example, 5.6 cm. Ask the children to use a straight edge to draw a line they think is about this length. Ask the children to measure their lines using a ruler to see how accurate their line was.
- For question 2, the children can work on their own to measure and draw the lines. Let them check each other's completed drawings for accuracy.

Answers for Pupil Book 4 page 41

<u>Think and share:</u> handspan, kilometre, litre and metre. Handspan is not a standard unit of measure of length. Kilometre and metre are too long to be on a ruler. Litre is a unit for measuring lengths.

Individual answers, for example: We use the word length when we measure in the horizontal. When we measure in the vertical we use the word height instead of length to differentiate between them.

1 a cm   b km   c mm   d cm
    e mm   f mm   g m   h m
    i km

2 a–c (lines drawn of the correct lengths)

Answers for Workbook 4 page 27

1 (basketball court) trundle wheel, metres; (pencil) ruler, cm; (path) trundle wheel, metre; (a belt) metre stick, cm; (your friend's height) metre tape, cm; (distance around a tree trunk) 5-metre tape, metres/cm; (width of a computer screen) metre stick, cm

Read and write lengths

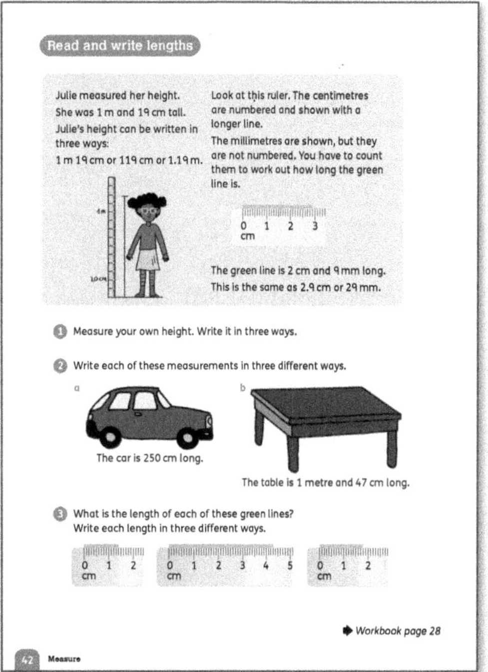

Materials
Metre sticks; rulers showing millimetres; tape measures; printed or photocopied sheets of rulers (actual size).

Warm-up
As a mental warm-up, choose any of the 'Calculation skills' activities on pages 28–29.

Focus
- Once the children have done some practical measuring tasks using metre sticks, rulers and tape measures, show them how lengths can be written in different ways (in metres and centimetres, in centimetres only or in metres and parts of metres using decimal notation). If necessary, remind them about the work they did in Unit 5 on decimals.
- At this level, the children will need to use different notations for measurements; this includes using a larger and smaller unit such as 3 kg and 250 grams and the equivalent decimal notation of 3.25 kg.
- When using digital notation, the children need to understand that the decimal part of the measurement is a fraction of the whole unit and is the same as a number of smaller units, for example, 0.6 m is $\frac{6}{10}$ of a metre, which is the same as 60 cm.
- Ask the children to read through the information on **Pupil Book 4 page 42**.
- The children have already measured their height in centimetres, so for question 1 they can either use the measurement they have or measure again using different measuring instruments.
- In question 2, the children read and write the measurements in different formats.
- Before the children complete question 3, make sure they can read the scale on a ruler and understand that the intervals are in millimetres.

Follow-up

Check that the children have a ruler showing millimetres before asking them to complete **Workbook 4 page 28**.

Challenge

Let the children work in pairs or groups to make a measurement 'Scavenger hunt' using items in the classroom. They find ten items (or any number you choose) and use their own rulers to measure and list just the lengths on a sheet. Afterwards, they give their sheet to another pair to find objects that match each length. This can also be done using a range (5–7 cm, less than 1 m and so on).

Support

- If any children need support to measure in millimetres and to draw accurate lengths, provide them with a sheet of printed rulers similar to the one below. You can make sheets of rulers that are actual size by photocopying real rulers or you can find examples on the Internet.

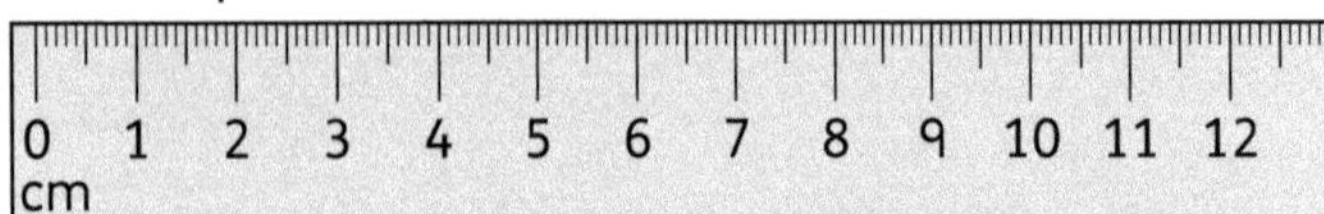

- Give the children lengths (for example, 5 cm and 6.8 cm) and ask them to colour the ruler to show each length. This helps them to treat the ruler like a number line and allows them to focus on the task rather than worrying about measuring or drawing.

Interesting mistakes

Some children may be confused about how many centimetres are equal to 1 m, and how many millimetres are equal to 1 cm. In their work on mass and *capacity* they learn that there are 1000 g in 1 kg, and 1000 ml in 1 l respectively, so they may think that there are 1000 cm in 1 m, or 100 mm in 1 cm.

The correct abbreviation for centimetre is cm and not Cm, cM or CM. Similarly, the correct abbreviation for kilometre is km and not Km, kM or KM. The prefixes *kilo-*, *centi-* and *milli-* come up in a variety of contexts in mathematics and the sciences and it is worth getting the children into the habit of writing these symbols correctly now to avoid future problems.

Answers for Pupil Book 4 page 42

1 Individual answers

2 a 250 cm, 2.5 m, 2500 mm

b 1.47 m, 147 cm, 1470 mm

3 14 mm, 1.4 cm, 0.014 m

43 mm, 4.3 cm; 0.043 m

18 mm, 1.8 cm, 0.018 m

Answers for Workbook 4 page 28

1 a 2 cm 9 mm, 2.9 cm   b 3 cm 8 mm, 3.8 cm

c 6 cm 2 mm, 6.2 cm   d 5 cm 4 mm, 5.4 cm

e 2 cm 1 mm, 2.1 cm   f a line 3.9 cm long

Litres and millilitres

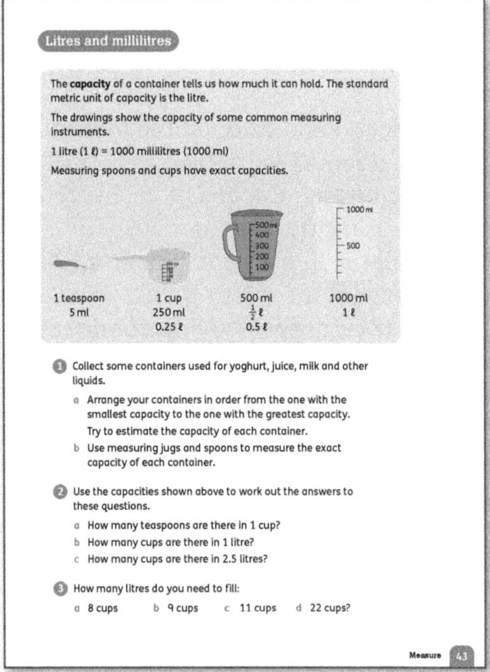

The children learnt to measure and compare capacity at Level 3. This year they will build on those skills and *estimate*, measure and calculate using *litres* and *millilitres*.

Materials

Containers with various capacities (such as jugs, spoons, bottles, cups); empty and clean containers from household products (such as yoghurt pots, milk bottles, juice bottles, hand soap and detergent bottles); 10 ml measure (such as a syringe or eyedropper); 5 ml spoon; two calibrated measuring containers (such as test tubes or small jugs); set of measurements for display (see 'Warm-up').

Warm-up

As a mental warm-up, display a set of measurements like the ones shown below. Remind the children that 1 litre = 1000 millilitres and ask them to jot down as many ways as they can to make 1 litre using any number of measurements from those given.

Let the children compare and check each other's answers.

800 ml	200 ml	700 ml	½ litre
500 ml	300 ml	750 ml	350 ml
¼ litre	600 ml	400 ml	250 ml
900 ml	150 ml	250 ml	150 ml
100 ml	100 ml	¼ litre	300 ml

Focus

- Start this lesson with some practical activities to remind the children of what they learnt at Level 3 and to develop their skills further.
- Show the children a range of containers. Hold up pairs of containers and ask them which holds more. Show them a litre container or the mark on a calibrated measuring jug. Remind them that there are 1000 millilitres in a litre.
- Ask them to estimate the capacity of each container before filling it with water and pouring the contents into the measuring jug to measure the capacity.
- Show the children how to record capacities in millilitres and litres, for example, 1200 ml or 1 litre 200 ml. Also demonstrate how to record capacity as a decimal of a litre, such as 1.2 ℓ.
- Discuss the fact that 1 ml is one-thousandth of a litre, that 100 ml is one-tenth of a litre and so on.
- Fill a standard litre measure with water and ask the children how you could share the litre equally into two identical, unmarked containers.
- Once they have discussed their ideas, give each group a litre of water and any container that can hold more than half a litre.
- Let them pour what they estimate to be half the liquid into the other container.
- The children should then use a calibrated measure (measuring jug or other container labelled in millilitres) to check how accurate they were at pouring half a litre.
- Turn to **Pupil Book 4 page 43**. Let the children work in pairs or small groups to read the information and complete the questions.
- For question 1, make sure that each group has a range of different containers.
- Remind the children that they can skip count in groups to help them work out question 2 if they need to.
- For question 3, remind the children that they have already worked out that 4 cups = 1 litre in question 2.

Challenge

Give the children a number of different containers and a bottle of water. Ask them to pour 100 ml of water into each of the four different containers. When they have done this, they can use a measuring jug to check their estimates. Repeat for 250 ml and/or other amounts.

Answers for Pupil Book 4 page 43

1 **a** and **b** Individual answers

2 **a** 50 teaspoons   **b** 4 cups   **c** 10 cups

3 **a** 2 litres   **b** 2.25 litres
   c 2.75 litres   **d** 5.5 litres

Estimate and calculate amounts of liquid

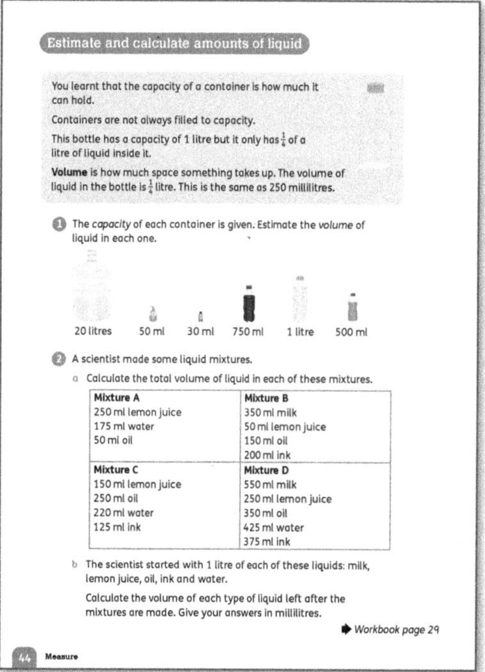

Materials

Containers with various capacities (such as jugs, spoons, bottles, cups); empty and clean containers from household products (such as yoghurt pots, milk bottles, juice bottles, hand soap and detergent bottles); 10 ml measure (such as a syringe or eyedropper); 5 ml spoon; two calibrated measuring containers (such as test tubes or small jugs); sticky notes; set of measurements from previous lesson.

Warm-up

Display the measurements from the 'Warm-up' activity from the previous lesson. Ask one child at a time to choose one of the measurements in millilitres and write it in litres on a sticky note, then stick it on the display.

Focus

- Show the class a measuring jug (or any container you have available) with a clearly marked scale and pour some liquid into it without filling it. Ask the class:
  - *What is the capacity of this container?*
  - *How much liquid is in the container at the moment?*
  - *How much more liquid can the container hold? How can we work this out?*
  - *What does it mean if we say something is 'filled to capacity'?*
- Repeat this with a different container or amount if necessary.
- Read through the information on **Pupil Book 4 page 44**, referring back to your practical demonstration.

The correct word for the amount of liquid in the container is *volume*, but the children are more likely to talk about 'how much' is in the container. Model the correct vocabulary but allow the children to talk about the liquid in containers in a way that makes sense to them.

- For question 1, the children need to visually estimate how much is in each container using their knowledge of fractions and applying it to the overall capacity. Allow them to answer using either fractions of amounts or millilitres.
- Question 2 involves working with addition of amounts in millilitres and subtraction from 1000 ml.

Follow-up

To consolidate the work on reading measuring scales and calculating in litres and millilitres, ask the children to complete **Workbook 4 page 29**.

Answers for Pupil Book 4 page 44

1 Possible answers: 8 litres, 25 ml, 10 ml, 200 ml, 900 ml, 375 ml

2 a **A** 475 ml **B** 750 ml **C** 745 ml **D** 1950 ml or 1.95 litres

b milk: 100 ml   lemon juice: 300 ml   oil: 200 ml
ink: 300 ml   water: 180 ml

Answers for Workbook 4 page 29

1 a 950 ml, 0.95 ℓ   b 225 ml, 0.225 ℓ
c 850 ml, 0.85 ℓ   d 400 ml, 0.4 ℓ
e 575 ml, 0.575 ℓ   f 500 ml, 0.5 ℓ

Calculate with measures

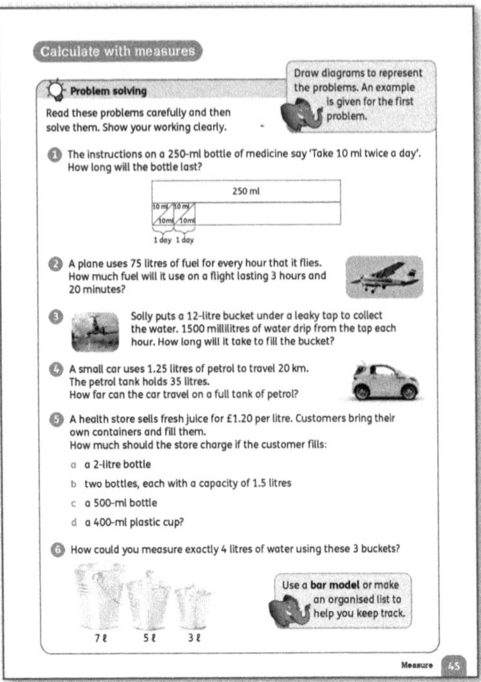

Materials

Measuring jugs; range of (clean) containers; a sheet of drawings of 1-litre measuring jugs.

Warm-up

- As a mental warm-up, ask the children to write down a capacity greater than 1 litre, but less than 2 litres.
- Ask a child to say their amount and then ask another child to work out how much more liquid is needed to make 2 litres.
- Repeat, asking different children to contribute.

Focus

- Problem solving: **Pupil Book 4 page 45** is a problem-solving page that involves measures (including time).
- Discuss strategies the children could use to solve the problems and allow some time to check answers and discuss methods that the children used.
- Allow them to work in pairs if necessary to complete question 1–6.

Challenge

Let the children explore the volume of stones or other irregular solids by working out how much water is displaced when the item is put into a measuring jug. This is an important scientific concept and helps the children understand that 1 cm^3 is equivalent to 1 ml. The children can investigate this in pairs or small groups and feed back to the class about what they discovered.

Support

- Continue with practical work on capacity. The children can work in groups to collect a range of (clean) containers. They can then use a measuring jug to fill each container to work out its capacity, then list the containers and let other groups estimate the capacity of each. They can give each other a score out of 5 for accurately guessing the capacity.
- You can also prepare a sheet of drawings of empty measuring jugs with a scale to 1 litre marked on each and let the children colour these to show the capacity of several small containers.
- If any children need support to work out what to do when they have to calculate amounts in litres and millilitres, encourage them to use diagrams or bar models (see page 22). Here are three examples:

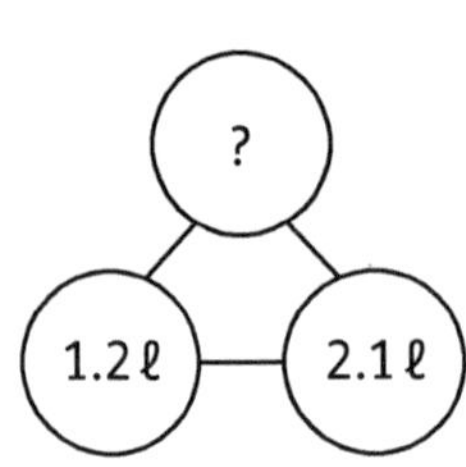

2 ℓ 450 ml	
?	1 ℓ 200 ml

4.85 ℓ	
4.2 ℓ	?

Interesting mistakes

The children may get confused about the distinction between the terms *capacity* and *volume* as they are often used interchangeably in real-life contexts.

Capacity means the amount of space inside a container. In other words, it is a term we use for containers that can be filled. Make sure the children understand that the capacity of a container is how much it can hold (for example, 1 litre); it may hold less than that if it is not filled.

Volume is the amount of space that a solid or liquid takes up. So we can talk about the volume of liquid inside a container – this volume might be less than the capacity of the container if the container is not full. When a container is not filled to capacity, we talk about the volume of liquid in it.

Answers for Pupil Book 4 page 45

1 12.5 days

2 250 litres

3 8 hours

4 560 km

5 a £2.40 b £3.60 c £0.60 d £0.48

6 Possible answer: Fill the 7-litre bucket, then fill the 3-litre bucket from the 7-litre bucket. There will be 4 litres left in the 7-litre bucket.

Mass

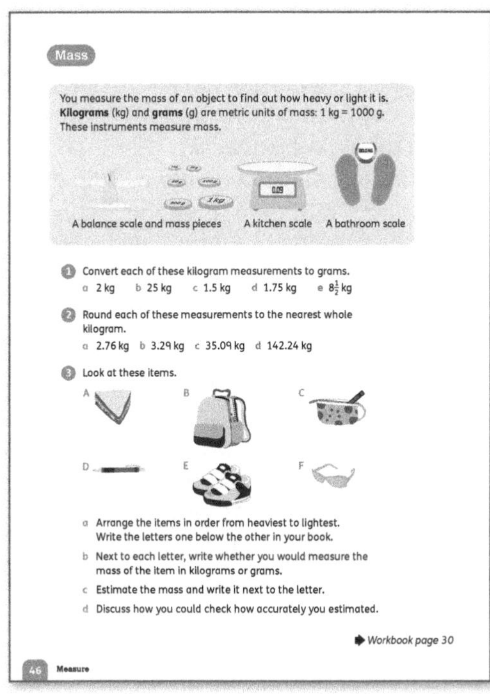

Materials

Objects for weighing (such as a brick, vegetables, books, stationery, coins); balance scales and weights; bathroom and kitchen scales; a 1-litre plastic bottle.

Warm-up

Choose any suitable 'Mental maths' activities from pages 27–29.

Focus

- Ask the children what they remember about *mass* and weighing objects from previous years.
- Tell the class that we often use the word weight when we really mean mass. For example, we say things like *I weigh 50 kilograms*, rather than *I have a mass of 50 kilograms*.

In science, mass and weight are not the same things. The weight of an object depends on gravity, so if you were on the moon, where there is less gravity, you would weigh less. The mass of any object does not change, it is a measure of how much matter is in the object. As the children advance, they should begin to talk about the mass of objects as this is the correct terminology and will be used in science lessons as well.

- Remind the children that 1 *kilogram* = 1000 *grams*, $\frac{1}{2}$ kilogram = 500 grams and $\frac{1}{4}$ kilogram = 250 grams, showing these on balance scales if possible. Make sure that they know the abbreviations for kilograms (kg) and grams (g).
- Show the class some objects. Ask them to suggest things with a mass that is similar to, lighter than and heavier than each object.
- Ask the children to suggest objects that would be weighed in kilograms or grams.
- Name various objects and ask the children to suggest a sensible unit to measure them in.
- Turn to **Pupil Book 4 page 46**. If possible, show the children balance scales, a kitchen scale and a bathroom scale (digital is fine). Let the class share their ideas about how each scale works. Correct any misconceptions.
- Let the children work in pairs to complete the measuring tasks on **Workbook 4 page 30** to make sure that they can estimate and measure mass.
- Demonstrate how to convert 0.5 kg into grams, by reminding the children that 0.5 is one-half, and one-half of 1000 g is 500 g. Repeat for 0.25 kg, which is one-quarter of 1000 g. Bar models may be useful here.
- The children can work in pairs to convert the measurements in question 1.
- Before the children do question 2, remind the class how to round decimals to the nearest whole number using number lines if necessary.
- Let the children work in pairs to complete question 3.

Answers for Pupil Book 4 page 46

1 a 2000 g   b 25 000 g   c 1500 g
   d 1750 g   e 8500 g

2 a 3 kg   b 3 kg   c 35 kg   d 142 kg

3 a B, E, C, A, F, D

   b **A** grams **B** kilograms **C** grams **D** grams **E** grams **F** grams

   c Individual answers, for example:
   A 250 grams **B** 4–5 kilograms **C** 400 grams **D** 5–10 grams **E** 900 grams **F** 50 grams

   d Individual answers

Read measuring scales

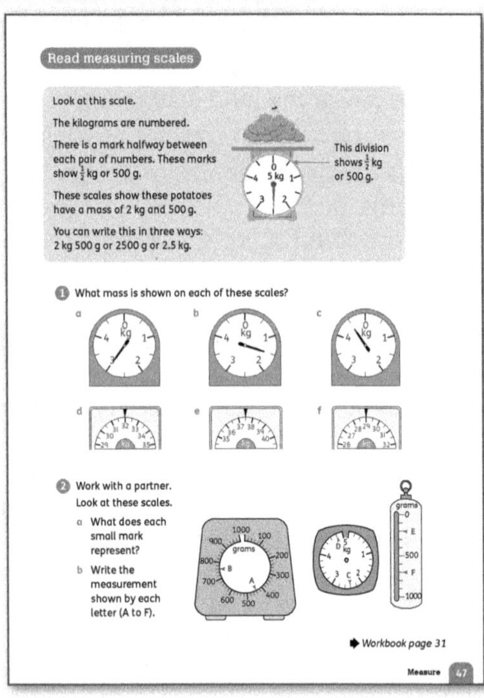

Materials
Balance scales; bathroom and kitchen scales; pictures of everyday items or real objects with known masses (see 'Support').

Warm-up
Use any of the 'Comparing numbers' activities on page 25 as a mental warm-up.

Focus
- Remind the children that measuring scales are similar to number lines and that they have to work out what interval is shown by each division to make sense of them.
- Spend some time working through the example on **Pupil Book 4 page 47**. Show the children examples of different measuring scales to reinforce the concepts.
- Do question 1 orally with the class, making sure that the children are able to read and make sense of the scales.
- Let the children work in pairs to discuss and complete question 2. Assist the children to read the scales where necessary.

Follow-up
Use **Workbook 4 page 31** for additional practice and consolidation.

Challenge
Make a set of task cards and encourage the children to complete as many as they can. Here are some possible tasks:
- *List five things that are small but that have a mass of more than 1 kilogram.*
- *Find five things that are less than 1 gram.*
- *Put this set of objects in size order: a cricket ball, a tennis ball, a basketball, a hockey ball, a beach ball, a table-tennis ball. How would the order change if you put them in order from lightest to heaviest? Why?*

Support
Let the children work in pairs to make equivalent units of mass using pictures of everyday items or real objects. For example, the children could show that a 1 kg bag of sugar is equal to two $\frac{1}{2}$ kg (500 g) bags by pasting pictures of these next to each other in their books. This could also be done using < or > to compare sets of objects.

Interesting mistakes
- The most common mistakes occur when converting between units of measurement. Many errors are due to misunderstanding place value and how to use the decimal point. Give the children plenty of experience in ordering measurements, saying conversions out loud and writing measurements in a place-value table (see page 22).
- Most measuring scales have only certain intervals marked on them. Kitchen scales may have masses marked in 10- or 100-gram intervals, or in $\frac{1}{2}$ kg intervals; bathroom scales may be marked in kilograms and $\frac{1}{4}$ kilograms; measuring tapes and rulers are generally marked in centimetres, with millimetres shown but not labelled. The children need to know how to count the intervals, work out how much each one represents and use this information to read a measurement between the marked units accurately. As with all measuring topics, lots of exposure to practical examples and real-life measuring equipment will help the children develop this skill.

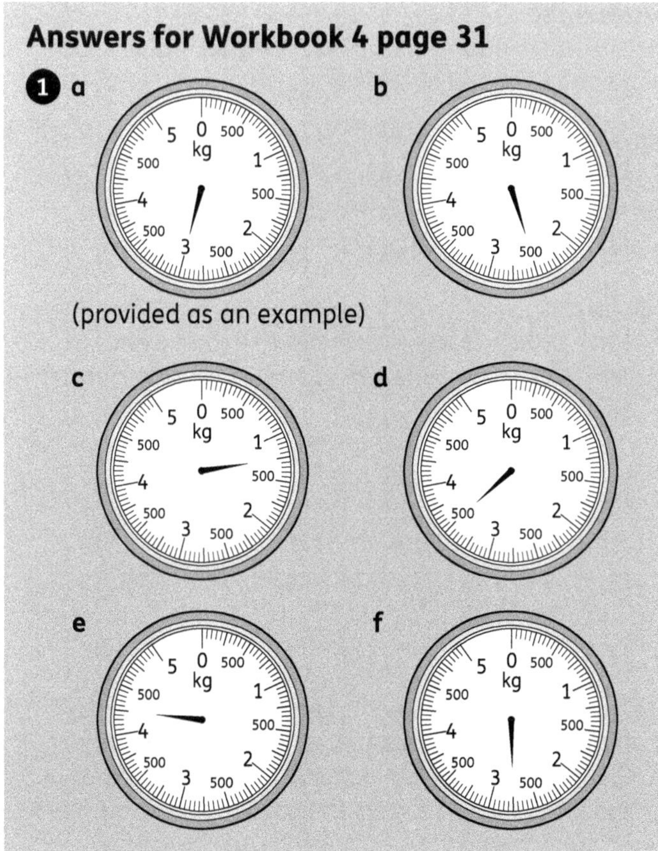

Answers for Workbook 4 page 31

1 a, b

(provided as an example)

c, d

e, f

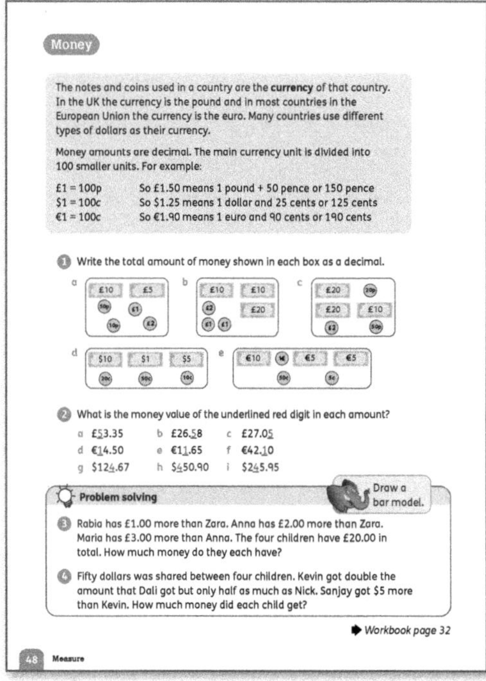

Money

• Tell the children to write 50p (or 50 small units of the local *currency*) more than each of their written prices.
• Finally, ask them to work out how much they have left over from 50p if they subtract each of their original prices from 50p.

Focus

• Ask the children to work in groups to make a list of the *coins* and *notes* used in your country.
• List these on the board and check whether they have included them all. If not, ask where you could find out about any other coins or notes (*from the Internet or the Central Bank*).
• Explain that every country uses money. Money comes in the form of coins (made from metal) and notes (made from paper, and increasingly from strong plastic polymers).
• Explain that this book uses mainly UK currency and that that is made up of pence and pounds.
• The UK currency items in circulation (in 2022) are:
  ○ coins: 1p, 2p, 5p, 10p, 20p, 50p, £1 and £2
  ○ notes: £5, £10, £20 and £50.
• Write some amounts of money up to £10 (or local currency) on the board, for example £8.65, £5.09, £9.85, £0.95, £9.08, 89p, £1. Ask the children to help you order them from smallest to greatest. (*89p, £0.95, £1, £5.09, £8.65, £9.08, £9.85*)
• Read through the examples on **Pupil Book 4 page 48** with the class to revise how to write money amounts as decimals. Money is written using decimals in virtually all currencies.
• The children can do question 1 using pounds, but you could ask them to use local currency if you prefer.
• For question 2, direct the children to look at the currency signs before they give the values of the underlined digits.
• Let the children work in pairs or on their own to complete **Workbook 4 page 32**. Observe as they do this to make sure that they are confident and able to work with money amounts. Most children have a fairly sophisticated concept of money, but as we move towards cards and cashless societies, they may not be as familiar with all the coins and notes.
• <u>Problem solving</u>: Read question 3 and question 4 on **Pupil Book 4 page 48** and let the children suggest strategies for solving each problem. Point out that they can use initials to represent the children rather than spend time writing out the full names.

Materials

Pictures or real examples of UK currency; coins and notes from your local currency.

Warm-up

• As a mental warm-up, ask the children to write down three 2-digit prices between 10 and 50 in a row in their books.

Interesting mistakes

Children sometimes use the £ sign and p sign in the same amount, for example, they might write £3.45p. remind them that the .45 part of the number is a fraction of the £ part, and therefore is covered by the £ sign. If you just write '45p' you are not using fractions, so you use the p sign. Show them real prices and talk about what they mean to help their understanding.

Answers for Pupil Book 4 page 48

1 a £18.60 b £44 c £52.70
d $16.80 e €21.55

2 a £50 b 50p c 5p
d €10 e €1 f €0.10 or 10c
g $4 h $400 i $40

3 Zara £3, Rabia £4, Anna £5, Maria £8

4 Kevin $10, Dali $5, Nick $20, Sanjay $15

Answers for Workbook 4 page 32

1

Pounds (£)	Pence (p)	Total in pounds and pence as a decimal
3	40	£3.40
13	5	£13.05
12	35	£12.35
5	99	£5.99
7	9	£7.09
14	0	£14.00

2 £1.88 = 1p + 2p + 5p + 10p + 20p + £1 + 50p
(provided as an example)

£2.99 = 50p + 50p + 50p + £1 + 20p + 20p + 5p + 2p + 2p

£4.00 = 20p + 20p + £2 + 50p + 10p + £1

£5.25 = £2 + £2 + 50p + 50p + 5p + 10p + 10p

£61.50 = £50 + £5 + £5 + £1 + 20p + 20p + 10p

3 a £39.50 £39.05 £35.90 £35.09 £30.95 £30.59

b £36 £40 £31 £35 £39 £31

c £212 d £424

Calculate money amounts

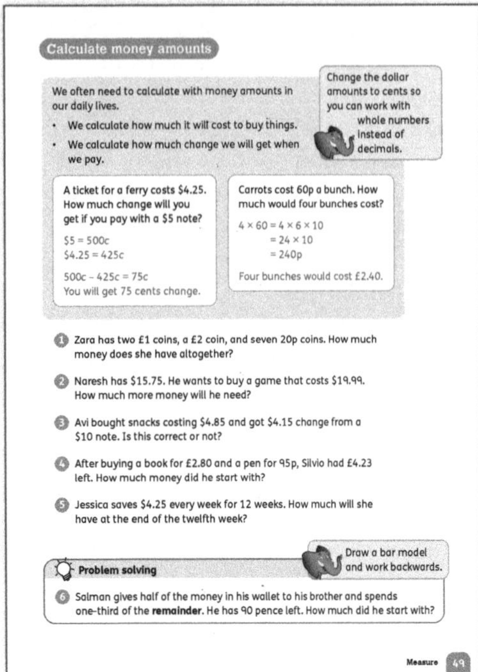

Materials

Pictures or real examples of UK currency; coins and notes from your local currency; play money; calculators.

Warm-up

As a mental warm-up, use the 'Counting in given steps' activity on page 23. Alternatively, use the 'Money problems' activity on page 27.

Focus

- Have a class discussion about how and when we calculate with money in our daily lives. Ask questions such as:
  - *When do we use and work with money amounts?*
  - *How do we have to think mathematically about money?*
  - *What calculations do we do with money?*
  - *How do you know how much to pay when you buy a few items?*
  - *How do you know if you have been given the right change?*
- The children may say things such as 'The cashier rings up the amounts and tells you what to pay.' Explore statements like these further to make sure that the children realise the cash register acts like a calculator to find the total and the change.

> In some places, amounts are rounded to avoid using small denomination coins. Note that money amounts are always rounded down, and not up so the rounding is in the customer's favour. If appropriate, discuss how informal traders work with money when they don't have electronic cash registers.

- Turn to **Pupil Book 4 page 49** and work through the examples with the class. Encourage them to suggest other methods of getting the same results.
- Remind the children that they can round money amounts to the nearest whole pound and calculate an estimate that they can use to check their answer is sensible. For example, to work out the total cost of two items costing £3.85 and £7.20 they can work out £4 + £7 = £11 to estimate the total.
- For question 1, the children can draw the money or use play money if necessary.
- Encourage the children to use a number line to help them work out question 2.
- Let the children discuss question 3 and share their methods of deciding.
- Encourage the children to change amounts to pence or pounds to help them work out question 4.
- The children may not know how to multiply decimals so encourage them to share strategies for working out question 5.

- <u>Problem solving</u>: The children may say that it is not possible to solve question 6 as they don't know the amounts. Ask them to imagine half a cake and remind them that even if they don't know the cake's size, shape or type, they can still think of half of it. So, they can think of Salman's money as the 'whole cake'. If they draw *bar models* (see page 22) using fractions, they will be able to work backwards to find the solution.
  For example:

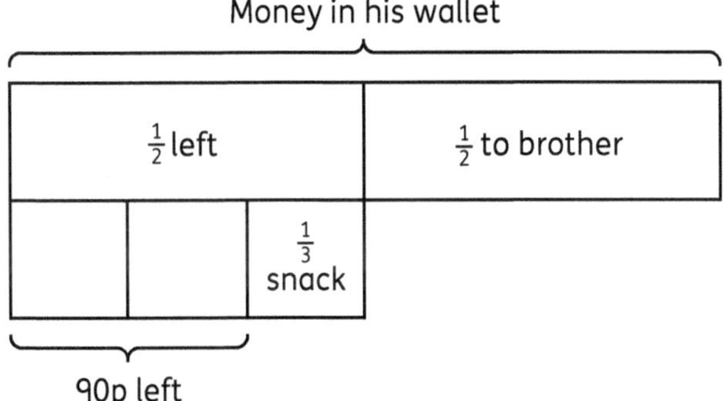

Support

Give the children a set of prices in pounds or local currency (such as £12.50, £7.25, £19.20). If you use local currency, choose appropriate amounts. Ask the children to work in pairs, with or without a calculator, to work out how much change a person would get if they paid for each of these items with a £20 note.

Interesting mistakes

The children may not always treat 0 correctly in money amounts. For example, they might write £5.50, £5.05 or £0.55 incorrectly. If they need support with this, make the equivalent of a place-value table for money like this:

Pounds (£)	.	Tens (pence)	Ones (pence)

This allows them to check amounts and helps them to work out why 'five pounds and five pence' is written as £5.05.

Answers for Pupil Book 4 page 49

1 £5.40 **2** $4.24

3 No, he should have received $5.15 in change.

4 £7.98 **5** €51.00 **6** £2.70

End-of-unit check

In addition to ongoing practical measuring tasks, use these questions and activities to assess how well the children have grasped, and can explain, the concepts in this unit.

- *What units would you use to measure the length of a kitten/a desk/the side of a triangle in your maths book? (centimetres for a kitten, metres and centimetres for a desk, millimetres or centimetres for a triangle)*

- *How many millimetres is 4 cm/6 cm/15 cm? (40 mm/60 mm/150 mm)*
- *How many centimetres is 20 mm/30 mm/84 mm? (2 cm/3 cm/8.4 cm)*
- Show the children two containers.
  - *Which of these two containers holds more?*
  - *How much more fits into this container than into that one?*
- *How can you measure how much a container holds?*
- Ask the children to find three containers of different shapes that all have the same capacity. (*for example, three 1 litre containers; three 250 ml containers*)
- *How many millilitres are there in a litre/half a litre/quarter of a litre? (1000 ml/500 ml/250 ml)*
- Show the children an object.
  - *What units would you use to weigh this object?*
  - *Do you think this object is heavier or lighter than a kilogram weight?*
- Ask the children to name three objects that they would weigh in kilograms (or grams). (*for example: kilograms: chair, dog, car; grams: book, mouse, apple*)
- *A chair weighs 2 kg and an exercise book weighs 50 g. How could you add these two masses? (convert 2 kg to 2000 g and add to make 2050 g or 2.05 kg)*
- *If I subtract 100 g from 0.5 kg I get 0.49 kg. Is that correct? (no, 0.5 kg – 100 g = 0.5 kg – 0.1 kg = 0.4 kg) Where did I make a mistake? (You converted 100 g to 0.01 kg instead of 0.1 kg.)*

To assess problem-solving skills, you can give the children word problems.

You could also develop some task cards involving equivalent measures. In the card below, the first balance shows that three triangles are equal in mass to one square. The children have to look at the other balances and decide whether they need to find something heavier or lighter, and then choose a suitable set from the ones given (or you can ask them to draw their own selections of squares and/or triangles).

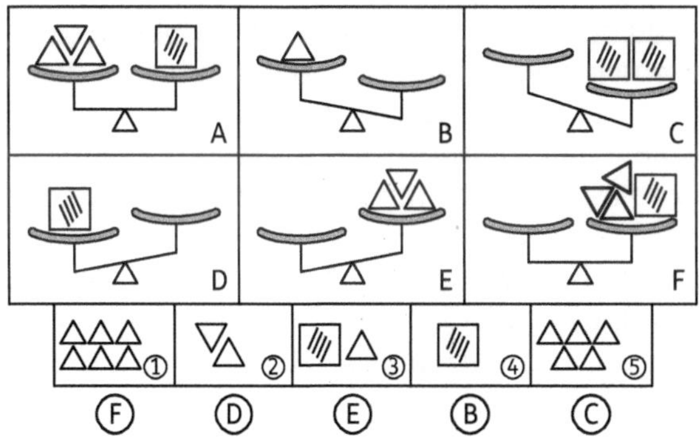

Count and calculate

• Expand the task by challenging the children to make their own unique patterns for different skip counts. They can use colour and shape, but they can also just combine sets of dots without shape outlines, for example, using curves, waves, flower petals and any other ideas they have.
• Encourage the children to be creative. Use the completed paintings to count the number of dots in any design using skip counting. If the designs are large and complex, you could just count the dots in a section of the work.
• Explain to the class that they have made counting patterns and that in this unit they are going to look in more detail at how patterns work and how we use patterns and place value to help us to calculate quickly and efficiently.

Learning objectives

• Recognise and extend number sequences
• Count on and back in steps of constant size
• Count in multiples of 6, 7, 9, 25 and 1000
• Apply place-value knowledge to known additive and multiplicative facts
• Reinforce and extend mental strategies for addition

Key words

count on   count back   skip count   compensate
partition   place value   bridge (tens, for example)
sum   total   multiple   model   mentally
method   sequence

Unit introduction

Materials

Pictures of pointillist paintings; poster paints and cotton buds or thick marker pens.

Teaching guidance

• Pointillism is a type of painting, using tiny dots instead of brushstrokes. It can be used as a fun way to introduce children to *skip counting* in different intervals. The children can make the dots using poster paint and cotton buds, or thick marker pens.
• Start by showing children some pictures of pointillist paintings. ('Beach at Heist' by Georges Lemmen is a good example, and you will find lots of other examples on the Internet.)
• If you have time, you could ask the children to research what this type of art is and to find their own examples before starting the activity.
• Explain that they are going to produce their own pointillist maths pictures using groups of dots (skip counting).
• The children will already be familiar with skip counting in twos, fives and tens so you can start with any of those numbers. Here are some examples showing skip counting in tens.

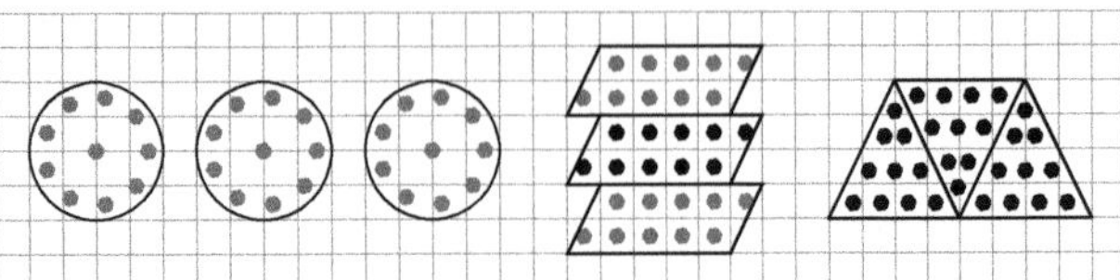

• Ask the children to make their own patterns using a given number and any shapes they like. Use the patterns to estimate and then count how many dots they used.

Count on and back

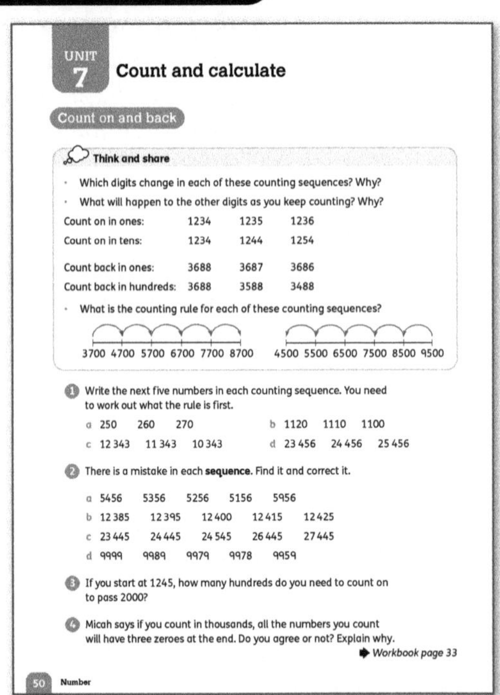

Materials

A large place-value table (see page 22) to thousands (for display); number grids (10–1000 and 100–10 000, see 'Focus'); calculators.

Warm-up

Display the place-value table and revise the names of the places (ones, tens, hundreds and thousands) as well as the term 'digit'. Then select one or more of the 'Place value and number sense' activities from pages 23–27 as a mental warm-up.

Focus

• <u>Think and share:</u> The children have already begun to explore skip counting patterns, so they can work through the questions on **Pupil Book 4 page 50** in groups before discussing their answers and ideas as a class.

- Allow different groups to share their ideas, particularly about how the other digits change when you continue counting in steps.
- Make sure the children understand that if you *count on* in tens (for example), only the tens digit changes unless you reach 9 tens, then the hundreds digit will increase by 1 (hundred). Similarly, if you *count back* in tens, the hundreds will decrease by 1 when you reach 0 in the tens place.

Display a 10–1000 grid like the one below (you could also reproduce this for each child if you prefer).

10	20	30	40	50	60	70	80	90	100
110	120	130	140	150	160	170	180	190	200
210	220	230	240	250	260	270	280	290	300
310	320	330	340	350	360	370	380	390	400
410	420	430	440	450	460	470	480	490	500
510	520	530	540	550	560	570	580	590	600
610	620	630	640	650	660	670	680	690	700
710	720	730	740	750	760	770	780	790	800
810	820	830	840	850	860	870	880	890	900
910	920	930	940	950	960	970	980	990	1000

- Choose a number (for example, 330) and discuss the values of the numbers to the left and right of it (ten less, and ten more), then look at the values of the numbers above and below it (hundred less and hundred more).
- Spend some time looking at the digits that 'change' when you move left and right (that is, the digits in the tens place) as well as the digits that change when you move up or down the grid (the digits in the hundreds place).
- Play some games where you draw a 3 × 3 section of this grid and place one number in it (you can vary the position for interest). The children have to work out what the missing numbers are.

Once the children are confident with numbers to 1000, repeat these activities for counting in hundreds and thousands using a 100–10 000 grid in intervals of 100 like the one below.

100	200	300	400	500	600	700	800	900	1000
1100	1200	1300	1400	1500	1600	1700	1800	1900	2000
2100	2200	2300	2400	2500	2600	2700	2800	2900	3000
3100	3200	3300	3400	3500	3600	3700	3800	3900	4000
4100	4200	4300	4400	4500	4600	4700	4800	4900	5000
5100	5200	5300	5400	5500	5600	5700	5800	5900	6000
6100	6200	6300	6400	6500	6600	6700	6800	6900	7000
7100	7200	7300	7400	7500	7600	7700	7800	7900	8000
8100	8200	8300	8400	8500	8600	8700	8800	8900	9000
9100	9200	9300	9400	9500	9600	9700	9800	9900	10 000

- Let the children discuss the patterns in question 1 on **Pupil Book 4 page 50** and identify the rules orally before they write the next numbers in each *sequence*. Allow them to use a calculator to check their own answers.

- You can do question 2 orally with the class. Give the children time to think and ask them to show when they have an answer before asking different children to share their ideas.
- The children can refer back to the charts you displayed to work out question 3.
- Let the children think about question 4 on their own before asking them to share their ideas in pairs or groups.

Follow-up

Use **Workbook 4 page 33** to consolidate counting on and back in ones, tens, hundreds and thousands. Let the children complete the questions and then compare and check each other's answers using a calculator if necessary.

Interesting mistakes

Some children might have difficulty counting in tens when they have to cross the hundreds, and in hundreds when they have to cross the thousands. Use counting grids and charts to help them work out the answers and also to practise counting aloud in different steps to bridge the barriers.

Answers for Pupil Book 4 page 50

<u>Think and share:</u> When you count on in ones, the ones digit changes. When you reach 9 and for your next count, the tens digit also changes and goes up by 1 and the ones digit is zero. We have now counted up 10.

When you count on in tens, the tens digit changes. After you reach 9 tens, the hundreds digit also changes and goes up by 1 and the tens digit is zero. We have now counted on 100.

When you count back in ones, the ones digit goes down. After you reach the ones digit 0, the tens digit also changes and goes down by 1 and the ones digit is 9. We have now counted back 10.

When you count back in hundreds, the hundreds digit does down. The thousands digit goes down by 1 once you go past zero hundreds. We have counted back 1000.

Rule 1: Count on in thousands
Rule 2: Count back in thousands

1 a count on in tens; 280, 290, 300, 310, 320

   b count back in tens: 1090, 1080, 1070, 1060, 1050, 1040

   c count back in thousands: 9343, 8343, 7343, 6343, 5343

   d count on in thousands: 26 456, 27 456, 28 456, 29 456, 30 456, 31 456

2 a 5956 should be 5056

   b 12 400 should be 12 405

   c 24 545 should be 25 445

   d 9978 should be 9969

3 8 hundreds

4 No. Possible explanation: It depends which number you start at. For example, if you start at 27 and count on in thousands you will get 27, 1027, 2027, . . . – these numbers do not have 3 zeros. But if you start at 0 and count in thousands, you will get 1000, 2000, 3000, . . . – these numbers do have 3 zeros.

Answers for Workbook 4 page 33

1 a 8465, 8466, 8468, 8469

   b 2020, 2030, 2050, 2070

   c 3910, 4310, 4410

   d 3700, 4700, 7700, 8700

2 2643, 2743, 1743, 1742, 1732, 1832, 2832, 2833, 2733

3 a 5000   b 3430   c 4450   d 2080

   e 9999   f 1999

Count in multiples of 6, 7 and 9

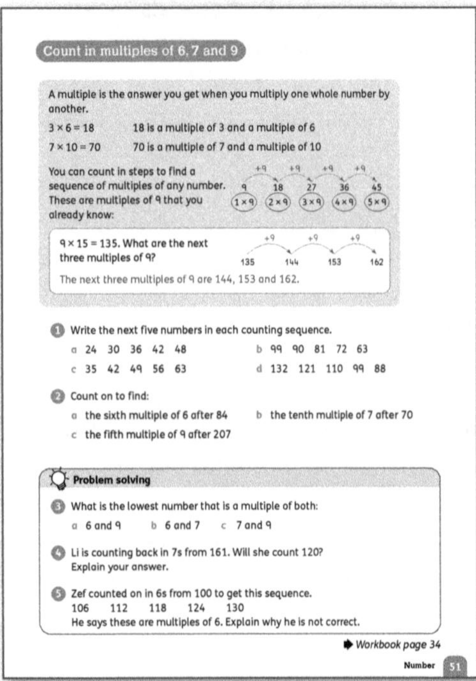

Materials

Interlocking cubes; skip counting game boards (see below).

Warm-up

Revise the 6, 7 and 9 times tables as a mental warm-up.

Focus

Read through the information on **Pupil Book 4 page 51** with the class. Make sure that they understand what a *multiple* is.

- Let the children work through the example in the white box on their own.
- Ask questions to make sure that they understand, for example: *Why are you told that 9 × 15 = 135?* (to have a multiple of 9 to start from) *Why are there three jumps of 9?* (*You have to find the next three multiples of 9, and it is easier to add nines than to multiply 16 × 9, 17 × 9 and 18 × 9.*) *How do you easily add 9?* (*Add 10 and subtract 1.*)
- Ask the children to complete the questions. They should be able to do question 1 using known facts.
- For question 2, the children can record the numbers they count or work *mentally*. Some children will realise that they can multiply and add to find the end number, which is mathematically correct.
- <u>Problem solving</u>: For question 3, you may need to show the children how to list multiples to find the lowest number that appears in both sets.
- The children can either count back to work out question 4, or find the difference between 120 and 161 and check if this is divisible by 7.
- Let the children discuss question 5 in pairs.

Follow-up

Turn to **Workbook 4 page 34** to combine skip counting in sixes, sevens and nines with finding rules for counting sequences. Let the children work through the questions in pairs and share the answers as a class.

Challenge

Let the children work in pairs to design games or activities to practise skip counting in multiples of 6, 7 and 9.

Support

If any children need support counting in multiples, they can use interlocking cubes. Design a set of skip counting game boards like the one below (or let the children design these and exchange with others). The children can then put interlocking cubes on the board to help them work out the missing numbers.

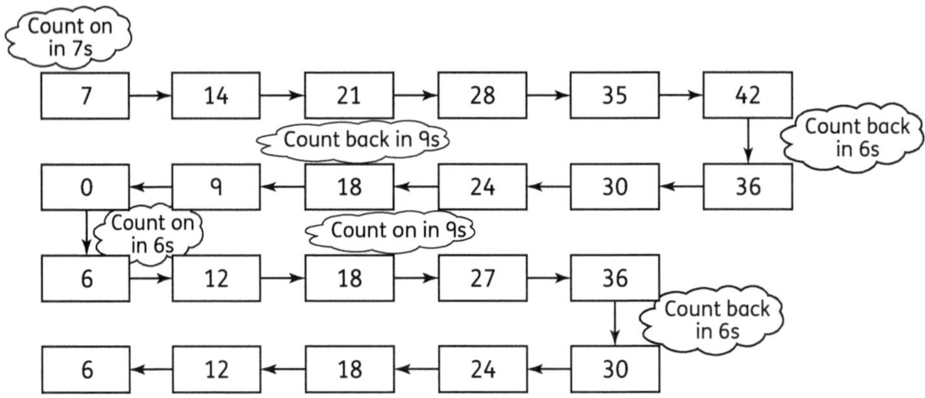

Answers for Pupil Book 4 page 51

1 a 54, 60, 66, 72, 78   b 54, 65, 36, 27, 18
c 70, 77, 84, 91, 98
d 26 456, 27 456, 28 456, 29 456, 30 456

2 a 120   b 140   c 252

3 a 18   b 42   c 63

4 No, 161 is a multiple of 7, but 120 is not a multiple of 7. She will count 119 instead.

5 106 is not a multiple of 6. If you do not start on a multiple of 6 when you count in sixes, then all the numbers in the sequence will not be multiples of 6 either.

Answers for Workbook 4 page 34

1 a 202, should be 204   b 281, should be 282
c 447, should be 448   d 170, should be 172

2 a 84, 105, 112
b 126, 120, 108 c 171, 153, 144
d 5076, 5085, 5094, 5103
e 1638, 1624, 1610

3 These numbers circled and ordered from smallest to greatest: 18, 27, 36, 63, 72, 99, 108

4 Individual answers

Count in 25s

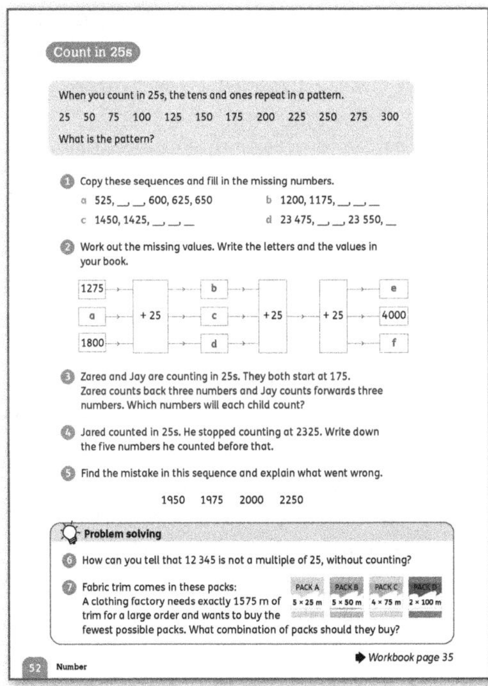

Materials

A chart with multiples of 25 from 25 to 1000 (for display, as shown in 'Focus'); sticky notes; calculators.

Warm-up

Use any suitable 'Place value and number sense' activity from pages 23–27 as a mental warm-up.

Focus

The idea of skip counting and number patterns is developed further in this lesson, where children learn to count in twenty-fives.

- Ask the children to look closely at the number sequence on **Pupil Book 4 page 52**.
- Give them some time to discuss the pattern. (The tens and ones digits repeat the 25, 50, 75, 00 pattern as you count up.) The children should recognise that they have seen similar patterns in multiples of 5.
- Let the children use a calculator to count up in twenty-fives to a higher and higher range to see that this pattern repeats. You could give them a starting number such as 3000, confirm that it's a multiple of 25 (as the tens and ones digits fit the pattern) and have them count up in twenty-fives from there.
- Display a large chart like the one below. Use sticky notes to cover some of the numbers. Point to a covered number and ask a child to say what it is. Remove the sticky note to confirm.

25	50	75	100
125	150	175	200
225	250	275	300
325	350	375	400
425	450	475	500
525	550	575	600
625	650	675	700
725	750	775	800
825	850	875	900
925	950	975	1000

- Leave the chart on display in the classroom for a while and let the children complete the questions on **Pupil Book 4 page 52**.
- The children can work on their own to complete question 1. Let them use a calculator to check their answers.
- Make sure that the children understand how the flow diagram works before they complete question 2. They should understand that the number on the left is the input, the +25 blocks are operations which give new numbers that form the input for the second part of the diagram. The end result is the output. When they have the output, they can work backwards and use inverse operations to find the input. Again, they can check their answers using a calculator.
- The children can use the chart shown above to work out question 3 if they need to.
- Remind the children that we can count on and back in twenty-fives. To answer question 4, they need to count back.
- Let the children do question 5 orally in groups or pairs.
- <u>Problem solving</u>: Question 6 is a reasoning activity. The children need to think about what they have learnt and explain how they can tell that this is not a multiple of 25.

- Remind the children to make an organised list and to keep track of the options they try as they aim to solve question 7.

Follow-up

Use **Workbook 4 page 35** to consolidate skip counting on and back in twenty-fives and thousands.

Give the children time to consider the patterns and to talk about three that they find interesting.

Answers for Pupil Book 4 page 52

1 a 550, 575 **b** 1150, 1125, 1100
   c 1400, 1375, 1350 **d** 23 500, 23 525, 23 575

2 a = 3925, b = 1300, c = 3950, d = 1825, e = 1350, f = 1875

3 Zarea: 150, 125, 100; Jay: 200, 225, 250

4 2200, 2225, 2250, 2275, 2300

5 2250 should be 2025

6 Possible answer: Any multiple of 25 should end in 25, 50, 75 or 00. The last two digits of 12 345 do not fit this pattern.

7 one of Pack A, one of pack B and four of Pack C

Answers for Workbook 4 page 35

1 Chart A

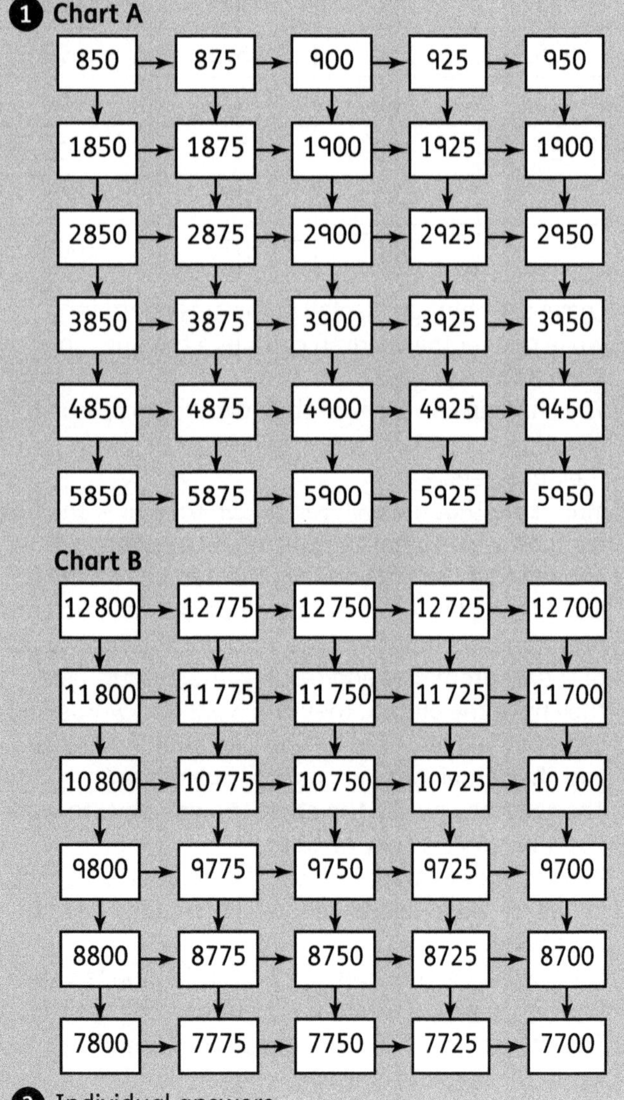

850	875	900	925	950
1850	1875	1900	1925	1900
2850	2875	2900	2925	2950
3850	3875	3900	3925	3950
4850	4875	4900	4925	9450
5850	5875	5900	5925	5950

Chart B

12 800	12 775	12 750	12 725	12 700
11 800	11 775	11 750	11 725	11 700
10 800	10 775	10 750	10 725	10 700
9800	9775	9750	9725	9700
8800	8775	8750	8725	8700
7800	7775	7750	7725	7700

2 Individual answers

Count in multiples to add or subtract

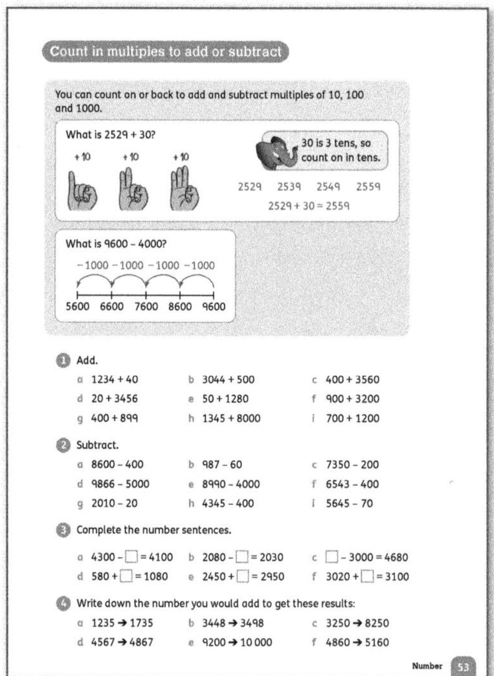

Count in multiples to add or subtract

You can count on or back to add and subtract multiples of 10, 100 and 1000.

What is 2529 + 30?

+10 +10 +10

30 is 3 tens, so count on in tens.

2529   2539   2549   2559

2529 + 30 = 2559

What is 9600 − 4000?

−1000 −1000 −1000 −1000

5600   6600   7600   8600   9600

1 Add.
   a 1234 + 40   b 3044 + 500   c 400 + 3560
   d 20 + 3456   e 50 + 1280   f 900 + 3200
   g 400 + 899   h 1345 + 8000   i 700 + 1200

2 Subtract.
   a 8600 − 400   b 987 − 60   c 7350 − 200
   d 9866 − 5000   e 8990 − 4000   f 6543 − 400
   g 2010 − 20   h 4345 − 400   i 5645 − 70

3 Complete the number sentences.
   a 4300 − ☐ = 4100   b 2080 − ☐ = 2030   c ☐ − 3000 = 4680
   d 580 + ☐ = 1080   e 2450 + ☐ = 2950   f 3020 + ☐ = 3100

4 Write down the number you would add to get these results:
   a 1235 → 1735   b 3448 → 3498   c 3250 → 8250
   d 4567 → 4867   e 9200 → 10 000 f 4860 → 5160

Number 53

Materials

10–1000 and 100–10 000 charts (see 'Focus' in 'Count on and back' on page 83); calculators (optional).

Warm-up

Use any suitable 'Counting backwards and forwards' activity from page 23 as a mental warm-up.

Focus

- Display the 10–1000 chart again. Have a number talk (see pages 17–18) to introduce the idea of finding a number that is a multiple of 10 or 100 more than a given number. For example, ask:
  - *How could you find the number that is 300 more than 390? (300 is three jumps of 100, so move down three blocks, or increase the hundreds digit by 3 to get 690.)*
  - *What is 80 more than 140? (80 is 8 tens, add 6 tens to get to 200 and then add another 2 tens to get to 220.)*
  - Ask more questions and let different children explain how they used the chart to find the answer.
- Work through and discuss the addition and subtraction strategies using counting on **Pupil Book 4 page 53**.
- Revise this using the 10–1000 and 100–10 000 charts if necessary, before asking the children to complete all the activities independently.
- Spend some time discussing their *methods* to make sure that they understand the ideas.
- The children can work through questions 1–4 independently, and then check their answers in pairs or using calculators.
- Discuss the strategies the children used to answer question 4.

Answers for Pupil Book 4 page 53

1 a 1274   b 3544   c 3960
   d 3476   e 1330   f 4100
   g 1299   h 9345   i 1900

2 a 8200   b 927   c 7150
   d 4866   e 4990   f 6143
   g 1990   h 3945   i 5575

3 a 200   b 50   c 7680
   d 500   e 500   f 80

4 a 500   b 50   c 5000
   d 300   e 800   f 300

Make 100

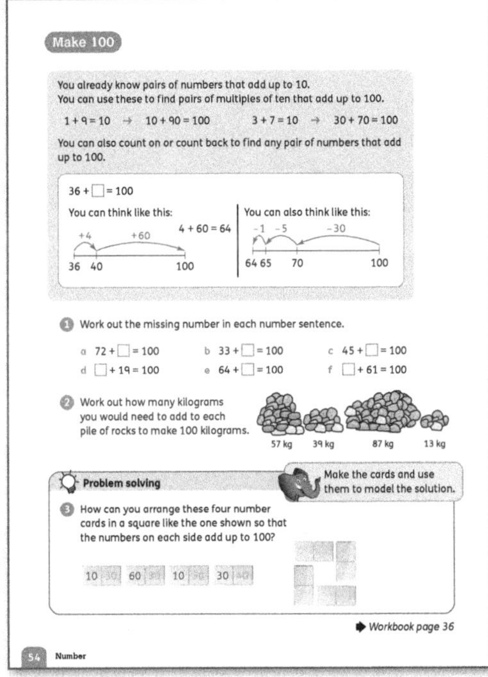

Materials

Sets of digit cards 1–9; blank cards; ruler marked in millimetres.

Warm-up

- As a mental warm-up, play a game to reinforce number bonds to 10. Give each child a digit card with one of the numbers from 1 to 9.
- Display a 2-digit number (for example, 47) and ask the children what they'd need to add to it to make the next multiple of 10. For example, for 47 the next 10 is 50, so the children with a card showing 3 need to hold it up.
- Repeat this for a number of different values, varying the number in the ones position to give everyone a chance to answer.

Focus

- Ask children to work in pairs to read through and discuss the examples on **Pupil Book 4 page 54**.
- Let the children work on their own to complete question 1 and question 2.
- <u>Problem solving</u>: For question 3, the children can make their own cards and physically move them around to help them *model* the solution. Most children will use some form of trial, error and improvement to solve problems like this one.

Follow-up

Use **Workbook 4 page 36** to consolidate bonds to 100 as well as inverse operations. Let the children compare answers and check each other's work.

Support

The children can use a ruler marked in millimetres to help them work out bonds to 100 (and 1000 if they can scale the answers by 10). This also helps reinforce work on estimating and measuring length. Show the children how to find an amount on the mm scale, for example, 46. They should read this from 0. They can then use the ruler to work out that they need 54 to get to 100 mm (jumping along the ruler much like a number line).

Answers for Pupil Book 4 page 54

1 a 28   b 67   c 55
   d 81   e 36   f 39

2 43 kg; 61 kg; 13 kg

3

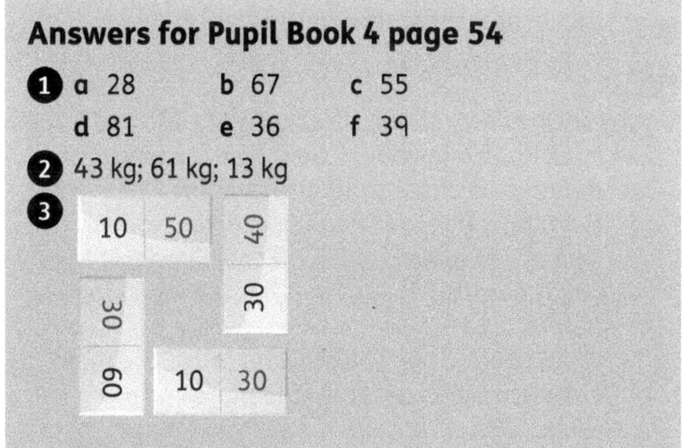

Answers for Workbook 4 page 36

1 55 cm

2 a 21 + 79 = 100   100 − 21 = 79   100 − 79 = 21
   b 25 + 75 = 100   100 − 25 = 75   100 − 75 = 25
   c 42 + 58 = 100   100 − 42 = 58   100 − 58 = 42
   d 51 + 49 = 100   100 − 51 = 49   100 − 49 = 51
   e 64 + 36 = 100   100 − 64 = 36   100 − 36 = 64
   f 10 + 90 = 100   100 − 10 = 90   100 − 90 = 10
   g 87 + 13 = 100   100 − 87 = 13   100 − 13 = 87
   h 52 + 48 = 100   100 − 52 = 48   100 − 48 = 52
   i 23 + 77 = 100   100 − 23 = 77   100 − 77 = 23

3 a 75p   b 87p   c 41p   d 28p

Make 1000

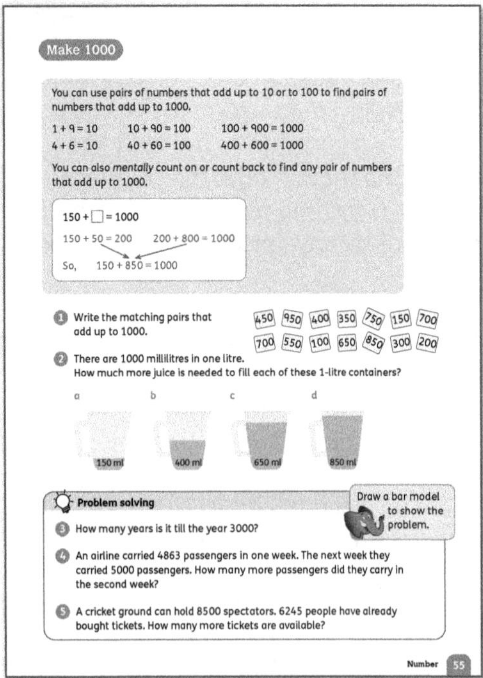

Warm-up

Use one of the 'Calculation skills' activities on pages 28–29 as a mental warm-up.

Focus

- Use the pairs of numbers that *total* 10, 100 and 1000 on **Pupil Book 4 page 55** to reinforce the idea that 100 is ten times 10 and that 1000 is ten times 100. Ask:
  - *What happens to the numbers in each addition when you scale up to 100? (Each number in the pair that totals 100 is ten times the matching pair that totals 10.)*
  - *What about when you scale up to 1000? (Again, each number is ten times the matching pair that totals 100.)*
- Display a few more number bonds to 10. Let the children write the matching pairs for 100 and 1000.
- Once they can do this, display a few number bonds to 100 that are not multiples of 10, for example, 72 + 28 = 100. Ask the children to predict the matching pair for 1000. They should work out that this is 720 + 280.
- Use the worked example to explain how you can count up to the next 10, 100, or any number you feel comfortable with, and then use known facts and scaling to find the matching number to make 1000 (or any multiple of 1000).
- Make sure the children realise that they need to use the values in question 1 to make pairs that total 1000. They will not use all the numbers. You can ask them to list the ones they don't use and to write what they would add to each one to make 1000.
- The children use number bonds and scaled facts to work out question 2.
- <u>Problem solving</u>: Encourage the children to draw simple bar models (see page 22) to show each problem in questions 3–5. They can use these to help them find the differences.

Interesting mistakes

- The children may attempt to count on instead of recalling addition facts to find pairs to 100 and 1000. You can address this by discussing methods and also by practising addition facts and combinations. Encourage the children to jot down addition facts because this often makes the connection clearer.
- The children may make place-value errors when they are asked to scale up addition facts. For example, if they know that 44 + 56 = 100, they may write that 404 + 506 = 100 instead of 440 + 560. Let them use a calculator to check their addition and they will see that they have made a mistake. Remind them that 1000 is ten times 1000, and that when you multiply whole numbers by 10 (to get the pair for 1000) both numbers will have a 0 in the ones place.

Answers for Pupil Book 4 page 55

1. 450 and 550; 350 and 650; 150 and 850; 700 and 300
2. **a** 850 ml **b** 600 ml **c** 350 ml **d** 150 ml
3. Individual answers
4. 137 passengers
5. 2255 tickets

Add multiples of 10

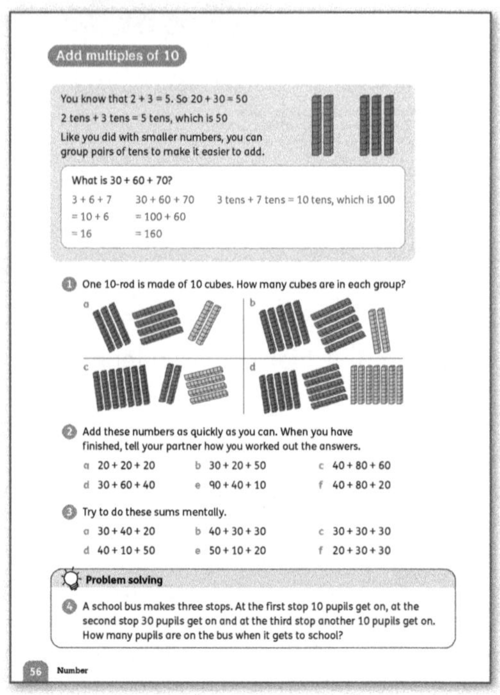

Warm-up

- As a mental warm-up, ask the children to write down as many ways as they can to make 15 by adding three 1-digit numbers.
- Write some children's suggestions on the board under the heading 'Makes 15'. For example, they might suggest:

Makes 15
1 + 5 + 9
3 + 4 + 8
2 + 7 + 6

Focus
- Add the heading 'Makes 150' to the right of the 'Makes 15' heading. Ask: *How could you make 150 using each of these additions as a starting point? (150 is 10 times 15, so you can scale up the calculation by multiplying each number by 10.)*
- Using the first number sentence, for example, you could make 10 + 50 + 90. Say this as: *1 ten plus 5 tens plus 9 tens, which is 15 tens*, to reinforce the idea of mental strategies using known number facts. Keep these lists on display for the next lesson.
- Turn to **Pupil Book 4 page 56**. Read through the example. Let the children work on their own to complete the calculations in questions 1–3. Give them time to discuss the strategies they use.
- <u>Problem solving</u>: Encourage the children to show their working clearly for question 4, either by using diagrams or by writing their calculations.

Answers for Pupil Book 4 page 56
1 a 90   b 120   c 130   d 170
2 a 60   b 100   c 180   d 130
   e 140   f 140
Possible answers: look for bonds to 100; use number bonds to ten and add the tens digits and then make your answer ten times greater
3 a 90   b 100   c 90   d 100
   e 80   f 80
4 50 pupils

Mental strategies for adding

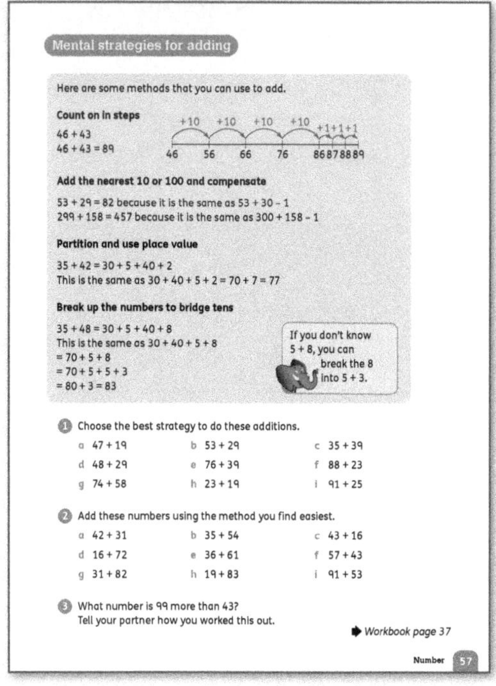

Materials
100 chart; sets of 10p coins or place-value counters (see **Pupil Book 4 page 8**) or base-ten 10-rods and ones cubes (see page 21).

Warm-up
- Draw a spider diagram with 8 legs on the board. Write a 2-digit number in the centre of the diagram. On the 8 legs write: + 100, + 10, + 9, + 99, + 11, + 8, + 98, + 12.
- Choose children to give the answer for the addition on each leg. Discuss their methods and encourage them to add the nearest 10 or 100 and compensate (for example, to add 99, they add 100 and subtract 1).
- Repeat with a 3-digit number written in the centre of the diagram.

Focus
- Turn to **Pupil Book 4 page 57**. The children have already used these mental strategies:
  - counting on in steps
  - adding the nearest 10 or 100 and compensating
  - *partitioning* and using *place value*
  - breaking up the numbers to *bridge tens*.
- Let them work in small groups or pairs to read through and talk about the examples.
- Then the children can complete questions 1–3 using the strategies they find most efficient.

Follow-up
Use **Workbook 4 page 37** for additional practice or as a homework task.

Challenge
Give the children a set of mixed 2-, 3- and 4-digit numbers and ask them to write additions involving two or three numbers to make each total.

Support
For children who are less confident adding multiples of 10, provide a 100 chart and show them how to move down the grid in columns. For example, start on 40 and add 30 by moving down three places in the table. It may also help to provide sets of 10p coins (or place-value counters or base-ten 10-rods and ones cubes) to let them group and add using the concrete apparatus until they no longer need it.

Interesting mistakes
When adding numbers that bridge multiples of ten, the children may forget to include the *sum* of the ones in the answer. For example, 36 + 34 = 70, but they may write 60. Talk about the methods they use, for example, ask:
- *What do you get when you add 6 to 4? (10)*
- *How many more tens do you have? (6 more tens, 30 + 30)*
- *How many tens does that make? (7 tens)*

Do the same for numbers such as 37 + 45, where you will get a ten and 2 ones. Use jottings to help children keep track.

Answers for Pupil Book 4 page 57

1   **a** 66   **b** 82   **c** 74   **d** 77
   e 115   **f** 111
   g 132   **h** 42   **i** 116

2   **a** 73 **b** 89   **c** 59   **d** 88   **e** 97
   f 100   **g** 113   **h** 102   **i** 144

3   142 Possible method: Add 100 and 43 and then take away 1.

Answers for Workbook 4 page 37

1   **a** 39 + 13 = 52 (provided as an example)
   b 41   **c** 26   **d** 85   **e** 56
   f 59   **g** 82   **h** 75   **i** 33
   j 111   **k** 41

End-of-unit check

Use some or all of these questions and activities to assess how well the children have understood the concepts in this unit.

- *What number is* (give a number) *more than* (give a number)?
- *How many more is 1234 than 1034?* (200) (and similar)
- *What must I add to 2454 to get 2554/2464/3454?* (100/10/1000) (and similar, including subtractions)
- *Increase/Decrease 4567 by 400.* (4967/4167) (and similar using multiples of 100)
- *I have 650 grams of sugar. How much more sugar do I need to make 1000 grams?* (350 g) (and similar, including capacity measures)
- Display a set of number cards and ask the children to find pairs that make 100 (or 1000). Include some numbers that do not make pairs.
- Have a number talk (see pages 17–18) with the class about mental strategies. Display a calculation and let the children describe their strategies for finding the answer. Discuss why these strategies are useful in mathematics and when the children might use them.

Mixed practice 1

You can use Mixed practice 1 on **Pupil Book 4 pages 58–59** to assess the children's confidence in the concepts from Units 1–7.

Answers for Mixed practice 1, Pupil Book 4 pages 58–59

1   **a** 4433   **b** 3003

   c

   d 4400 and 3000

2   Set B

3   **a** square   **b** pentagon
   c rhombus   **d** regular hexagon

4   **a** stop 137   **b** stop 132   **c** stop 136

5   **a** A = 0.2, B = 0.5, C = 0.9, D = 1.25, E = 1.8
   b 0.21, 0.51, 0.91, 1.26, 1.81
   c 0.05

6   c: 12.7 → 130 should be 12.7 → 13; d: 27.9 → 27 should be 27.9 → 28

7   **a** A = 0.5 kg, B = 1.3 kg, C = 1.7 kg, D = 2.9 kg
   b 1 kg, 1 kg, 2 kg, 3 kg
   c 2.4 kg   **d** 6.4 kg

8   **a** red tank   **b** 70.95 litres

9   500 g flour, 100 ml oil, 250 g cheese, 8 small tomatoes, 400 g chopped onions and peppers, 1 g chilli powder

10   £14.35

11   **a** 48, 42, 36   **b** 85, 95, 105
   c 135, 144, 153   **d** 450, 475, 500

Learning objectives

- Identify lines of symmetry in different shapes
- Complete a symmetrical shape using the line of symmetry
- Identify lines of symmetry on shapes and patterns

Key words

symmetrical   symmetry   line of symmetry
horizontal   vertical   diagonal   regular

Unit introduction

Materials

Counters in two colours; sheets of paper.

Teaching guidance

Place the children in groups. Give each group eight counters (four of one colour and four of another) and a sheet of paper with a dashed line drawn down the middle. Ask them to investigate how many different ways there are to arrange the counters in a row so that the arrangement is *symmetrical* about the line.

- Here is one way to get them started:

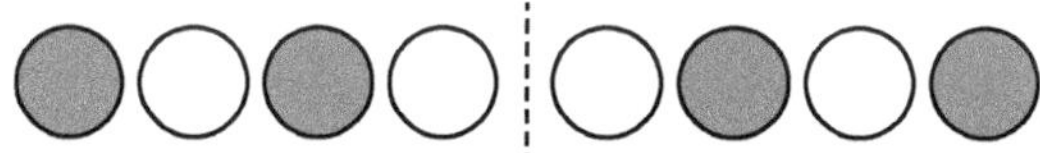

- Give the children some time to explore this and then ask them to share their answers. Ask them how they decided that they had all possible solutions.
- There are six possible ways of arranging the eight counters in a row symmetrically. If the children have more, they have repeated some; if they have fewer, they have not found them all. Here are the other five ways:

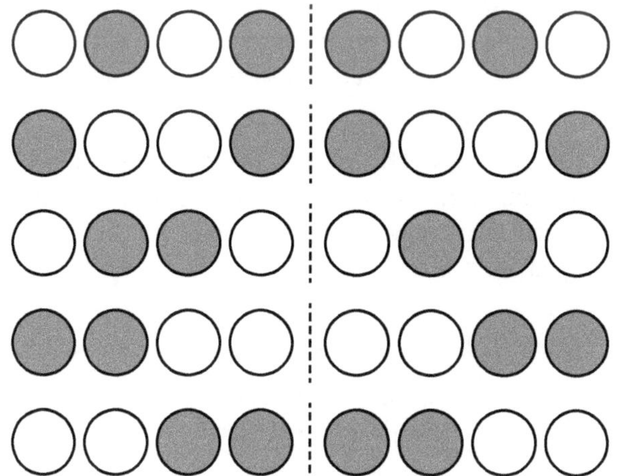

To extend this task, ask questions such as:

- *What happens if you add one more counter of each colour? (It is not possible to make symmetrical arrangements with five of each colour.)*

- *What if you have nine counters in total (four of one colour and five of the other)? Can you still make symmetrical arrangements? (It is only possible if you place the extra counter on the dashed line so that half of it is in each half of the arrangement. There are six possible ways to do this.).*

Is it symmetrical?

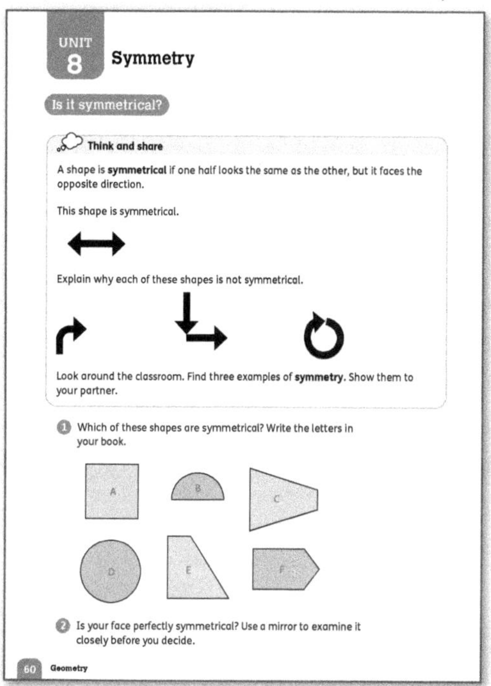

Materials

Large 2D paper shape (such as a circle or a rectangle); mirrors; access to digital cameras and photo-manipulation software (optional); thin paper for tracing.

Warm-up

- Revise the term and the concept of *symmetry* with the class, referring back to the activity with the counters from the unit introduction (see opposite). Explain that shapes are symmetrical when one half is an exact 'mirror image' of the other.
- Show the children a large symmetrical 2D shape cut out of paper, for example, a circle or a rectangle. Tell them that if you fold the shape in half, the term for the fold line is the *line of symmetry*.
- Hold a mirror along the fold line to show how half of the shape is reflected in the mirror to make the whole shape.

Focus

- <u>Think and share:</u> Read the information on **Pupil Book 4 page 60** with the class. Ask them to place a small mirror across the middle of the double arrow shape to see how the shape is symmetrical. Alternatively, do this as a class demonstration.

- Let the children talk about the other arrow shapes and then take feedback about why they are not symmetrical. Encourage the children to reason mathematically, using the words *half, identical* and *mirror image* in their explanations.
- Ask the children to look at shapes around the room (for example, doors, windows, floor tiles, brick patterns, ventilation grids and so on) and identify which are symmetrical and where the lines of symmetry lie. It is useful to have a mirror to demonstrate the symmetry if the children cannot see where the line of symmetry is.
- Let the children work in pairs to decide whether the shapes in question 1 are symmetrical or not. Discuss the answers with the class and let the children say how they decided.
- For question 2, in their pairs, the children decide whether their faces are symmetrical. Human faces are not symmetrical. If possible, let the children manipulate a photograph of their face using a computer. Snipping half of their face and flipping it to make two identical but opposite halves, will show them how different their face would look if it was completely symmetrical.

Answers for Pupil Book 4 page 60

<u>Think and share:</u> The shapes are not symmetrical because there is no way to divide the shapes in half so that the two halves look the same but face the opposite direction.

1 A, B, C, D, F

2 No (faces are never perfectly symmetrical).

Lines of symmetry

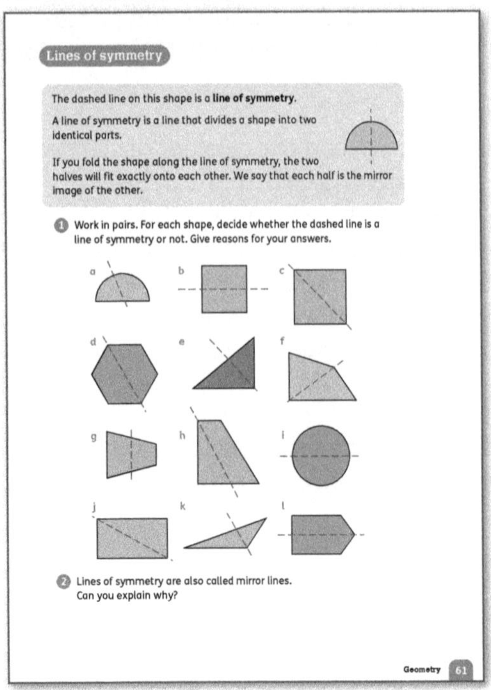

Materials
Semicircle cut out of paper or card; paper shapes with one or two lines of symmetry, that can be folded; small mirrors; rulers.

Warm-up
- Show the children the semicircle you have made. Fold it in half and point out that the fold makes a line. Remind them that this is called the line of symmetry. The two halves on either side of the line will fit exactly onto each other when you fold the shape along that line. Now fold in a different place, and show that this is not a line of symmetry.
- Give each group one or two paper shapes. Ask them to fold the shapes to make a line of symmetry. Have them mark the lines using a ruler.
- Let the groups show the class their shapes with the line(s) of symmetry marked.

Focus
- Turn to **Pupil Book 4 page 61**. For question 1, let the children work in pairs to decide whether the dashed lines on the shapes are lines of symmetry. Check the answers orally with the class.
- Discuss question 2 with the class.

Challenge
Let the children do a small project to find examples of symmetry in the local environment. If they have access to digital cameras, they could record these photographically and prepare a slide show. If not, they could draw the symmetrical patterns they find and make a poster to show their designs.

Support
Let the children use a small mirror to check whether the line is a line of symmetry. Alternatively, give the children thin paper and let them trace the shapes and fold them to check whether the line is a line of symmetry or not.

Interesting mistakes
The children may find it difficult to visualise what a shape will look like if folded. They may also think that if the two parts are identical, they are also symmetrical. This means that many children will say that rectangles and other parallelograms are symmetrical about their *diagonals* (because the diagonal divides the shape into two equal parts).

If the children make this mistake, encourage them to trace and fold shapes to check, or use a mirror to check whether the two halves fit onto each other (without rotating either one).

Answers for Pupil Book 4 page 61

1
a no	b yes	c yes	d yes
e no	f no	g no	h no
i yes	j no	k no	l yes

2 Possible answer: A line of symmetry is called a mirror line because if you put a mirror along the line of symmetry, half of the shape is reflected in the mirror so you see the whole shape.

Symmetrical shapes

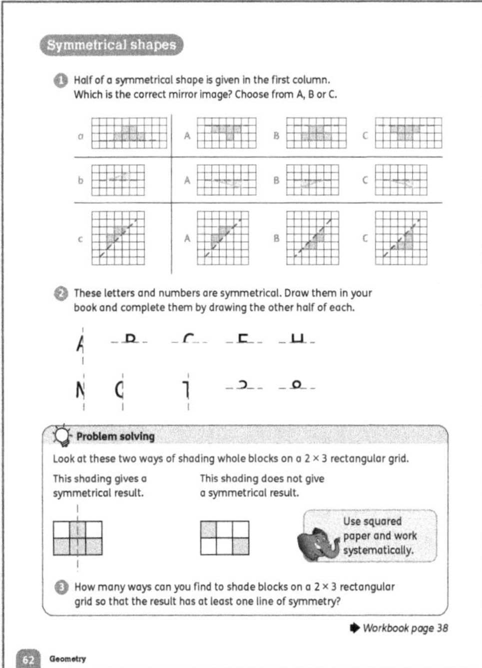

Materials
Small mirrors; squared paper; scissors.

Warm-up
Ask each child to draw their own initials, and decide whether each letter is symmetrical. Then ask them to repeat with all the letters in their names. The children then share their answers with the class.

Focus
The questions in this lesson will help the children to consolidate their ideas about symmetry and allow them to explore symmetry in patterns.

- Make sure that each group has a mirror. Use question 1 on **Pupil Book 4 page 62** to revise the idea that placing a mirror on the line of symmetry gives you a view of the other, symmetrical, half of the shape. Let the children use their mirrors to find the matching part of each shape.
- Let the children complete question 2 on their own. They can use a mirror to find the other half of each letter or number if they cannot visualise it.
- Problem solving: Hand out squared paper to each pair of children. Read through question 3 with the class. Make sure that the children understand the task before asking them to complete it. They can shade any number of blocks as long as the pattern is symmetrical. They can have a vertical or horizontal line of symmetry. (You may wish to demonstrate this.) When they have finished, share all the results as a class.

This problem-solving activity is a good opportunity to reinforce or teach children that they need to work systematically and keep track of what they have done to make sure that they don't repeat patterns or leave out any possible answers.

Follow-up
Let the children complete question 1 and question 2 on **Workbook 4 page 38** to consolidate the concept of completing a shape using a single, given line of symmetry.

Problem solving: Use questions 3 and 4 to informally assess whether the children are able to complete and draw shapes using two given lines of symmetry.

Challenge
If the children found the problem-solving activity interesting, you can extend this activity to a 3 × 3 square and ask them to find as many symmetrical patterns as possible. The square offers more options because it also has diagonal lines of symmetry.

Support
If any children have difficulty drawing shapes with two lines of symmetry, show them that they can complete the task in stages. Draw the matching half for one line of symmetry and then do the same for the other.

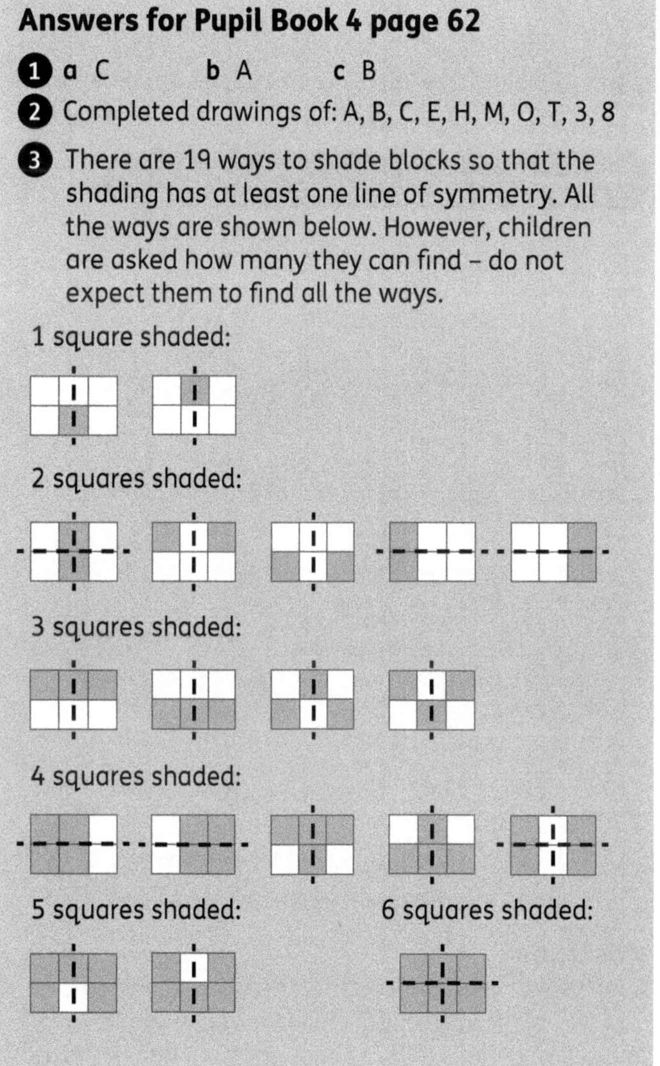

Answers for Pupil Book 4 page 62

1 a C   b A   c B

2 Completed drawings of: A, B, C, E, H, M, O, T, 3, 8

3 There are 19 ways to shade blocks so that the shading has at least one line of symmetry. All the ways are shown below. However, children are asked how many they can find – do not expect them to find all the ways.

1 square shaded:

2 squares shaded:

3 squares shaded:

4 squares shaded:

5 squares shaded: 6 squares shaded:

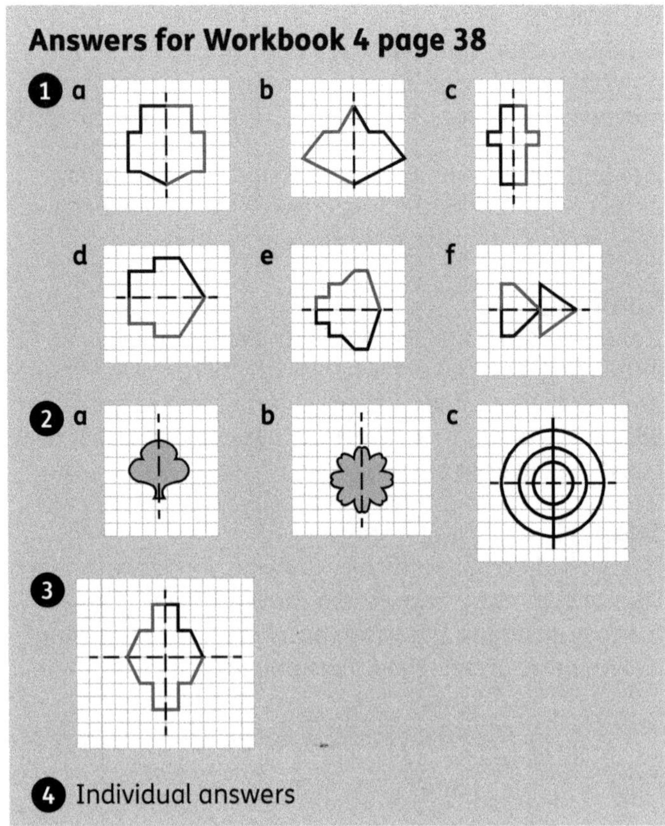

Answers for Workbook 4 page 38

1. a, b, c, d, e, f

2. a, b, c

3.

4. Individual answers

Investigating lines of symmetry

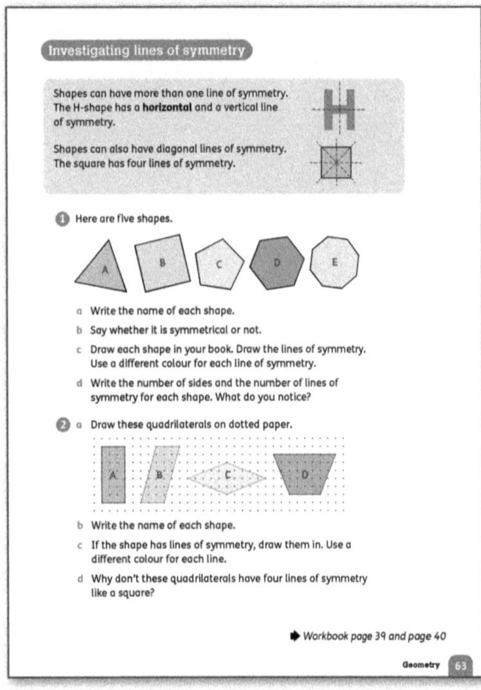

Materials
A mixture of regular polygons and other symmetrical and non-symmetrical 2D shapes made from plastic or thick card; some similar shapes cut from paper; dotted paper (see **Pupil Book 4 page 63**, question 2).

Warm-up
Revise *regular* polygons by asking the children to sort the shapes into two sets: regular polygons and irregular polygons. If necessary, remind them that regular polygons have all sides the same length and all angles the same size.

Focus
- Turn to **Pupil Book 4 page 63**. Explain that shapes can have more than one line of symmetry and point out the lines on the examples given.
- Tell the class that they are going to learn more about shapes and how many lines of symmetry they have.
- Give each group a set of shapes that cannot be folded (plastic or thick card) and ask them to classify them as symmetrical or non-symmetrical. (Check their sets to make sure that they are correct). Then ask them to sort the symmetrical shapes into those that have only one line of symmetry and those that have more than one.
- Draw three columns on the board with these headings: 'No line of symmetry', '1 line of symmetry' and '2 or more lines of symmetry'. Display various paper shapes and ask the children to say where each would fit into the table. Ask the children to come to the front of the class and fold the shapes to confirm their choices. Continue sorting shapes in this way, naming them and discussing their properties as you do so to revise those concepts.
- Let the children complete question 1 and discuss the answers as a class before moving on. Make sure they realise that these shapes are regular.

> Regular polygons have the same number of lines of symmetry as they have sides.

- Hand out the dotted paper and let the children work in pairs to investigate symmetry in the shapes in question 2. Have a class discussion about their findings. Use the terms *horizontal* and *vertical* to describe the lines of symmetry.

Follow-up
Ask the children to complete **Workbook 4 page 39** on their own. All the children should be able to find at least one of line of symmetry on each shape. Some children may find all the possible lines.

Challenge
Turn to **Workbook 4 page 40** and challenge the children to find all the possible lines of symmetry on the different shapes. Encourage them to convince each other that they have found all the possible lines of symmetry. This is a good way of getting them to reason mathematically and support their answers.

Interesting mistakes
The children may assume that all quadrilaterals have four lines of symmetry based on their answers to question 1 part d on **Pupil Book 4 page 63**. Let them work out for themselves that this is not the case and let them share their ideas about why. The reasoning and discussion will help them use mathematical vocabulary and formulate their ideas.

Answers for Pupil Book 4 page 63

1 a–d

A equilateral triangle
symmetrical
3 sides
3 lines of
symmetry

B square
symmetrical
4 sides
4 lines of
symmetry

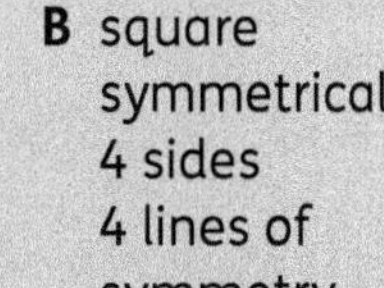

C regular pentagon
symmetrical
5 sides
5 lines of
symmetry

D regular hexagon
symmetrical
6 sides
6 lines of
symmetry

E regular octagon
symmetrical
8 sides
8 lines of
symmetry

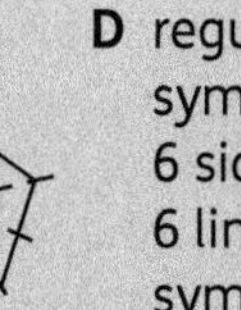

The number of sides is the same as the number of lines of symmetry.

2 a–c

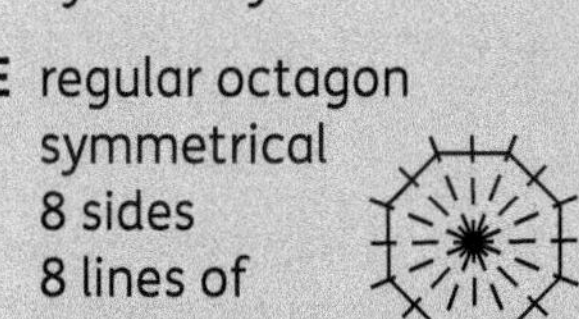

A rectangle **B** parallelogram **C** rhombus **D** trapezium

  d Possible answer: They are not regular polygons. The square is the only quadrilateral that is regular.

Answers for Workbook 4 page 39

1 Possible answers:

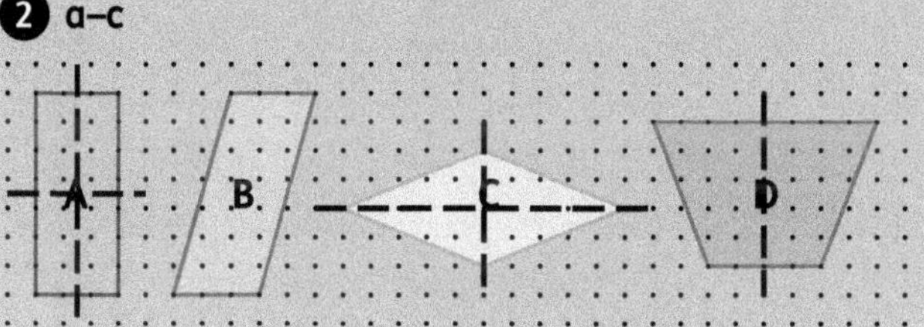

a equilateral triangle

b regular hexagon

c regular decagon

d square

e square

f regular pentagon

g regular octagon

h regular heptagon

Answers for Workbook 4 page 40

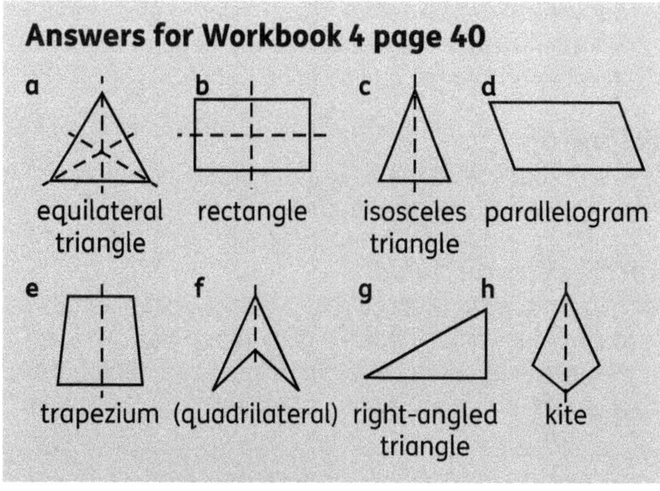

Symmetry around us

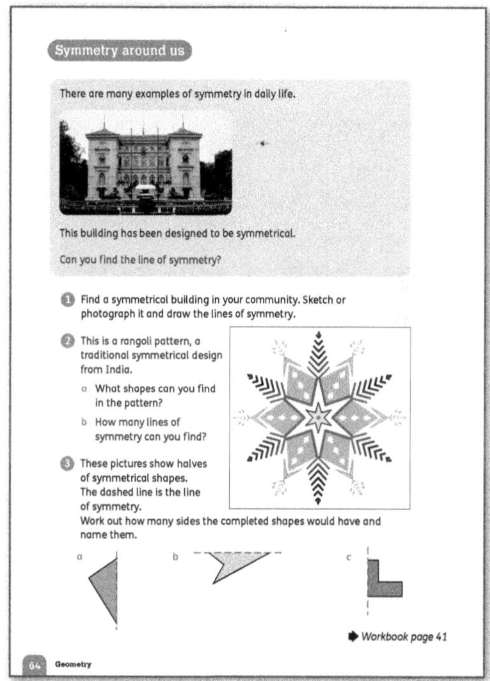

Materials
Digital cameras (such as smartphone cameras).

Warm-up
Use the 'Shapes around us' activity on page 30 as a mental warm-up.

Focus
- Use the photograph on **Pupil Book 4 page 64** to talk about symmetry in the environment. Remind the children that natural objects and living things may look symmetrical, but that they are not usually perfectly symmetrical. Buildings, bridges, rugs and clothing are often symmetrical. The building in the photograph has one vertical line of symmetry, down from the central flagpole.
- Set question 1 as homework or organise a class outing so that the children can look for examples of symmetry in the local area.
- Discuss the pattern in question 2 with the class. Let them find their own examples of symmetrical designs or patterns and share them.

- Let the children work out the answer to question 3 on their own. Some children might need to draw the completed shapes and count the sides.

Follow-up
Use **Workbook 4 page 41** for more practice in identifying and making symmetrical patterns.

Interesting mistakes
In question 3 part c, some children might count each line that continues across the line of symmetry as two sides. If they get an incorrect number of sides, encourage them to show the class how they worked it out.

Answers for Pupil Book 4 page 64
1 Individual answers
2 a Possible answers: kites, hexagons, triangles and 12-sided star (dodecagon)
  b 6
3 a 4 sides; kite
  b 6 sides; hexagon
  c 8 sides; octagon

Answers for Workbook 4 page 41
1 Lines of symmetry:

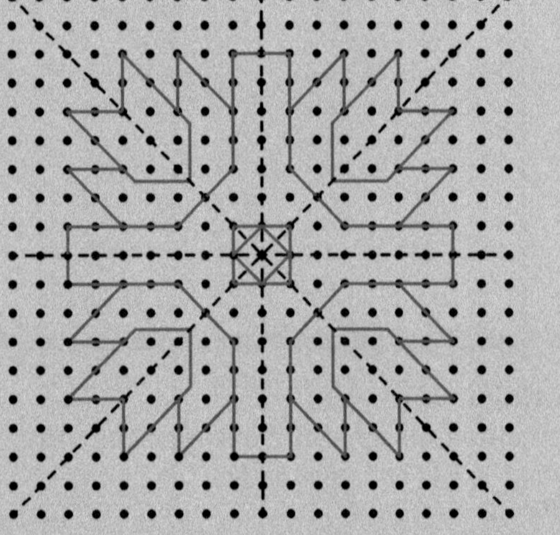

Individual answers for the colouring, which must be symmetrical.
2 Individual answers

Use some or all of these questions and activities to assess how well the children have understood the concepts in this unit.
- Show a shape. *Is this shape symmetrical? Why or why not?*
- *What does symmetry mean?* (If you fold a shape or picture or pattern along a mirror line, the two halves of the pattern fit exactly onto each other.)
- *What is a line of symmetry?* (a line that divides the shape into two identical parts)
- Show a symmetrical shape. *Where is the line of symmetry on this shape? Is there only one?*
- *What do you call this fold line that divides the shape into two equal halves that fit exactly onto each other?* (a line of symmetry or mirror line)
- *Give two examples of polygons with more than one line of symmetry.* (for example: square, rectangle, regular hexagon)
- *Draw a shape with more than one line of symmetry.*

Data and charts

Learning objectives

- Plan and carry out an investigation to answer a statistical question
- Collect, record and organise data using tallies and frequency tables
- Draw bar charts, pictograms and Venn diagrams to display data
- Solve comparison, sum and different problems using data presented graphically

Key words

data cycle   tally   tally mark   tally chart
frequency table   frequency   bar chart
axis/axes   pictogram   symbol   key
Venn diagram   properties

Unit introduction

Teaching guidance

- Display a list of questions that can be answered by collecting data. Include questions that involve measurement over time and questions where it is not clear what data you could collect to answer them. Here are some questions you could ask:
  - *Which subject do children find most interesting?*
  - *Which sport is least popular among our class?*
  - *Which TV show is most popular with kindergarten children?*
  - *Do children prefer to read fiction or non-fiction books?*
  - *How much water do bean plants use in a week?*
  - *What happens to the temperature of a cup of tea if you leave the cup standing on a table in your house?*
  - *Which of your textbooks is the easiest to read?*
  - *Do people listen properly when you read something to them?*
  - *What is the most common car colour?*
  - *What type of litter is the biggest problem in our area?*
- Explain that these are statistical questions and that we can collect data to answer them. Point out that we use the word 'record' to mean write down. When we collect data, we need to record what we find out and decide how to organise it so that it makes sense to anyone reading it.
- Show the class this simple diagram of the steps in the *data cycle*.

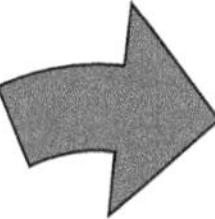

- Let the children work in groups to discuss how they could answer each question by collecting, organising and displaying data. Then ask the groups to share their ideas. Once a group has answered, avoid other groups repeating the same ideas by asking: *Who else has a similar method? Does any group have a different method?*
- If you want the children to do a statistical investigation, let the groups choose one of the questions, or make up their own question. Explain that they are going to carry out their investigation over the next few days. Let them write a plan for collecting, organising and displaying the data. Check that their plans are sensible and practical.
- Set some time aside over the next few lessons for the children to work on their investigations. Tell them what they need to present to you or the class to show the results of their investigation.

Tally charts

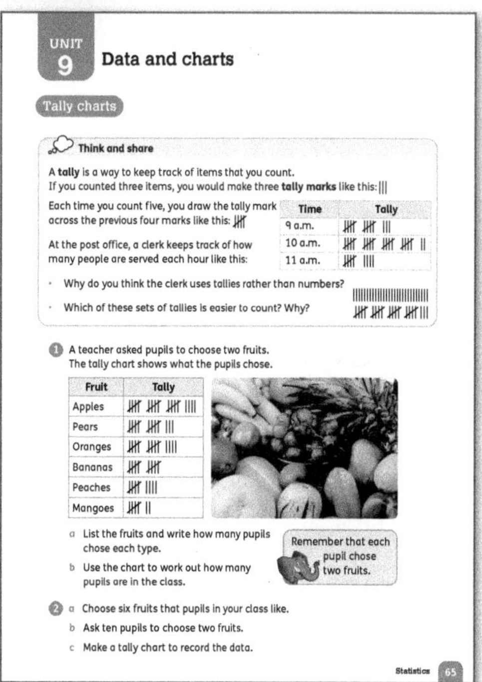

Time	Tally
9 a.m.	|||| |||| |||
10 a.m.	|||| |||| |||| |||| ||
11 a.m.	|||| ||||

Fruit	Tally
Apples	|||| |||| |||| ||||
Pears	|||| |||| |||
Oranges	|||| |||| ||||
Bananas	|||| ||||
Peaches	|||| ||||
Mangoes	|||| ||

Warm-up

- <u>Think and share:</u> Check that the children know what the term *tally* means and what a what a *tally mark* is before asking them to read through the information on **Pupil Book 4 page 65**. The children can work in groups or pairs to answer the questions. Take feedback as a class before moving on.

Focus

- Make sure the children realise that it is faster and more efficient to make a short tally mark than to write a number, especially if you are counting many items, for example, the number of adults and children who attend a large sporting event. The 'gate' that makes the fifth tally mark allows you to quickly count in fives to find the total number in any set of tally marks.
- The children can work independently to interpret the *tally chart* in question 1.
- For question 2, the children will need to move around the classroom in order to survey 10 children to collect the data for their tally chart. Let them compare their tally charts with others in their group and let them suggest why they might all get different results.

Answers for Pupil Book 4 page 65

<u>Think and share:</u> The clerk uses tallies rather than numbers because that way you can just add one tally each time instead of having to cross out your number each time. Counting groups of 5 tallies is easier because it is faster to count up in fives than to count up in ones.

1 a 19 apples; 13 pears; 14 oranges; 10 bananas; 9 peaches; 7 mangoes

   b 72

2 a–c Individual answers

Frequency tables

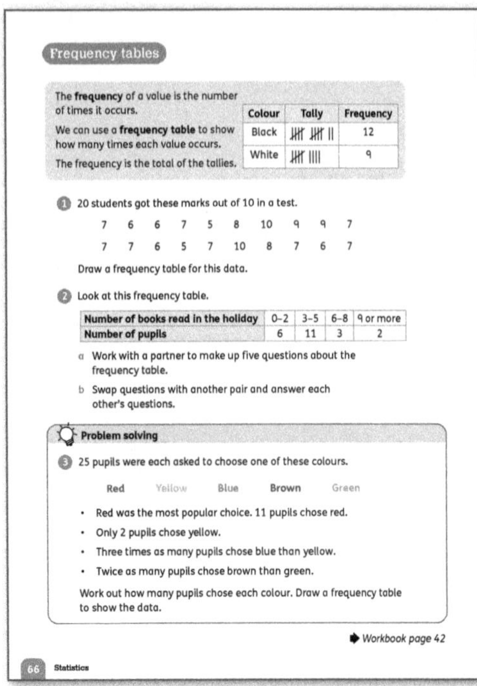

Materials

1–6 digit cards (for the Workbook investigations); examples of bar charts (if possible); highlighter pens.

Warm-up

Use any suitable 'Place value and number sense' activity (pages 23–27) or 'Mental maths' activity (pages 27–29) as a mental warm-up.

Focus

- Carry out a classroom survey to demonstrate creating a *frequency table* as follows.
- Write the names of four currently popular songs or TV shows (or any other things that you think all the children will know about) one below the other on the board.
- Ask the children to decide which one of these they like best. Point to the first item. Ask the children to raise their hands if that is the one they like best.
- Tally the results. Repeat this for each item.
- Draw a vertical line to the right of the tallies. Write 'Frequency' at the top of this column.
- Ask different children to add the tallies and write the total in the frequency column.
- Explain that the word *frequency* tells us how many people voted for each item (how frequently the item was chosen).
- Point out that the frequency table in the example at the top of **Pupil Book 4 page 66** has a column for tallies and for the frequency of each colour.
- Let the children work on their own to complete question 1. They need to create a table, tally the number of each mark and add up the tallies to find the frequency for each mark. Let them compare and check each other's completed tables.
- Next, look at the table in question 2. Point out that this is also a frequency table, but it does not contain tallies. Frequency tables like this one are shortened forms of tally charts.
  Set a time limit for pairs of children to make up five questions about the table. At the end of the time limit, the children can exchange their questions with another pair.
- <u>Problem solving:</u> For question 3, the children can list the colours and numbers in the table. They don't need tallies here.

Follow-up

- Turn to **Workbook 4 page 42**. Give each pair of children a set of digit cards 1–6.
- Ask them what numbers they think they will get if they pick a card 30 times. (They should replace the card each time so they are always picking from a complete set.)
- Ask them if they think any of the numbers is harder to pick than any other. Discuss their answers.
- Explain that each number has the same chance (is equally likely) as all the others of being picked. If you pick a card enough times, the frequency of each number will be more or less the same. (Statistically, if you picked a card 600 times, you could expect each

number to come up 100 times, but the reality is that you would probably only see equal frequencies after far more picks than that.)
- Let the children decide how they are going to work to tally the results of 30 picks. (They may take turns to pick a card 15 times and then to tally.) Then ask them to do the experiment and record the results.
- Discuss question 2 with the class. Ask them to suggest headings for the frequency table and make sure these are suitable before they do the counting part. Possible headings are:

Vowel	Tally	Frequency
a		
e		
i		
o		
u		

- Counting the vowels is quite tricky and the children may miss some as they read over them. Encourage them to use highlighters or different markings for each vowel and to mark them before they tally and count.

Challenge

After completing **Workbook 4 page 42**, let the children combine their card-picking results with those of other groups to see whether the frequency of each number being picked gets closer to being the same if they combine their results.

Answers for Pupil Book 4 page 66

1

Mark	Tally	Frequency
5	II	2
6	IIII	4
7	JHT III	8
8	II	2
9	II	2
10	II	2

2 **a** and **b** Individual answers

3

Colour	Frequency
Red	11
Yellow	2
Blue	6
Brown	4
Green	2

Answers for Workbook 4 page 42

1 Individual answers

2

Vowel	Tally	Frequency
a	JHT JHT I	11
e	JHT JHT JHT JHT JHT JHT IIII	34
i	JHT JHT JHT	15
o	JHT JHT JHT I	16
u	JHT JHT III	13

a e   **b** 21   **c** 89

Bar charts

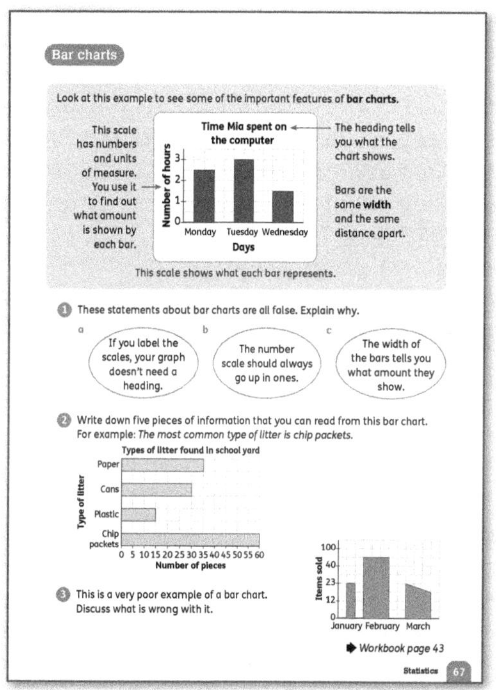

Materials

Large, simple bar chart; squared paper.

Warm-up

Choose a mental addition and/or subtraction activity as a warm-up, such as 'Find the operation' (page 27) or 'Addition and subtraction' (pages 27–28).

Focus

- Show the class a large, simple *bar chart* if you have one, or use the example on **Pupil Book 4 page 67**. Point out the features of the graph: the heading, the scale, the horizontal and vertical *axes*, the labels and the bars themselves.

Note that a bar chart has bars of the same width with equal spaces between them. Graphs where the bars touch are called histograms and the children will learn about these in later years. Histograms may also have bars of different widths.

- Discuss how the bar chart in the example on **Pupil Book 4 page 67** could be redrawn with horizontal bars. Let the children explain clearly what they would do to draw such a chart.
- Ask questions such as:
  - *Where would the heading go?* (*at the top of the graph*)
  - *Where would you put the numbered scale?* (*on the horizontal axis*)
  - *How would you decide how long each bar should be?* (*using the scale and data from the existing graph*)
  - *What would you write on the vertical axis?* (*the names of the days*)
  - *Does it matter if you make the bars a different width to the ones on the vertical chart?* (*no, as long as they are all the same width on the graph you draw*)
- You could give the children squared paper and ask them to redraw the graph based on the discussion.
- Display one or two more bar charts and ask the children to read them carefully. Ask questions, or let the children ask each other questions. Spend some time talking about how children found the answers to the questions.
- Do question 1 orally as a class.
- Question 2 could be a whole-class discussion or children could discuss the graph in groups and feed back their ideas.
- For question 3, let the children work in pairs to identify and list the information before sharing with the class.

Follow-up
- Explain to the children that they are going to draw a bar chart with all of the features they have learnt about. Ask them to turn to **Workbook 4 page 43**.
- Ask the children to complete the frequency table and check that their answers are correct before moving on.
- Let the children then work independently to draw their own bar charts in question 1. They can decide whether to draw a horizontal or vertical chart and choose their own interval on the scale.
- Then let the children compare and check each other's work.
- <u>Problem solving:</u> In question 2, the children need to think about the data in their graph and how likely it is that the spinner will land on each colour. Blue is the colour with the highest frequency, so it makes sense that the largest sector on the spinner would be blue. Note that the children may argue differently and it is important to let them explain their thoughts.

Challenge
Ask the children to design and test a spinner that is likely to land on yellow least often, on green most often, and more or less the same number of times on purple and black. They should do at least 30 trials and tally their results.

Support
Prepare a checklist with the children of the features that a bar chart should have. They can use this to make sure that any graphs they draw or interpret have these features.

Interesting mistakes
When graphs are drawn on squared paper or graph paper, the children may count the blocks and use that number as the amount for each bar. Keep reminding them that to interpret graphs we need to read the information on the frequency scale. It may help to draw three different bar charts showing the same data on differently scaled sets of axes. This will show the children that the length of the bar links to the scale and not the number of blocks on the paper.

Answers for Pupil Book 4 page 67
1 Possible answers:
  a All graphs need a heading so that there is no confusion about what you are showing.
  b As long at the scale is in steps of the same size, it can go up in other size steps.
  c The height/length of the bar shows the amount against the numbered scale but the width of the bar doesn't tell you anything

2 Possible answers: The children found more paper than cans; the least common type of litter is plastic; pupils only found four types of litter in the school yard. In total, the children found 140 pieces of litter.

3 Possible answers: no heading, bars different widths, scale incorrect, one bar with sloping top, label on the horizontal axis missing

Answers for Workbook 4 page 43
1 a (blue) 22, (green) 12, (yellow) 7, (red) 9, total number of spins = 50

  b Possible answer:

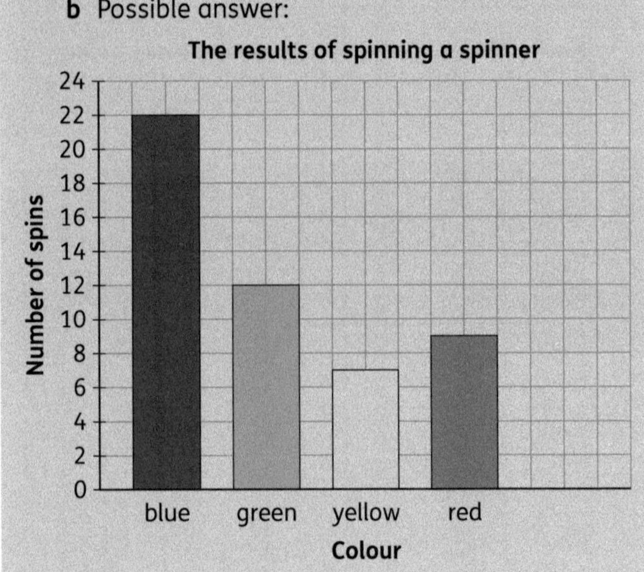

2 a Spinner coloured as follows: largest section (right): blue second largest (top left): green next largest (bottom left): red smallest section (left): yellow

b Possible answer: Blue had the highest frequency, the largest section of the spinner is probably blue. I coloured the other sections so the sizes matched the frequency of the spins.

Use bar charts to answer questions

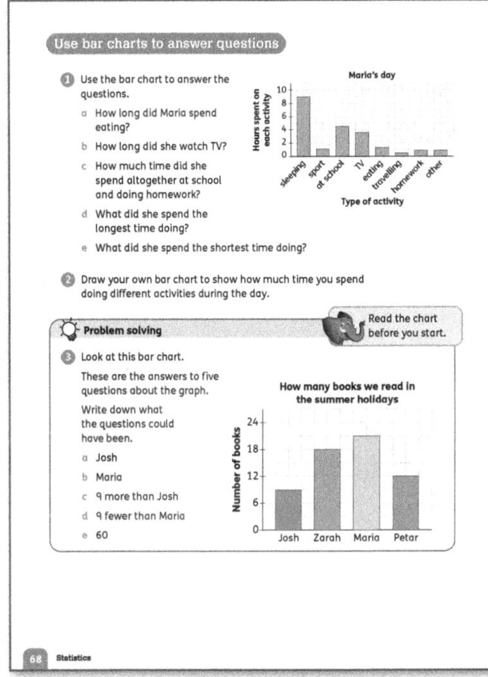

Warm-up

Use any suitable mental addition or subtraction activity from 'Calculation skills' on pages 28–29 as a warm-up.

Focus

- Let the children complete the questions on **Pupil Book 4 page 68** independently to consolidate their work on reading and interpreting graphs.
- Problem solving: When the children have written their questions for question 2, invite them to share them with the class. Ask: *Are everyone's questions the same for each answer?*

Answers for Pupil Book 4 page 68

1 a 1.5 hours **b** 3.5 hours **c** 5.5 hours
  d sleeping **e** travelling

2 Individual answers

3 Possible answers:
  a Who read the fewest books?
  b Who read the greatest number of books?
  c How many more books than Josh did Maria read?
  d How many fewer books did Petar read than Maria?
  e how many books did the children read in total?

More bar charts

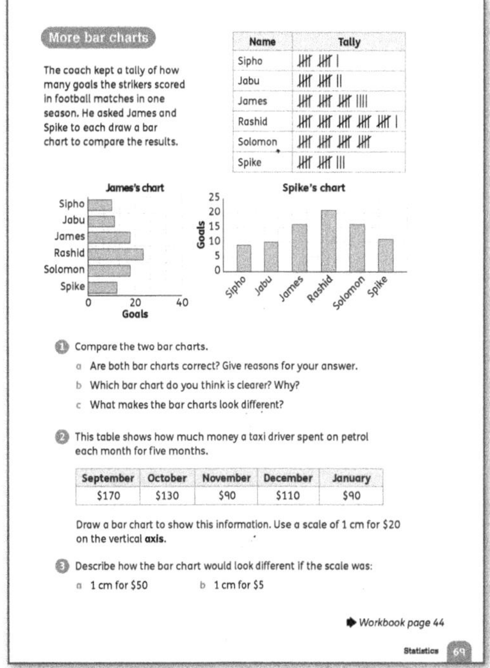

Materials

Squared paper; small counters.

Warm-up

Choose any 'Place value and number sense' activity from pages 23–27 as a mental warm-up.

Focus

- Discuss the two different bar charts on **Pupil Book 4 page 69** as a class. Make sure the children understand that both charts show the same data and that they are both technically correct.
- Let the children work in small groups to discuss question 1. Take feedback when they have done this.
- For question 2, the children can discuss how they will draw the graph before they draw their own.
- Discuss question 3 orally with the class. If any children need help to work out how the bar charts will change as the scale changes, you can show them bar charts drawn to each scale, or give them the scales and let them draw their own charts so they can see the difference.

Follow-up

Let the children work independently to complete the questions on **Workbook 4 page 44**. Discuss their answers to question 3 as a class. Let the children suggest why they have all drawn different bar charts. (The results are based on chance and different children will toss the counter onto different shapes and therefore get different outcomes.)

Answers for Pupil Book 4 page 69

1 **a** Possible answer: No, neither chart has a title. The bars in Spike's graph aren't the correct height

   b Individual answers, for example: Spike's bar chart is clearer because the scale for the number of goals shows 5 goals for every small square. James's shows 20 ÷ 3 = 6.7 goals for every small square, which is not a whole number of goals. That makes it very difficult to read! Spike's scale for the number of goals goes up to 25, which is more sensible than 40; it uses the space better. The gap between the bars is larger on Spike's graph, making it easier to see which bar is for which player.

   c The bars charts look different because one is horizontal and the other is vertical. The scales are different for the number of goals scored.

2

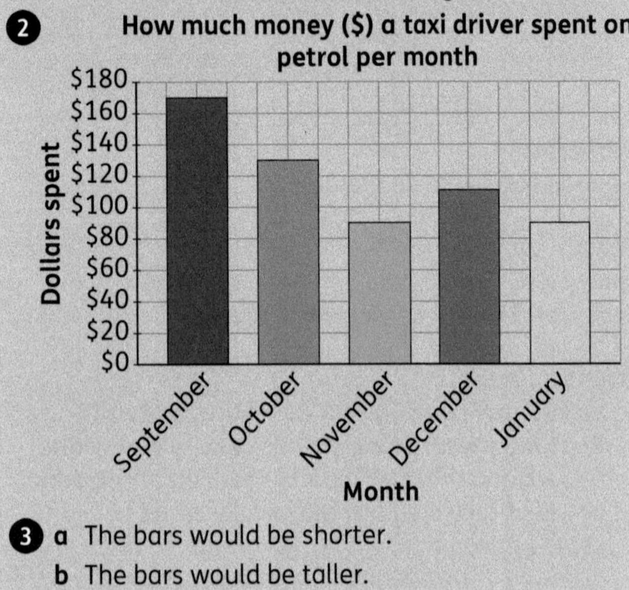

3 **a** The bars would be shorter.

   b The bars would be taller.

Answers for Workbook 4 page 44

1–**3** Individual answers. Possible title for the bar chart: Shapes landed on in a game; possible label for the vertical axis: Shape

Pictograms

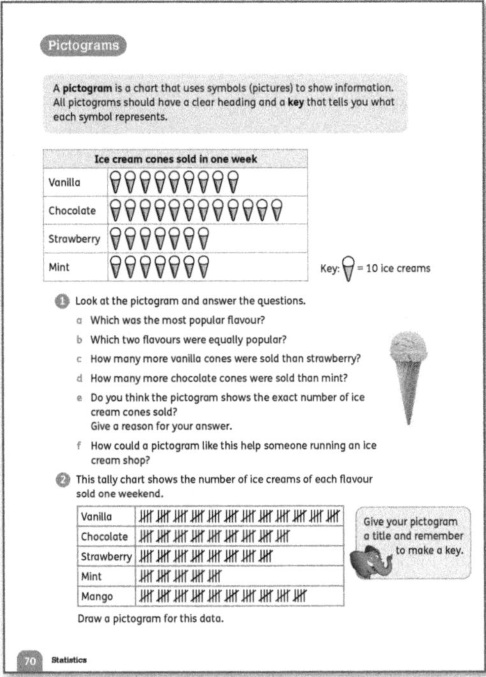

Warm-up

Select any activity from 'Place-value activities' (pages 24–25) or 'Rounding and estimating' (pages 26–27) to use as a mental warm-up.

Focus

The children have worked with *pictograms* in earlier years.

- Use **Pupil Book 4 page 70** to revise the concept of a pictogram. Make sure the children realise that one ice-cream cone *symbol* represents 10 ice creams. Discuss why a *key* is important in this type of chart.
- Let the children read and discuss the questions in question 1 in groups.
- Take feedback from the class and ask the children to explain how they reached their answers.
- For question 2, the children can copy the format of the ice cream pictogram at the top of the page. They could use the same key and symbol (1 cone = 10 ice creams) and show, for example, 45 ice creams using 4 full cones and 1 half cone. Some children may prefer to use the key 1 cone = 5 ice creams, so they do not need to draw half cones.

Answers for Pupil Book 4 page 70

1 **a** chocolate

   b strawberry and mint

   c 20 d 50

   e No, they are all multiples of 10.

   f It could help them know when they need to order or make more ice cream.

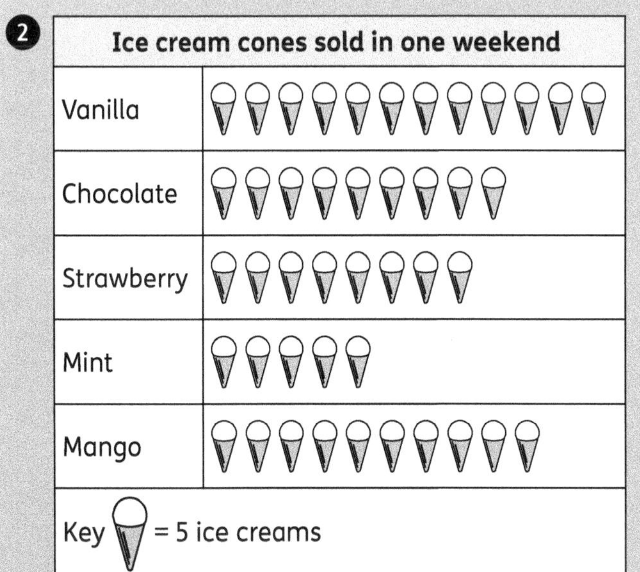

More pictograms

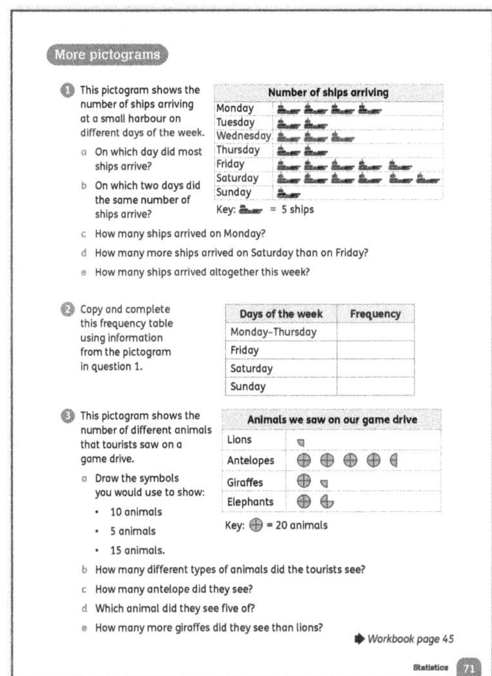

Warm-up

Use any 'Counting backwards and forwards' activity from page 23 as a mental warm-up.

Focus

Let the children work independently to complete questions 1–3 on **Pupil Book 4 page 71**. Discuss and compare answers as a class.

Follow-up

Use **Workbook 4 page 45** as an informal assessment task to check that the children can complete a frequency table and use the data to draw a pictogram. Check that they write a title and two clear sentences based on the graph.

Challenge

Ask the children to draw a pictogram showing how many moons each planet in our solar system has. You can ask the children to find this information (updated regularly on the NASA website) or you could also give them a table of information. These numbers were accurate in 2020, but new moons are discovered and confirmed regularly. (Pluto has been left out as it is technically a dwarf planet rather than a planet, but if you want to include it, it has five moons.)

Planet	Moons
Mercury	0
Venus	0
Earth	1
Mars	2
Jupiter	79
Saturn	62
Uranus	27
Neptune	14

The challenge in this activity includes developing a suitable scale for the chart and symbols that can be divided easily.

The children can decorate their charts and display these in the classroom.

Interesting mistakes

- The children may misinterpret graphs if they do not look at the key carefully. They need to be aware that one symbol can represent more than one item of data (many-to-one correspondence).
- The children may think that they need to draw detailed and accurate pictures rather than simple symbols. Remind them that simple shape symbols are more useful because they are much quicker to draw and can be divided easily to show partial totals. Squares, rectangles and circles are all useful shapes for pictogram symbols.

Answers for Pupil Book 4 page 71

1 a Saturday   b Tuesday and Thursday
   c 20 d 5   e 115

2

Days of the week	Frequency
Monday–Thursday	55
Friday	25
Saturday	30
Sunday	5

3 a 10 animals =

    5 animals =

    15 animals =

   b 4   c 90   d lions   e 20

Answers for Workbook 4 page 45

1 **a** Frequencies: Monday 17 Tuesday 13
Wednesday 21 Thursday 15 Friday 21
Saturday 32 Sunday 34

b The number of tourists who took a bus tour on each day of the week

Monday	⊕⊕⊕◔
Tuesday	⊕⊕◒
Wednesday	⊕⊕⊕⊕▽
Thursday	⊕⊕⊕
Friday	⊕⊕⊕⊕▽
Saturday	⊕⊕⊕⊕⊕⊕◔
Sunday	⊕⊕⊕⊕⊕⊕⊕

Key: ⊕ = 5 people

c Individual answers

Venn diagrams

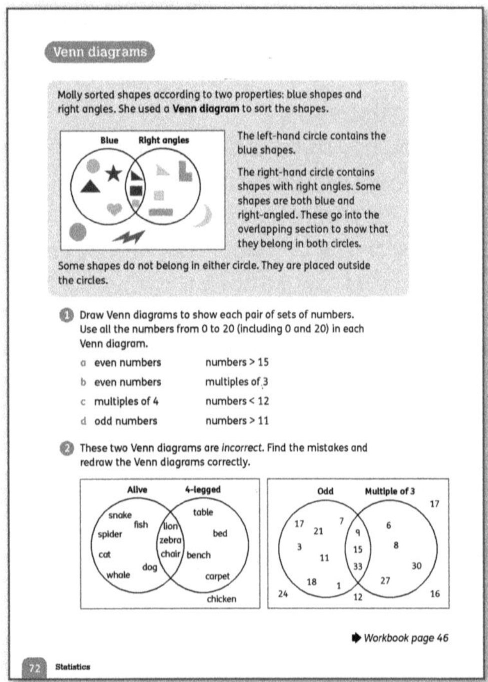

Materials

Large sheets of paper; chalk; rope; objects to sort (such as blocks in different colours or cut-outs of familiar 2D shapes in different colours); equipment for drawing circles (such as compasses, round lids or stencils); printed sheets with four blank Venn diagrams (optional).

Warm-up

Choose any suitable counting activity (such as 'Counting in given steps' on page 23) or multiples activity (such as 'Multiplication tables' on page 28).

Focus

- Set up a large *Venn diagram* by using chalk, hoops or rope to make two large overlapping circles on the floor. Explain that you have some shapes and you are going to make two sets.
- Sort the shapes into the circles or middle section using particular *properties* such as: *red* and *small*, or *right-angled* and *red*. (You will need to use criteria that fit the objects you are using.)
- Let the children work out what criteria you are using to sort the shapes, and discuss which shapes should go into the overlapping part of the circles (for example, both red and small).
- Let the children draw their own diagrams to represent the sorting.
- Turn to **Pupil Book 4 page 72**. Ask the children to look at the Venn diagram in the example and tell them to cover over the information next to it.
- Ask them to make statements about the Venn diagram and what it shows. Then let them uncover the information and check that they are correct. Read through the text with the class as they find each related part of the Venn diagram.
- Turn to **Workbook 4 page 46**. Let the children work in pairs to discuss what could go into each section of the Venn diagrams. Then ask them to complete them individually.
- Check this work and discuss any mistakes that the children made and why before going back to **Pupil Book 4 page 72**.
- For question 1, make sure the Venn diagrams that the children draw are large enough. If this is too difficult or is taking too long, support them by providing a printed sheet with four blank diagrams.
- Let the children discuss the diagrams in question 2 and find the mistakes before they redraw them correctly. Ask them to use colour to show any items that have changed from one part of the diagram to another. Again, supply printed diagrams if the children are struggling to draw their own.

Challenge

- Draw a Venn diagram with three overlapping circles. Write criteria on each circle, for example 'Long hair', 'Curly hair' and 'Black hair'. Discuss what will go into each overlap stressing that the central overlap has to be long, curly, black hair. Let individual children decide where they belong in the diagram.
- Another fun Venn diagram can be made by sorting film stars or singers and choosing some categories that could overlap (for example, wearing a hat, wearing glasses, wearing a necklace). The children can then find photographs of film stars and decide where they go in this Venn diagram. Anyone who is wearing a hat, glasses and a necklace goes in the central overlap.

Interesting mistakes

When you overlap circles in a Venn diagram, you get an area where the objects or data are common to both the sets. This can be confusing, as can the idea that some data does not fit into either circle. Let the children use concrete objects (such as plastic shapes in different colours or plastic numbers) to create their own Venn diagrams. Using concrete apparatus makes it easier for them to move items around if they are placed in the wrong groups.

> Venn diagrams are used at higher levels in set theory and also to show outcomes and answer questions about probability. A basic understanding of how they work can help children when they reach these stages.

Answers for Pupil Book 4 page 72

1 a
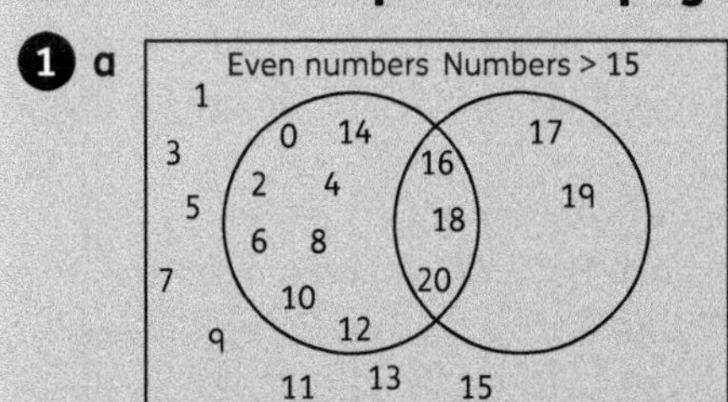

b
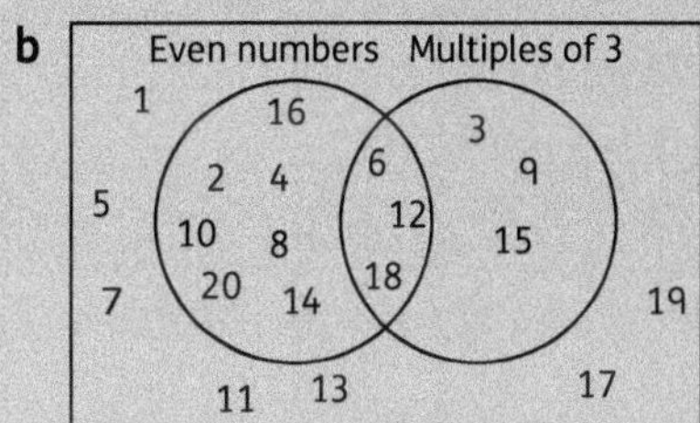

c
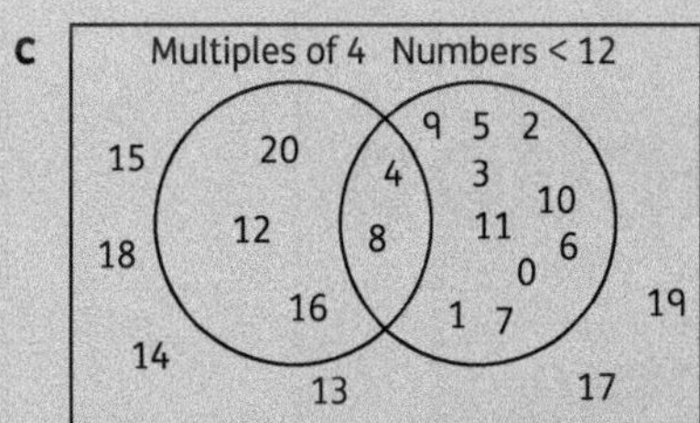

d
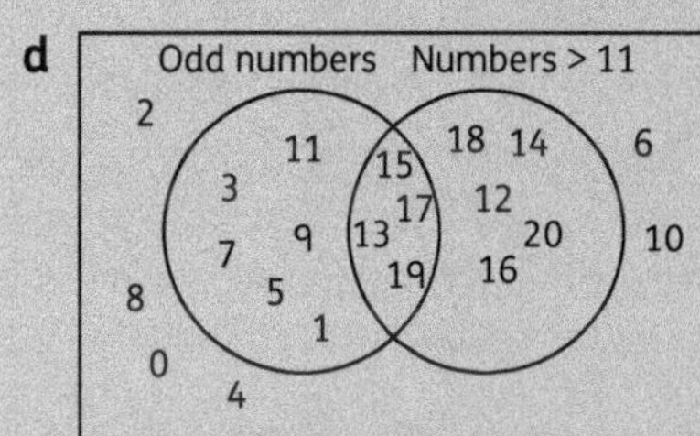

2
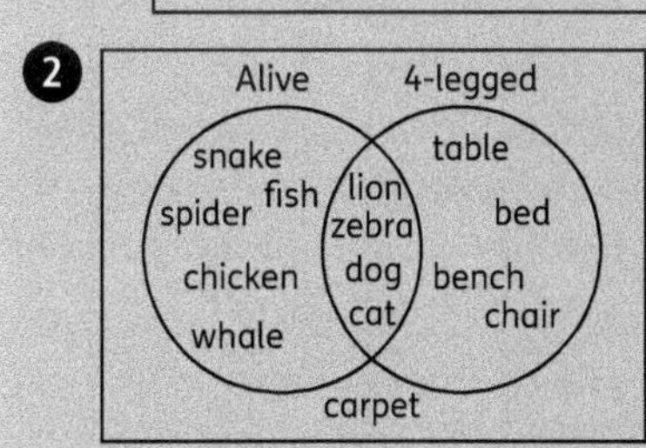

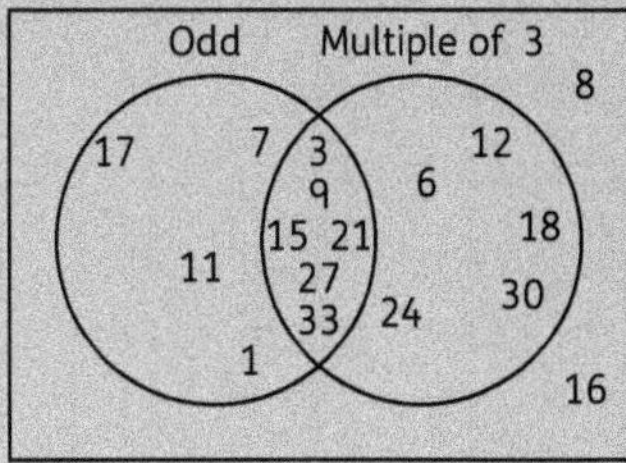

Answers for Workbook 4 page 46

1 a and **b** Individual answers to fit the sorting criteria given

2
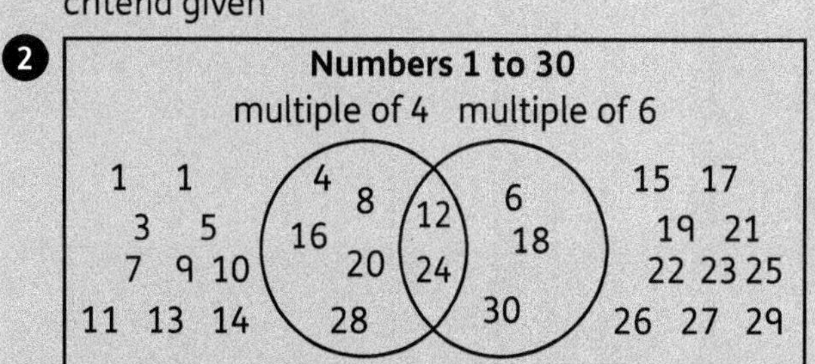

Solve problems using charts

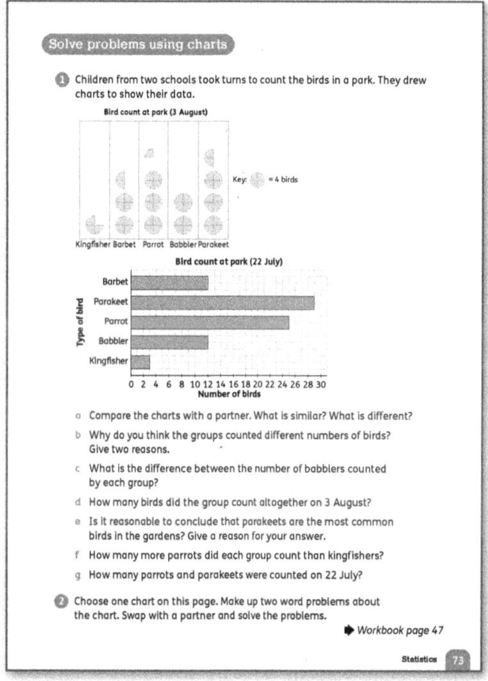

Warm-up

Choose any mental addition and/or subtraction activity from 'Calculation skills' on pages 28–29 as a warm-up.

Focus

- Turn to **Pupil Book 4 page 73**. There are no new concepts in this lesson.
- For question 1, the children are expected to apply what they know about graphs to solve comparison, sum and difference problems using the data.
- For question 2, the children should work on their own to write word problems about one of the charts. They can exchange their problems with the person next to them to solve, or you can collect all the problems and hand them out randomly.

Follow-up

- Use **Workbook 4 page 47** to check that the children can interpret and analyse information shown in a bar chart.
- The children can answer questions 1 and 2 on their own. Discuss how the children can compare the bars on the chart in the same way as they use bar models (see page 22). For example, to work out how much faster/slower one animal is than another, they need to find the bars giving the speed of each of the two animals and work out the difference.

Answers for Pupil Book 4 page 73

1 a Individual answers

  b Possible answers: They saw different numbers of birds. They counted in different parts of the park. They counted on different days.

  c 4

  d 48

  e Possible answers: No, there are a similar number of parrots. Yes, because this was the most common bird both groups saw.

  f 10 and 22

  g 29 parakeets and 25 parrots for a total of 54

2 Individual answers

Answers for Workbook 4 page 47

1 Ostrich: bar drawn on chart to 72
Giraffe: bar drawn on chart to 60

2 a 88 km/h   **b** cheetah
  c wildebeest, lions, cheetahs and springboks
  d 40 km/h   **e** 60 km/h   **f** 82.42 km/h

Ask some or all of these questions to assess how well the children have understood the concepts in this unit.

- Show a bar chart and ask:
  - *What does each bar represent?*
  - *What does the length of the bars represent?*
  - *What does 1 block/1 centimetre on the bar chart show?*
- *Why do we use four tallies with a line through to represent five? (So we can count them in fives, to get a total quickly.)*
- Show frequency tables, bar charts and pictograms and ask specific, directed questions to make sure the children can read and interpret the data.
- Show two bar charts that present the same data, but have different scales. *Can you explain why the bars on this graph are much longer than the bars on this graph?*
- Make up sum and difference questions using any of the bar charts from the unit to make sure the children can compare bars and solve problems using the data.
- Display a Venn diagram used to sort shapes and ask:
  - *What is this diagram called? (a Venn diagram)*
  - *Which shapes should go here?*
  - *Why is this shape here?*
  - *Which shapes go in this circle, but not this one?*
  - *Why are some shapes outside the circles?*

UNIT 10 Addition and subtraction

Learning objectives

- Estimate and use inverse operations to check calculations
- Add and subtract numbers with up to 4 digits using formal written methods
- Decide which operation and methods to use to solve problems and explain choices
- Solve two-step addition and subtraction problems in context

Key words

estimate   round   inverse operation
column method   column addition   difference
column subtraction   two-step problem   area

Unit introduction

The concepts of addition and subtraction are familiar to the children, but they are now expected to work in a higher number range and to increasingly use formal, written methods to add and subtract efficiently.

They are also expected to apply what they have learnt about rounding numbers to help them estimate the answers to calculations, so they can check that their own answers are reasonable.

Teaching guidance

A yohaku is a number puzzle where you have to work out which numbers fit into the grid and give the sum (or product) shown for each row and column. The simplest yohaku are 2 × 2 puzzles. For example:

		10
		10
7	13	+

This puzzle has no restrictions, so there are many possible solutions. Two solutions using only whole numbers are:

5	5	10
2	8	10
7	13	+

6	4	10
1	9	10
7	13	+

Most yohaku have some restrictions, which are given below the puzzle. Restrictions might include: use only odd numbers; use four consecutive even numbers; do not use multiples of 10.

- Either explain the concept to the class or ask them to find out about yohaku puzzles and how they work. Then give them a number of 2 × 2 puzzles to solve of different levels of difficulty (simple, intermediate and advanced). You can find yohaku puzzles online.
- Once the children have solved the 2 × 2 puzzles, ask them to make up some simple and some challenging 3 × 3 puzzles with restrictions.

Estimate and check

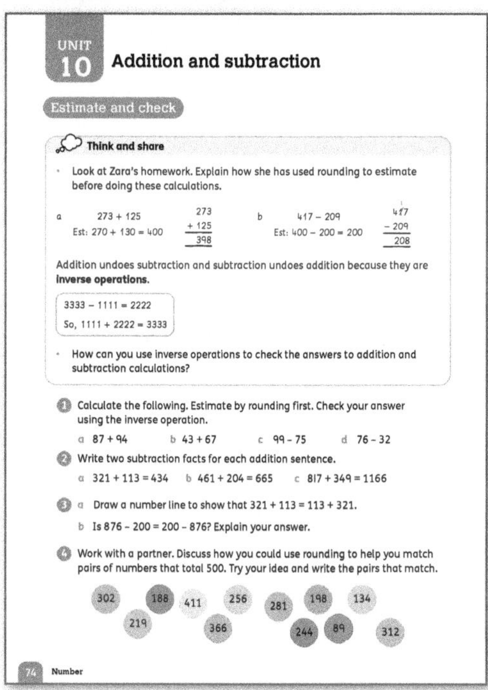

Warm-up

Choose a suitable activity from 'Counting backwards and forwards' (page 23) or 'Place-value activities' (pages 24–25) as a mental warm-up.

Focus

- <u>Think and share:</u> Turn to **Pupil Book 4 page 74**. Ask the children to work in groups to discuss the examples and answer the questions.
- Once they have done this, work through these questions systematically with the class to make sure they understand that an *estimate* is an approximate answer, and that good estimates are close to the real values.
- Discuss how to *round* to an appropriate place to make the calculation easier. Ask the class how they would round the numbers if they had to add 347 and 1289.

- Make sure they realise that they cannot round to thousands in this example, and that they will get the best estimate rounding to 10. Also, be clear that a reasonable answer is one that is close to the estimated answer, so it may or may not be the correct answer.
- Let the children explain how they could use *inverse operations* to check the answer to both addition and subtraction calculations. Ask them to give examples that are different from the ones in the book to show what they mean.
- Tell the children to write down their estimated answers for question 1, as well as their accurate answers. They should also show the inverse calculations they use to check their work.
- The children can work independently on question 2. For question 3, they can work in pairs and discuss their answer to part b.
- For question 4, the children work in pairs to develop and test a strategy for making pairs. Let them share their methods with the class.

Answers for Pupil Book 4 page 74

<u>Think and share:</u> If I have my answer to an addition, I can use subtraction to check my answer. Subtraction undoes addition. If the numbers are all the same when I do both the subtraction and addition, then my answer should be correct.

1 a 181   b 110   c 24   d 44
2 a 434 – 321 = 113; 434 – 113 = 321
   b 665 – 461 = 204; 665 – 204 = 461
   c 1166 – 817 = 349; 1166 – 349 = 817
3 a Individual answers
   b No, the order matters in subtraction
4 Individual answers

Explore adding and subtracting larger numbers

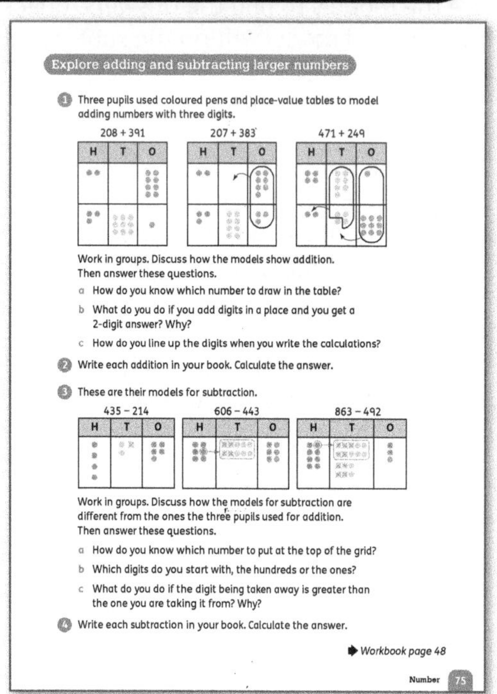

Materials

Sheets of place-value tables (see page 22); coloured marker pens.

Warm-up

- Choose a suitable activity from 'Place-value activities' (pages 24–25) as a mental warm-up.

Focus

- Turn to **Pupil Book 4 page 75**. The aim of this page is for the children to discuss the different approaches so that they begin to understand the process of adding and subtracting in columns and that they know how this links to place value. Let the children spend time talking about the models in question 1 in their groups.
- Take feedback when they have done this and work through each part of question 1 in turn.
- Give each child a sheet of place-value tables and make sure that they have coloured marker pens. Explain that they are going to do a number of additions and subtractions and that they can use a place-value table and marker pens to help them find the answers if they need to.
- Ask the children how they could model adding two 4-digit numbers. (*Use a place-value table with an extra column for thousands.*)
- For question 2, ask the children to write the calculations and the answers. Note that they might not write these in columns at this stage because they can use the models to find the answers. Check the answers, and model writing the sums in columns as you do so.
- For question 3, tell the children that they are now going to consider subtraction. Let the children spend time talking about the models in their groups. Take feedback afterwards and work through each question in turn.
- Again, for question 4, the children might not write the sums in columns at this stage because they can use the models to find the answers. Check the answers and model writing the subtractions in columns as you do so.

Follow-up

Turn to **Workbook 4 page 48** and use the first function machine diagram to explain what they need to do. Tell them that they may have to use inverse operations to work out some of the missing numbers.

Support

Allow the children to continue using place-value tables and marker pens to model the numbers and calculations if they need to.

Answers for Pupil Book 4 page 75

1 a Possible answer: Draw the first number being added in the top row. Draw the second number being added in the bottom row.

  b Possible answer: Take 10 of those dots and exchange for one 1 dot and move it one place to the left. You can only have 9 or less in a place.

  c Possible answer: The digits should be lined up by what place they are in.

2 599; 590; 720

3 a Possible answer: You put the number you are subtracting from.

  b the ones

  c Possible answer: You cross one dot out in the column to the left, and move 10 dots to the column you are working in.

4 221; 163; 371

Answers for Workbook 4 page 48

1 a 117

  b (right, top to bottom) 88; 106; 33

  c (right, top to bottom) 5; 20; 68

  d (left, middle) 206; (right, top) 143; (right, bottom) 586 e 748; 419; 598

  f (right, top to bottom) 554; 418; 213

  g (left, middle) 351; (left, bottom) –84;

  h (right, top) 75; (left, middle) 551; (left, bottom) 662

Use written methods to add

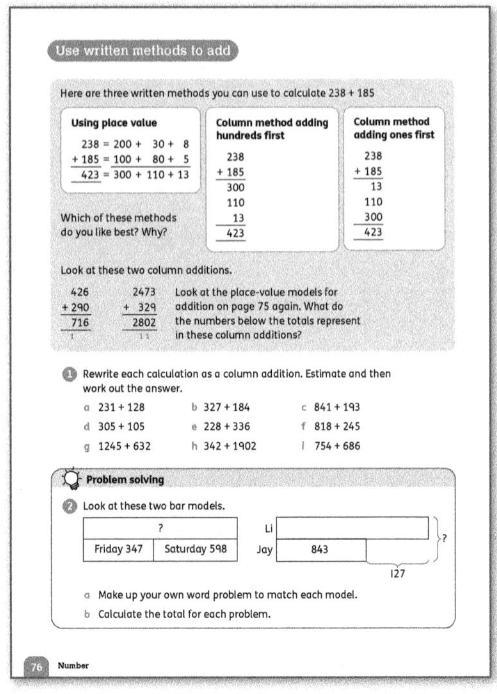

Materials

Place-value tables (see page 22); squared paper.

Warm-up

Choose a suitable activity from 'Counting backwards and forwards' (page 23) or 'Place-value activities' (pages 24–25) as a mental warm-up.

Focus

- Work slowly and carefully through the different addition strategies shown at the top of **Pupil Book 4 page 76** with the class, allowing time to discuss each one, and allowing the children to say whether they find it easy to use or not (and why).
- Demonstrate and discuss the *column methods* of adding. Make sure that the children understand that the numbers below the total are the groups of ten that are carried over. If necessary, draw place-value models like the ones on **Pupil Book 4 page 75** to show this practically.
- For question 1, tell the children to use rounding to estimate, and remind them to write down their estimate before they do the calculation. They should check their calculated answer is reasonable by comparing it to their estimate.
- <u>Problem solving</u>: The children can work on their own or in pairs to make up their own word problems for question 2. Share some of the word problems with the class.

Challenge

Ask the children to make up five *column additions* with mistakes. Let them exchange these with other children to find and correct the mistakes.

Support

The children should not be forced to move to a column method until they are confident and ready. Let them choose the method they prefer to do the calculations and spend time discussing how they worked once they have finished.

If any children have difficulty lining up the digits correctly in columns, it may help to work on squared paper.

Answers for Pupil Book 4 page 76

1 a 359   b 511   c 1034
    d 410   e 564   f 1063
    g 1877   h 2244   i 1440

2 a Possible answers: A website had 347 visitors on Friday and 598 visitors on Saturday. What was the total number of visitors?

    Jay has 843 stickers. Li has 127 more stickers than Jay. How many stickers do they have in total?

    b 945; 1813

Use written methods to subtract

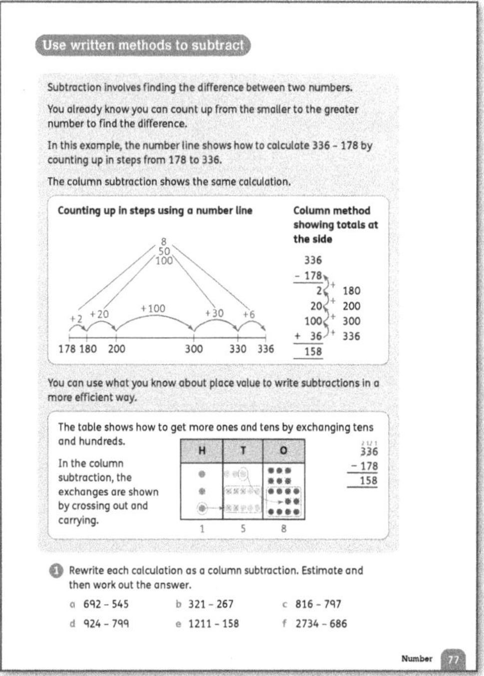

Materials

Squared paper; place-value tables (see page 22); coloured marker pens; online column subtraction games (optional).

Warm-up

Choose a suitable activity from 'Place-value activities' (pages 24–25) or 'Calculation skills' (pages 28–29) as a mental warm-up.

Focus

- Discuss and explain each method of finding the *difference* between two numbers shown on **Pupil Book 4 page 77**. Encourage the children to say which methods they prefer and why.
- Make sure that the children understand how to 'borrow' from the place to the left if they need to. The place-value table helps with this and the children can continue to use place-value tables and coloured marker pens to model the subtractions until they are confident carrying from one place to the next.
- Let the children work on their own to complete question 1. If any children have difficulty aligning the digits in *column subtractions*, provide squared paper for them to work on. Make sure that they estimate and write their estimates before they do the calculations.

Challenge

Provide a number of subtractions with some digits missing and ask the children to work out what they are. Here are some examples using 3-digit numbers:

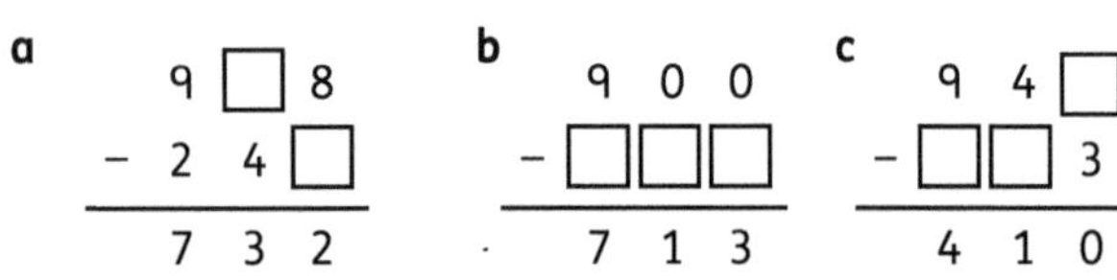

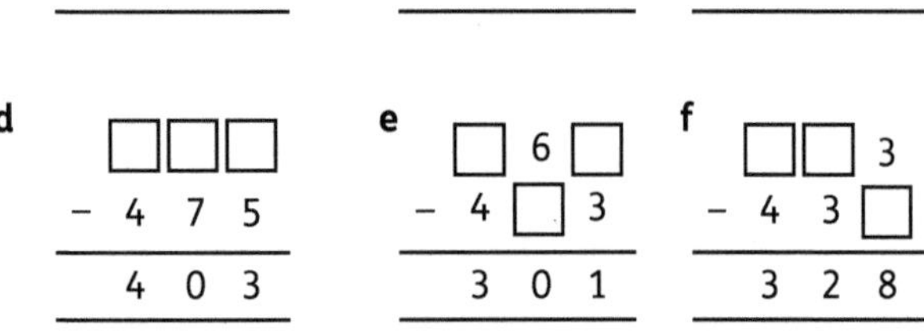

Once the children have solved these, ask them to make up some challenging subtractions of their own.

Support

There are many different programmes and activities available online to support children who find formal methods difficult. Search for 'column subtraction games' to find suitable activities for your children.

Interesting mistakes

Some children resist estimation because they don't like answers that are not 'correct'. To make sure that they estimate, encourage them to write down their estimates before they calculate. Also, provide calculations where you ask them for an estimate only, and not an actual calculation.

Answers for Pupil Book 4 page 77

1 Accept any reasonable estimates. Worked-out answers:

a 147 b 54 c 19 d 125
e 1053 f 2048

Subtraction problems

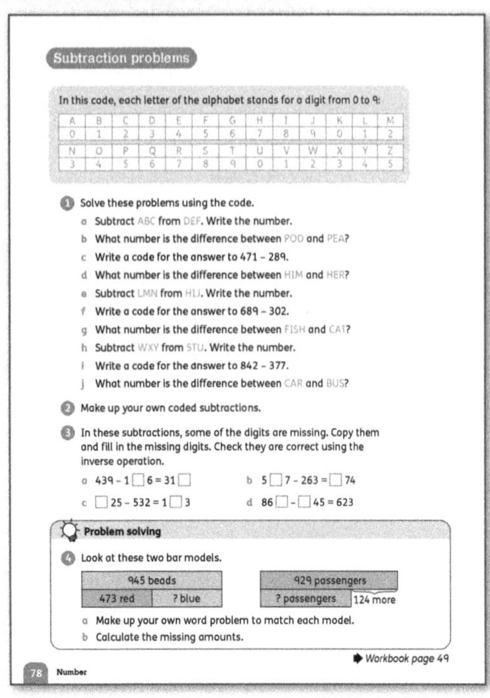

Materials

Squared paper; calculators.

Warm-up

Choose a suitable activity from 'Place-value activities' (pages 24–25) or 'Calculation skills' (pages 28–29) as a mental warm-up.

Focus

- Turn to **Pupil Book 4 page 78**. Ask the children to use the code letters to write the number for their first names. Check that they can interpret the code and that they realise that all the digits represent more than one letter.
- The problems in question 1 use different ways of talking about subtraction, for example: 'Subtract . . . from . . .' and 'What number is the difference between . . .'. Some children may find this a little confusing and initially write, for example, ABC above DEF in part a. Just remind them that making mistakes and resolving them is a valuable learning experience.
- For question 2, you can collect in their coded subtractions and give them to other children to solve.
- To make question 3 easier, give the children squared paper that they can use to write the calculations in columns.
- Problem solving: For question 4, the children can work in pairs to develop written problems to match the bar models. Share some of these with the class.

Follow-up

Use **Workbook 4 page 49** to consolidate addition and subtraction. Make sure the children realise that some pyramids involve addition while others involve subtraction. Challenge them to make a really difficult pyramid of their own for question 2 (but limit them to 3-digit numbers) and let them explain what makes it difficult.

Challenge

Ask the children to choose a three-letter code, for example, CAT. Then have them write five different subtractions that will give this result.

Interesting mistakes

- Some children may not realise that they need to subtract the smaller number from the larger number when problems are expressed in ways other than 234 – 156. This can normally be addressed by modelling solutions using base-ten blocks (see page 21) or talking through the calculation.
- The children may also say that you cannot subtract a larger number from a smaller one. This is not mathematically true, and you can let them explore this using a calculator. They will soon see that $9 - 5 = 4$ but that $5 - 9$ is also possible and that you get a result of -4. They will deal with negative numbers in more detail later.

Answers for Pupil Book 4 page 78

1 **a** 333   **b** 3

    c Possible answer: EHL – CIJ

    d 35   **e** 666

    f Possible answer: GIJ – DAC

    g 5678   **h** 656

    i Possible answer: IEC – DHH

    j 99

2 Individual answers

3 **a** 439 – 126 = 313   **b** 537 – 263 = 274

    c 725 – 532 = 193   **d** 868 – 245 = 623

4 **a** Individual answers

    b 472 (blue beads); 505 (passengers)

Answers for Workbook 4 page 49

1 **a** Possible answer: **b** Possible answer:

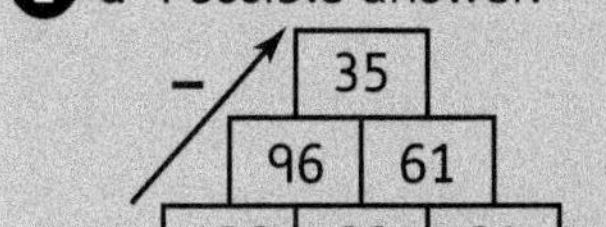
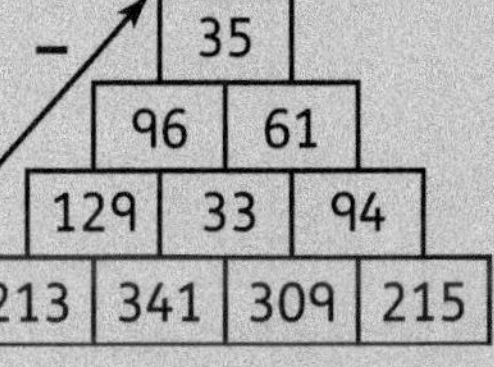
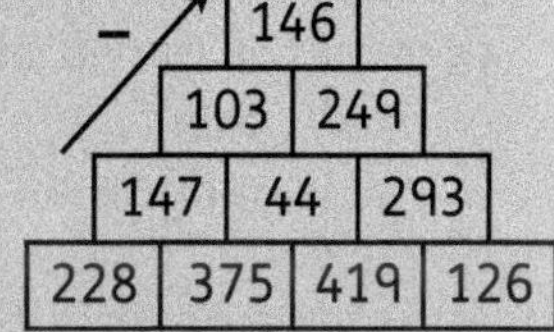

c
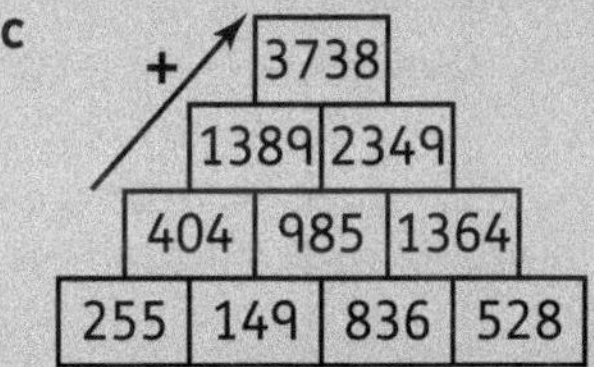

d
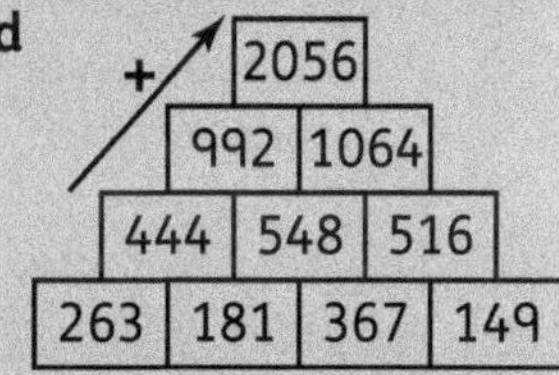

e
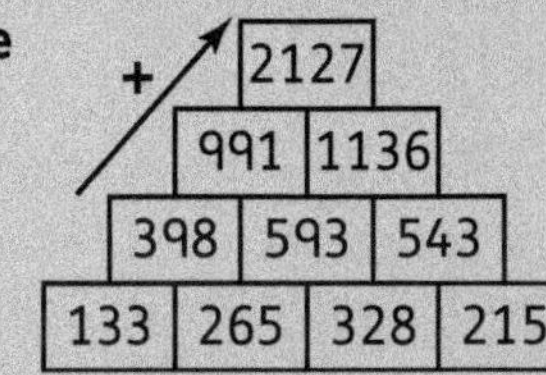

2 Individual answers

Do you add or subtract?

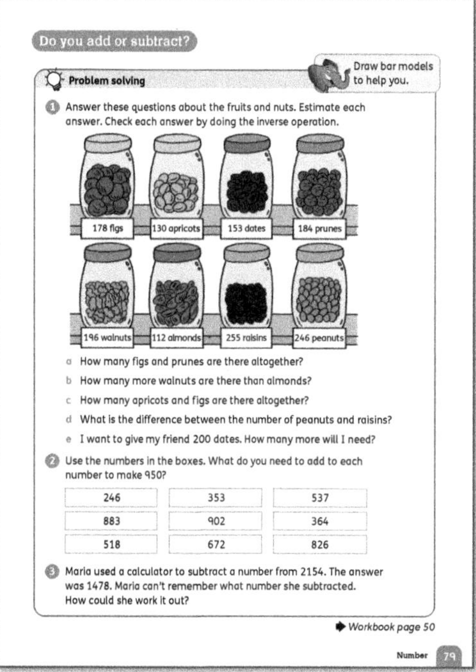

Materials
A large sheet of card for each group; coloured marker pens.

Warm-up
- Discuss words or phrases that tell us to add or subtract, for example more, altogether, in total, sum, difference, take away, less than. Record the children's suggestions on the board, accepting all contributions at this stage.
- Then go through the suggested words to check that they do indicate addition or subtraction. Circle those that do (rather than crossing out contributions that don't).
- Next, use different colours to sort the words and phrases into two groups – one group for addition and one for subtraction.

Focus
- Give each group a sheet of card and coloured marker pens. Ask each group to design a poster to teach other children which words mean add and which mean subtract.
- <u>Problem solving</u>: Turn to **Pupil Book 4 page 79** and let the children work independently to complete questions 1–3.

Follow-up
Ask the children to complete the target diagrams on **Workbook 4 page 50**. These involve 3-digit numbers and will help to consolidate the concepts from this lesson.

Interesting mistakes

Sometimes children assume that the word *add* always means that they have to add. **Pupil Book 4 page 79** question 2 is a good example of where a question includes the word 'add', but you have to subtract to find the answer. To find out what you need to add, you have to work out the difference.

Answers for Pupil Book 4 page 79

1 a 362   b 84   c 308
   d 9   e 47

2 246 + 704   353 + 597   537 + 413
    883 + 67   902 + 48   364 + 586
    518 + 432   672 + 278   826 + 124

3 She can calculate 2154 − 1478 to work out that the number she subtracted was 676.

Answers for Workbook 4 page 50

1 The missing numbers in each target diagram, clockwise from top right:
a 76, 69, 96, 88, 32, 38, 80
b 179, 129, 173, 210, 148, 152, 49, 232
c 218, 207, 215, 133, 135, 114, 223, 120
d 126, 128, 107, 63, 108, 131, 112, 99
e 425, 41, 280, 210, 500, 150, 453, 242
f 636, 590, 684, 402, 210, 450, 773, 30

Practise adding and subtracting

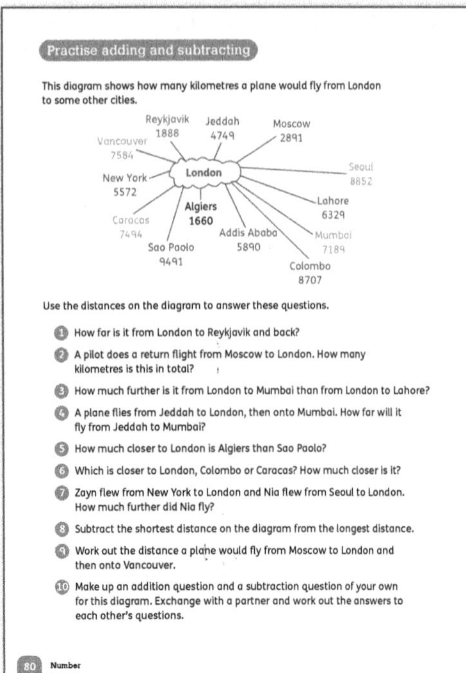

Materials

A large world map showing countries and major cities (for display) or an atlas for each group.

Warm-up

- Display the world map or ask the children to find a map of the world in their atlases that shows countries and major cities. Turn to **Pupil Book 4 page 80** and ask the children which cities they know and have them use the map to locate these.
- Find London on the map. Then ask the children to find the other cities. Note that the directions on the diagram can be used to help them do this, as can the index of the atlas. Let the children take turns to locate cities and to say in which country each one is found.

Focus

- Explain that the diagram at the top of **Pupil Book 4 page 80** shows the flight distances from London to different places. (These are correct, but they may vary from airline to airline and depending on routes.)
- Encourage the children to trace a few routes to check that they understand how to use the diagram. For example, ask:
  - *Put your finger on London, now move to Seoul. How far is that?*
  - *Start on London. Move to New York. Now move back to London. How far is that?* (The children don't need to work out the answer – they can just say, for example, '5572 and another 5572'.)
  - *Put your finger on Colombo, now move to London and onto Moscow. What distance is each part of the journey?*
  - Let the children work with a partner to do the calculations in questions 1–10.

Challenge

Let the children do their own research to find the distance from your country's main international airport to other places. (Be aware that these distances may be more than 10 000 km.) The children can then make their own diagram like the one at the top of **Pupil Book 4 page 80** and use it to make up some questions to test each other.

Answers for Pupil Book 4 page 80

1 3776 km   **2** 5782 km
3 860 km   **4** 11 938 km
5 7831 km   **6** Caracas; 1213 km
7 3280 km   **8** 7831 km
9 10 475 km   **10** Individual answers

Solve two-step problems

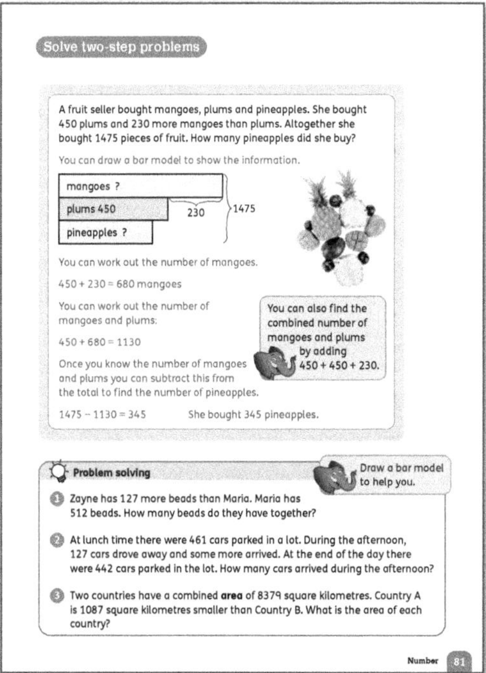

Materials
Graphic organisers (see 'Support').

Warm-up
Use one of the 'Rounding and estimating' activities (pages 26–27) as a warm-up.

Focus
- The example on **Pupil Book 4 page 81** demonstrates to the children that to solve a *two-step problem*, they have to calculate one value (the total of mangoes and plums) so that they can use that number in a second step to work out the answer: total (mangoes and plums) = number of pineapples.
- Work through the steps with the class, encouraging them to think of other methods of working by asking questions such as: *Could you draw a different bar model? Show me. Is there a different way to calculate this? Show me.*
- Problem solving: Once you have shown the children the process, let them work independently to solve the problems in questions 1–3. If necessary, remind them that the *area* of a country is how much land it covers.
- Afterwards, spend some time talking about how different children solved the problems and let the children describe and show their workings to the class.

Challenge
Give the children blank cards and let them make a set of five themed two-step problems of their own. They should write the solutions on the back. Collect the cards and hand them out to different children to extend and consolidate work on two-step problems.

Support
There are a number of different styles of graphic organisers to help children structure their thinking and workings to solve two-step problems. These can offer structure and support by getting the children to break down the steps, saying what they need to find and drawing a diagram to show it. They can also write an equation for each step. For example:

Step 1: Find . . .	Step 2: Find . . .
Bar model	Bar model
Calculation	Calculation

You can find examples of these online or work with the children to develop a frame or organiser that works for them.

Answers for Pupil Book 4 page 81
1. 1151 beads 2. 108 cars
3. Country A: 3646 km and Country B: 4733 km

End-of-unit check

Assess the children's addition and subtraction skills by asking them to complete calculations involving 3- and 4-digit numbers.

You may also like to ask some or all of the following questions to check their understanding.
- *What is the difference between 366 and 178? (188) . . . 941 and 199? (742) . . . 253 and 187? (66)*
- *How do you decide whether to add or subtract to find an answer? (For example: Words such as total, sum, altogether may mean add. Words such as difference, less than, how much bigger may mean subtract.)*
- *What is the total of 472 and 129? (601) . . . 588 and 259? (847) . . . 1376 and 3269? (4645)*
- *How much greater is (give a number) than (give a smaller number)?*
- *What do you need to add to (give a number) to get 2400?*
- *The total is 514. What two 3-digit numbers could you add to get this total? (for example: 213 + 301)*
- *Show how you would add/subtract (give two numbers).*

Angles and triangles

Learning objectives

- Revisit right angles and understand that an angle is a measure of turn
- Identify and classify angles as acute, right or obtuse
- Compare and classify triangles based on angle size and side lengths
- Understand that equilateral triangles are regular polygons

Key words

angle   turn   degree   right angle   acute angle
obtuse angle   polygon   classify
acute-angled triangle   right-angled triangle
obtuse-angled triangle   scalene   isosceles
equilateral

Unit introduction

The children have already worked with angles, including right angles, and they know that triangles are three-sided *polygons*. In this unit they are going to learn the names of acute and obtuse angles and that there are different types of triangles, which we *classify* mathematically using the properties of their angles and sides.

There is quite a lot of technical vocabulary involved here and you may need to develop some activities to help children learn the terms *acute* and *obtuse* angles and *scalene*, *right-angled* and *isosceles* triangles.

Materials

Sheets of card in different colours (or acetate sheets); scissors; compasses or round objects that children can draw around to make circles; geostrips (or strips of card or plastic); split pins.

Teaching guidance

- Tell the children that they are going to learn more about *angles* in this unit.
- Ask them to tell you a few things that they already know about angles. Many children find it difficult to articulate what an angle is but they should be able to tell you that it is a measure of *turn* between two arms.

- Show the children the following three methods of making simple equipment to show and measure angles (by comparison, not measuring angles in *degrees*). They will use these angle measures in the lessons in this unit.

Method 1
- Cut out two circles in different colours.
- Make a cut from the outer edge to the centre of each one.
- Slot them together so they overlap.
- Turn the outer circle to make an angle.

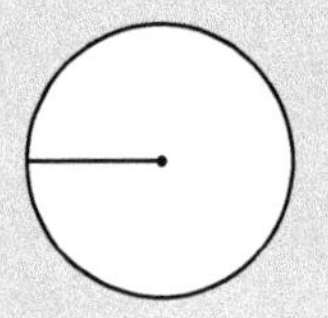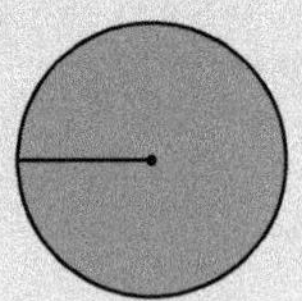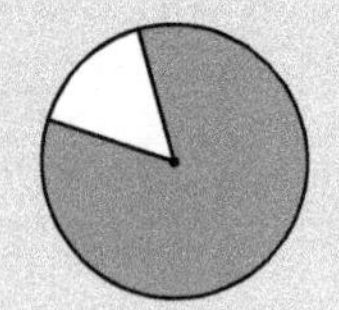

Method 2
- Use a square of card and make a slot in it as shown.
- Cut out a circle and make a cut from the outer edge to the centre.
- Slot the circle into the cut in the square.
- Turn the circle to make angles.

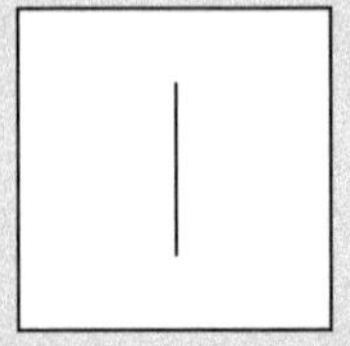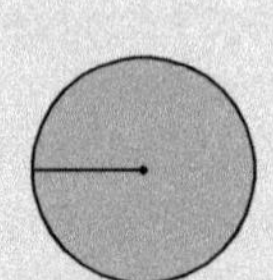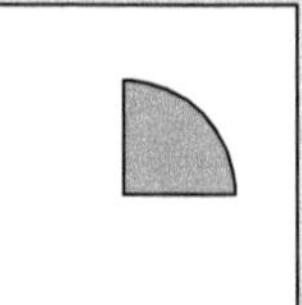

Method 3
- Connect two geostrips (or other strips) at one end using a split pin.
- The strips can be any length and different lengths.

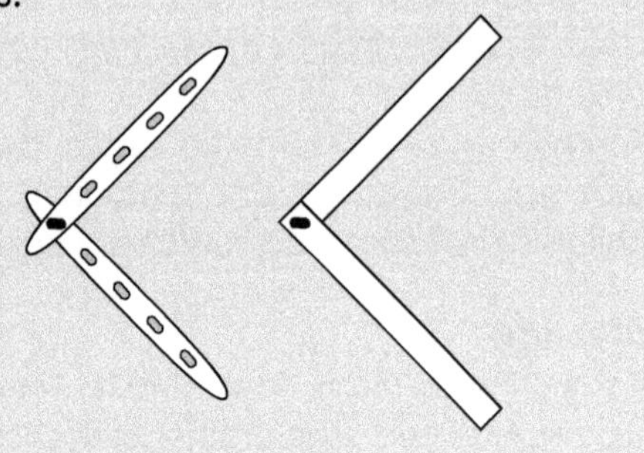

- Ask groups to choose a method. Let them collect the equipment they need and use it to make an angle measure. Ideally, each child should make their own. At the least, they need one per pair of children.

- Don't tell the children what size circles to draw or what length their strips should be. Provide a range of different-sized measuring tools to help the children notice that the size of the angle is not affected by the length of the arms.
- Once the children have made their measuring equipment, ask them to use it to make a *right angle*.
- Then let them find five right angles in the classroom. Include angles where you cannot see one of the arms. For example, an open window or door can make a right angle with the 'imaginary' line made by the door surround or the bottom of the window.
- Repeat this for angles that are smaller than a right angle and angles that are larger than one right angle but smaller than two right angles.
- Discuss how the children identified angles and used the equipment to check the sizes.
- Point out or elicit that the different equipment gave the same results. Show this by asking all the children to make a right angle again and compare these.

Types of angles

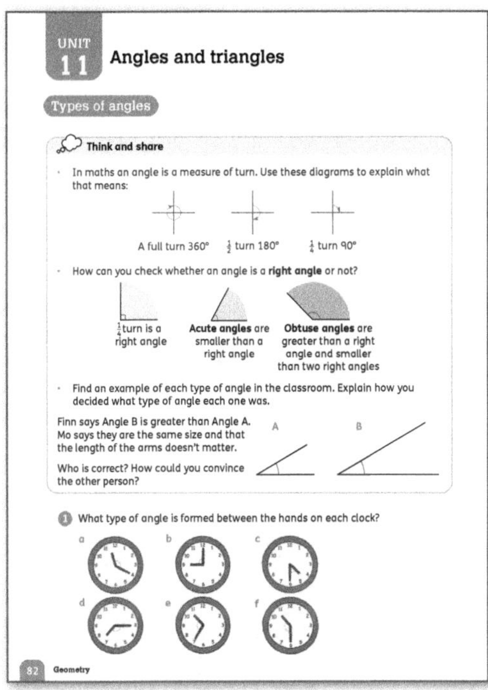

Materials
Angle measures (see 'Unit introduction'); clock face with moveable hands; two different-sized pairs of scissors.

Warm-up
Choose suitable activities from the 'Activity bank' on pages 23–30 as a warm-up. As there are no calculations in this unit, select activities that reinforce number facts and mental calculation skills.

Focus
- Think and share: Turn to **Pupil Book 4 page 82**. Ask the children to work in pairs or small groups to find examples of the angles in the classroom.
- Take feedback, making sure that the children can properly describe an angle and that they know they can use a corner to check whether an angle is a right angle.
- Ask them what the terms *acute angle* and *obtuse angle* mean and let them show examples of these angles using their angle measures and describe them in their own words.
- Share some of the examples they found in the classroom.
- Have the children share their ideas about the two angles A and B and their size. If they are not convinced, ask them to draw an angle of their own and use the angle measure to show its size. Then have them use a ruler to extend the length of the arms.
- Ask: *Has the angle changed at all?* (*No, and they can prove this by comparing it to the measure again.*)
- The children need to recognise that the size of an angle is related to the amount of turn between the arms. (They should also realise that the length of the arms has no impact on the size of an angle.)
- Clock faces are useful for demonstrating this. You can also use the two different-sized pairs of scissors. Show the class that you can make two angles the same size and the length of the cutting parts doesn't affect this.
- Ask the children to complete question 1. They can use their angle measures to check if they are not sure of the answers.

Interesting mistakes
Children sometimes find it difficult to recognise right angles in different orientations. Suggest that they prepare a set of ten right angles, all facing different directions, to help them see what the right angles look like.

Answers for Pupil Book 4 page 82
Think and share: Possible answer: You can tell whether an angle is a right angle or not by finding an object that has straight edges and meets at a corner like a book. If you can line your angle up with the corner of your object, it is a right angle.

Mo is correct. Possible answer: An angle is a measure of a turn, not length so the length of the arms is not important. You could trace the one angle and place it over the other to show how the angles are the same size.

1 a obtuse   b right angle   c acute
   d obtuse   e obtuse   f obtuse

Compare and order angles

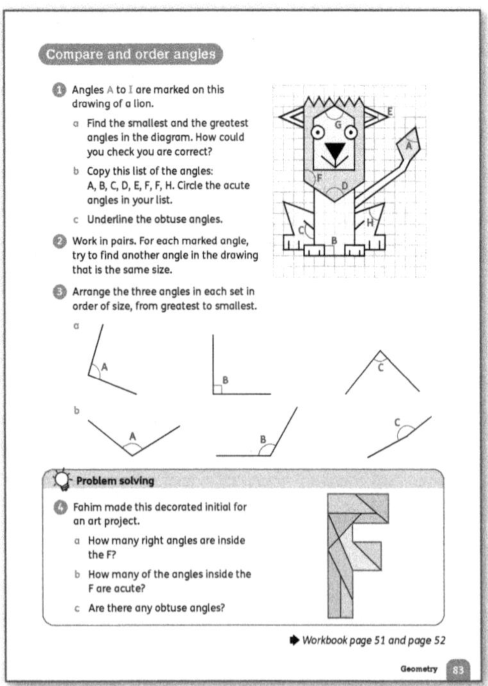

Materials

Angle measures (see 'Unit introduction'); cut-out angles of different sizes (acute, obtuse and right).

Warm-up

Choose suitable activities from the 'Activity bank' on pages 23–30 as a warm-up. As there are no calculations in this unit, select activities that reinforce number facts and mental calculation skills.

Focus

- Ask the children to turn to **Workbook 4 page 51**. Let them independently colour the angles to show which are acute, which are right angles and which are obtuse. They then sort the angles into the table in question 2. Check their answers before moving on.
- Once the children are able to identify the types of angles correctly, turn to **Pupil Book 4 page 83**.
- The children can work on their own or in pairs to find and list the different kinds of angles in question 1.
- Remind the children that they can use an angle measure to compare the angles in question 2. They can also use their knowledge of symmetry.
- Ask the children to try to do question 3 by sight before they measure and check.
- <u>Problem solving</u>: Let the children work in pairs to complete question 4. Ask them to think about how to keep track of the angles they count and how they could convince someone else that they were correct if that person got a different answer.

Follow-up

Use **Workbook 4 page 52** to consolidate work on identifying angles and to revise using tallies and frequency tables.

Challenge

The children can find star constellation maps, with lines drawn between the stars, online (Hercules and Orion are good examples). Challenge them to print or copy these and to mark the angles as acute, right angles or obtuse.

Support

- Provide a set of cut-out angles for the children to compare and sort by type. Physically handling the angles makes it easier for them to compare the angle sizes against a right angle to classify them.
- Ask the children to take one angle each and then to arrange themselves in a line showing their angles in order of size. Encourage the children to talk about how they compared the angles.
- Hold up one of the angles. Ask the children to draw a smaller/larger angle.

Interesting mistakes

The children may get confused between measuring length and measuring angles. Keep demonstrating angles using real-life examples, such as the hands on a clock, the covers of a book (open a book to view side on), arms out straight and at varying angles.

Emphasise that an angle is only created when two lines meet at a point – the angle is the amount of turn from one line towards the other.

Answers for Pupil Book 4 page 83

1 a A, H, B, C, F, D, G   b A, E, H   c C, D, F, G
2 Individual answers
3 a A, B, C   b C, B, A
4 a 8   b 18   c yes

Answers for Workbook 4 page 51

1 and **2** Angles coloured and sorted into the table as follows:
  Acute angles: a, d, f, i, l, n, o
  Right angles: g
  Obtuse angles: b, c, e, h, j, k, m

Answers for Workbook 4 page 52

1 a Individual answers   b Individual answers

Types of triangles

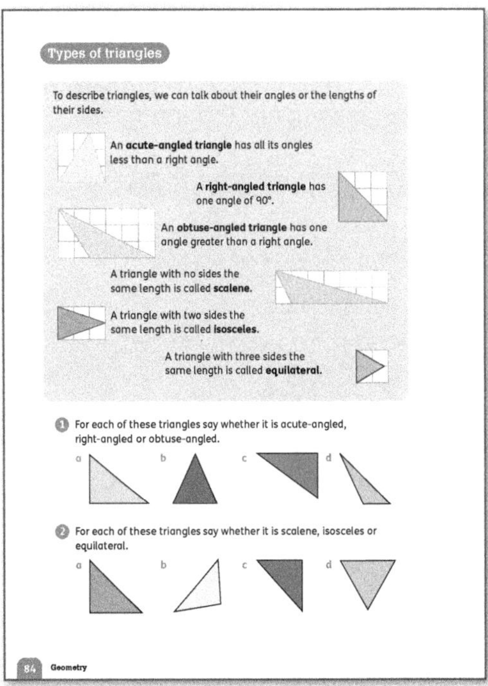

Materials
Geoboards; squared dotted paper; squared paper.

Warm-up
Choose suitable activities from the 'Activity bank' on pages 23–30 as a warm-up. As there are no calculations in this unit, select activities that reinforce number facts and mental calculation skills.

Focus
- Work through the types of triangle on **Pupil Book 4 page 84** with the class: *acute-angled*, right-angled, *obtuse-angled*, scalene, isosceles and *equilateral* triangles. Explain that the children can use the grid squares to work out whether angles are:
  - less than a right angle (acute angles form inside the corner of a square)
  - right-angled (right angles fit the corner of a square exactly)
  - more than a right angle (obtuse angles are larger than the corner of a square on the grid).
- Similarly, they can use the lengths of sides of the squares to work out whether the sides of a triangle are the same length or not. Reinforce the vocabulary and encourage the children to use it.
- The children can do question 1 orally in pairs.
- Do question 2 orally with the class, asking the children to say why they classified each triangle as they did.
- Ask the children: *Is an equilateral triangle a regular polygon?* You may need to remind them that a regular polygon has all sides the same length and all angles the same size (**Pupil Book 4 page 20**). Give them time to measure and compare side lengths and angles, and take their suggestions. Confirm that an equilateral triangle is a regular polygon.

- You could also ask them why, for example, a scalene or isosceles triangle is not a regular polygon. (*Its sides are not all the same length.*)

Follow-up
To consolidate understanding, ask the children to work with geoboards and/or squared dotted paper to make triangles according to your instructions. For example, you could ask:

- *How many different isosceles triangles can you make on a 4 × 4 square section of geoboard? Draw them on dotted paper.*
- *How many different right-angled triangles can you make on a 4 × 4 square geoboard? Draw them on dotted paper.*
- *Can you draw a regular three-sided polygon on this squared grid? If so, explain how.*

Challenge
Let the children explore the use of triangles to stabilise other shapes. They can do this by finding examples in real life, or by making quadrilaterals (and other polygons) by joining strips of card with split pins or using construction straws. They can try to push these over to see whether they are rigid or not. Adding a diagonal splits the quadrilateral into two triangles and this stabilises the shape.

Interesting mistakes
Children sometimes think that angles in a larger triangle are greater than the same angles in a smaller triangle. For example, they may say that Triangle B has larger angles than Triangle A.

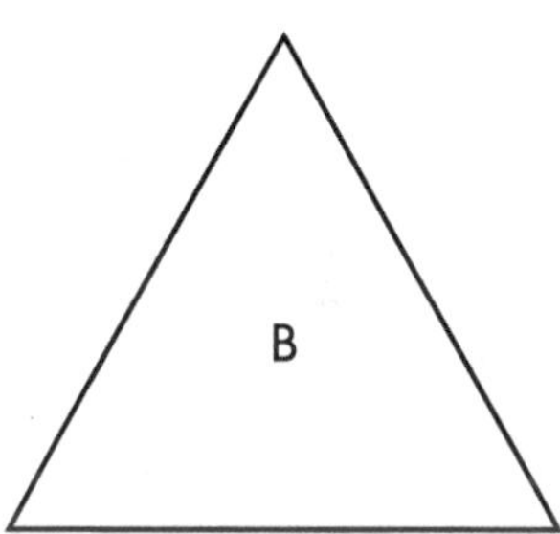

As with angles, let the children measure (by comparing) to check that the angles are the same size. Explain that the amount of space between angles doesn't change their size. Demonstrate this using two cut-out angles and moving them apart along a line.

> ### Answers for Pupil Book 4 page 84
> **1** a right-angled   b acute-angled
>   c right-angled   d obtuse-angled
> **2** a isosceles   b scalene
>   c isosceles   d equilateral

Classify triangles

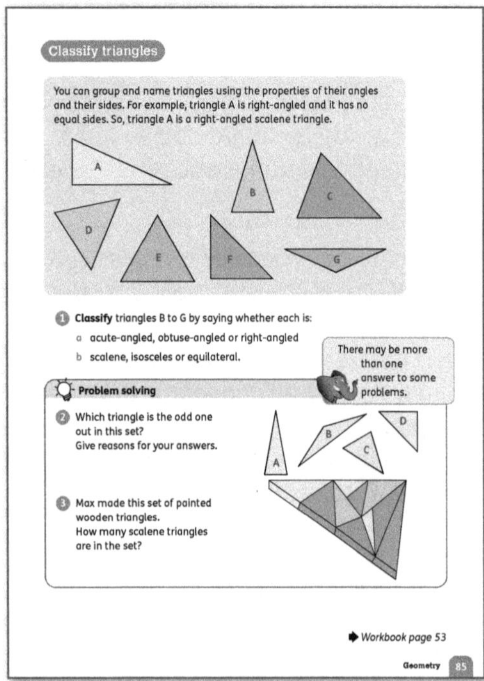

Materials

Geoboards and/or squared dotted paper.

Warm-up

Choose suitable activities from the 'Activity bank' on pages 23–30 as a warm-up. As there are no calculations in this unit, select activities that reinforce number facts and mental calculation skills.

Focus

- Revise the names and types of triangles using this practical activity.
- Let the children work on geoboards or dotted paper. Tell them that you are going to describe some triangles. As you describe each triangle, they should make it on the geoboard (or draw it on dotted paper). Here are some examples:
  - *This triangle is obtuse and one pair of sides are equal.*
  - *This triangle has a right angle. It also has a line of symmetry.*
  - *This triangle is acute-angled with no lines of symmetry.*
  - *This is an isosceles triangle with one side twice as long as one of the other sides.*
- Once you are sure that the children know the correct names for different types of triangles, let them work on their own to classify the triangles in question 1 on **Pupil Book 4 page 85**.
- Check the children's answers to question 1 before using **Workbook 4 page 53** to assess how well they understand the concepts in this lesson.
- Problem solving: Turn back to **Pupil Book 4 page 85**.
- For question 2, the children can work in pairs to share their ideas and reasons for choosing a particular triangle.
- Let the children work on their own to solve question 3.

Challenge

Explore combining different types of triangles to make quadrilaterals. Ask the children to explore what shapes they can make using two, three or four triangles and let them describe the properties of the shapes they make using what they know about the triangles.

Answers for Pupil Book 4 page 85

1 **a** right-angled: F
acute-angled: B, C, D, E
obtuse-angled: G
  b scalene: C, F
isosceles: B, D, G
equilateral: E

2 Possible answers: C, because it is not an isosceles triangle; B, because A, C and D are acute-angled and B is obtuse-angled.

3 4

Answers for Workbook 4 page 53

1 **a** the third triangle: It is the only triangle that is not a right-angled triangle.
  b the second triangle: It is an equilateral triangle; the others are all scalene.
  c the third triangle: It is the only scalene triangle; the others are all isosceles.
OR: It is obtuse-angled; the others are all acute-angled triangles
  d the second triangle: It is a an acute-angled triangle; the others are all obtuse-angled.

End-of-unit check

Use some or all of these questions to assess the children's understanding of angles and triangles.

- *What is an angle? (a measure of a turn)*
- *Is it true that an angle is the area between two lines? Explain why or why not. (No. It is the amount of turn from one line to the other.)*
- *Make a right angle using your fingers/arms. (The children show a right angle with their thumb and first finger or with one arm pointing vertically up and the other arm out to the side.)*
- *What important tips would you give someone to help them use their measuring tool to compare angles?*
- Show some acute, right and obtuse angles. *Label each angle as acute, right or obtuse.*
- Show a triangle. *What is the correct name for this triangle?*
- Draw an angle. *What type of angle is this?*
- *How can you tell whether a triangle is isosceles? (An isosceles triangle has two sides the same length.)*
- *Is this triangle right-angled? How can you tell? (If one angle is a right angle, the triangle is right-angled.)*
- *Is it possible to have a right-angled equilateral triangle? (No. In an equilateral triangle, all the angles are acute.)*

UNIT 12 Multiplication and division facts

Learning objectives

- Recall multiplication and division facts for multiplication tables up to 12×12
- Understand properties of multiplication (and division), including multiplying by 1 and 0 and dividing by 1
- Multiply three numbers, changing the order as needed
- Manipulate multiplication and division equations and understand and use the commutative property
- Understand and use the distributive property
- Use factor pairs and commutativity in mental calculations

Key words

factor   array   multiplication fact   product
times table   division fact   division
multiplication   inverse   multiply   fact family

Unit introduction

The concepts in this unit are mostly familiar to the children. They should already know their times tables facts for the 2, 3, 4, 5, 8 and 10 times tables.

They will now extend these to learn the 6, 7, 9, 11, and 12 times tables and associated *division facts*. They will also need to begin to use the patterns from tables in their work with multiples.

The children should also remember that *division* is the *inverse* of *multiplication*. However, they need to extend their understanding of this concept to realise that once they know multiplication facts, they can use these to derive two related division facts.

It is important that you continue to use practical and concrete methods to reinforce multiplication and division facts alongside the questions in the Pupil Book and Workbook.

Some concepts involving properties of multiplication and division that may be unfamiliar to the children are introduced later in this unit.

Materials

Squared paper; graph paper (optional).

Teaching guidance

- Give each child a sheet of squared paper and some crayons or marker pens.
- Choose some numbers less than 100 that have a number of *factors* (12, 24, 30, 36, 42, 48 and 60 are all good examples, although if you are using the larger numbers you could give the children graph paper to work on, so they can fit more diagrams onto a sheet).
- Ask the children to draw *arrays* or rectangles on the grid to show each number. Ask them to find as many ways as they can. For example, 12 could be shown as:

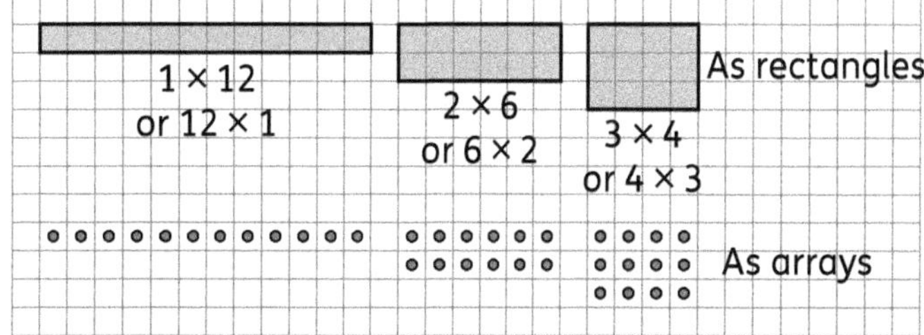

- Have the children write a *multiplication fact* for each diagram. Discuss how they do this (as $1 \times 12 = 12$ or $12 \times 1 = 12$) and whether or not these are the same. (They both give the same *product*.)
- Repeat for a few numbers to revise *times tables* that the children learnt in previous levels. They should know facts for 2, 3, 4, 5, 8 and 10 times tables.
- Provide lots of opportunity for the children to verbalise the facts so that they recognise and learn sound patterns to help memorise their tables. Saying *three times four is twelve* helps them to develop fluency and recall.

Revise multiplication facts

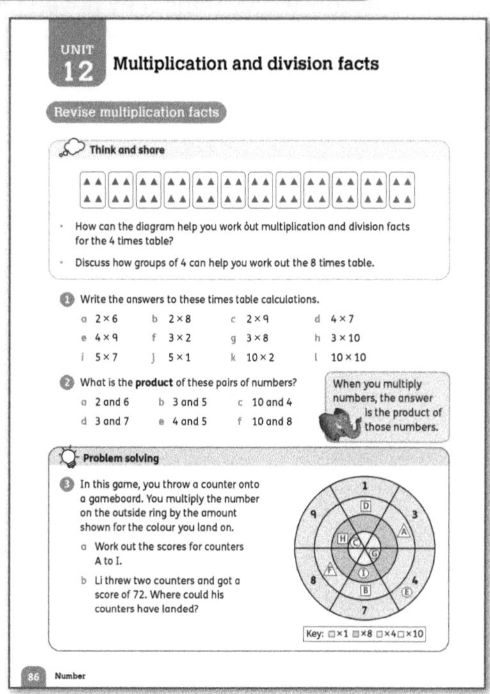

Warm-up

- Choose a suitable 'Mental maths' activity from pages 27–29 or spend the time on activities that will help the children learn their times tables.
- Other suitable mental warm-ups include:
  - giving the children function machines (see **Workbook 4 page 48**) with operators such as × 2 or × 4, with missing numbers for the children to complete
  - giving the children a number and asking the children to write down multiplications that give that number as the answer
  - displaying some arrays and asking the children to give the multiplication and division facts.

Focus

- <u>Think and share</u>: Turn to **Pupil Book 4 page 86**. Put the children in pairs and ask them to read through and answer the questions about the ten groups of 4. Check the answers and have a class discussion about how groups of 4 can help us work out facts for the 8 times table.
- The children write the answers to question 1 independently. Let them check each other's work.
- Before the children do the calculations in question 2, make sure they understand that the term product means the answer to a multiplication.
- <u>Problem solving</u>: Work out the score for counter A with the class to show them what to do in question 3 part a. Let them find counter A and ask them how to work out the score. Make sure they can see that the score for A is 3 × 4 = 12 points. Discuss how you want the children to record their answers to the rest of part a. Finally, let them work in pairs to complete part b.

Interesting mistakes

The children may know that multiplication is commutative but have difficulty applying this property to make multiplication easier. For example, they may know that 4 × 7 = 28 but not know what to do when faced with 7 × 4 because they think they haven't learnt the 7 times table yet.

Continuing to work with arrays and grids will help them to become more comfortable and familiar with this property.

More tables

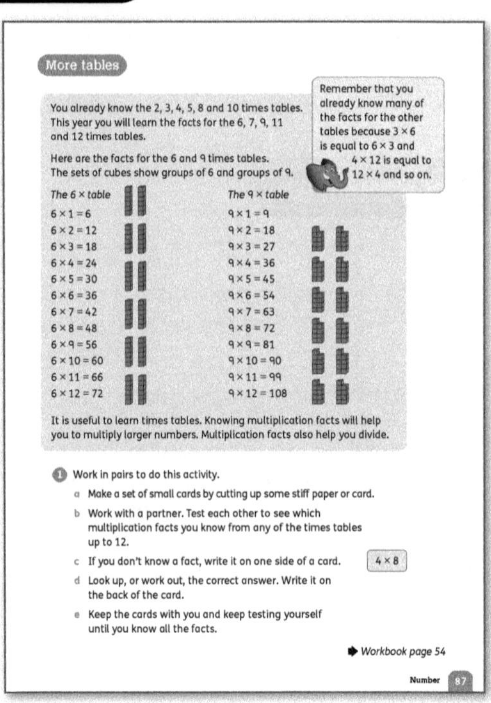

Materials

Sets of blank cards and coloured marker pens (for each child); counters.

Warm-up

Use a suitable 'Mental maths' activity from pages 27–29 or do some activities that will help the children learn their times tables.

Focus

- Spend enough time working through **Pupil Book 4 page 87** to make sure that the children know their 6 and 9 times tables and that they can *multiply* single-digit numbers as well as derive division facts and recognise multiples of these numbers.
- Have a number talk (see pages 17–18) with the class about why we don't need to learn the 11 and 12 times tables if we already know the facts for the other tables.
- Let the children share their ideas and use them to make a list of the facts for the 11 and 12 times tables based on what they already know. For example: *We know 2 × 11 = 22 and 3 × 11 = 33, so 11 × 1 = 11, 11 × 2 = 22 and 11 × 3 = 33.*
- Once you've listed the 11 times table, focus on the patterns in the products and ask the children to describe them. Repeat this for the 12 times table, taking contributions from different children as you work through the facts.
- The children may need to count on to find the facts for 11 × 11, 11 × 12 and 12 × 12 (once they have 11 × 12 they also have 12 × 11).
- Give each child a set of blank cards. Ask the children to follow the steps in question 1 to help them learn all the times tables facts up to 12 × 12. Encourage them to use their cards on a regular basis.

Follow-up

Use **Workbook 4 page 54** to consolidate times tables up to 12 × 12. Make sure the children realise that once they have completed this tables square, they can refer to it to help them find products if they forget a fact.

Challenge

Hold a times table competition. Divide the children into teams of four or five and ask them to write questions to test the other teams on the times tables.

Support

- Ask the children to make ten groups of 6 using counters. Ask them how many counters are in one group of 6, two groups of 6 and so on. Write the total on the board each time. Ask the children whether they can see a pattern.
- Reinforce the pattern obtained by counting in sixes using a number line. Show the children that, starting at zero and jumping 6 each time, they get the same pattern of numbers as when they added the sets of 6 together.
- Summarise the 6 times table in a whole-class session. Repeat this for the other tables as necessary.

Arrays

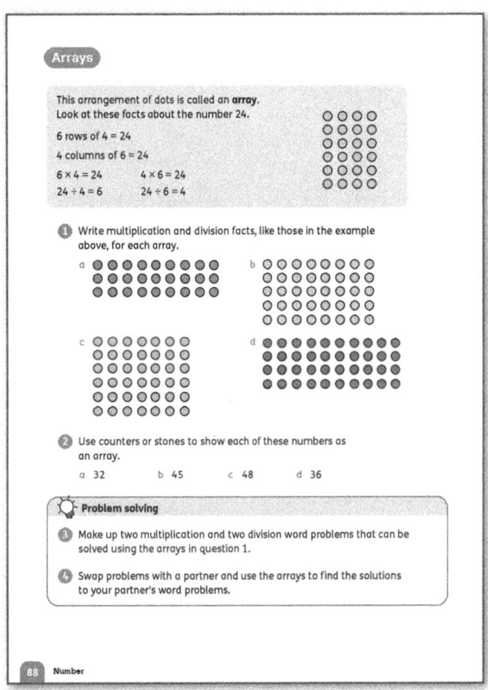

Materials

Stones, counters or Numicon number frames (see page 21); squared paper.

Warm-up

Continue to revise multiplication tables to 12 × 12 as a mental warm-up. The children could use their cards from the previous lesson to do this.

Focus

- Work through the example on **Pupil Book 4 page 88** with the class to make sure they know that division is the inverse of multiplication and that they can derive division facts from multiplication arrays.
- Do question 1 orally with the class before asking the children to write the related facts.
- Give the children stones, counters or number frames to make the arrays in question 2. Alternatively, they can draw the arrays on squared paper.
- Problem solving: For question 3, the children make up four word problems. Ask them to write the solutions to their own problems in their books.
- To complete question 4, the children swap their problems with a partner and solve each other's

problems. They then check the partner's answers. Resolve any disagreements as a class.

Interesting mistakes

We often deal with multiplication separately from division, so some children might see the two operations as different and separate and not realise that they are inverse operations. Working with arrays and physically dividing them into rows and columns may help the children to connect the two operations.

Answers for Pupil Book 4 page 88

1 a $3 \times 9 = 27$; $9 \times 3 = 27$; $27 \div 9 = 3$; $27 \div 3 = 9$
   b $5 \times 8 = 40$; $8 \times 5 = 40$; $40 \div 5 = 8$; $40 \div 8 = 5$
   c $6 \times 7 = 42$; $7 \times 6 = 42$; $42 \div 6 = 7$; $42 \div 7 = 6$
   d $4 \times 10 = 40$; $10 \times 4 = 40$; $40 \div 4 = 10$; $40 \div 40 = 4$

2 a Possible answers: 1×32, 32×1, 2×16, 16×2, 4×8, 8×4
   b Possible answers: 1×45, 45×1, 3×15, 15×3, 5×9, 9×5
   c Possible answers: 1×48, 48×1, 2×24, 24×2, 3×16, 16×3, 4×12, 12×4, 6×8, 8×6
   d Possible answers: 1×36, 36×1, 2×18, 18×2, 3×12, 12×3, 4×9, 9×4, 6×6

3 Individual answers. For example:

Possible multiplication word problem for array **a**: The children in the choir stand in 3 rows. There are 9 children in each row. How many children are there in total?

Possible division word problem for array **b**: There are 40 bananas in a box. The shopkeeper puts them into bags, with 5 in each bag. How many bags does she fill?

4 Individual answers

Grids

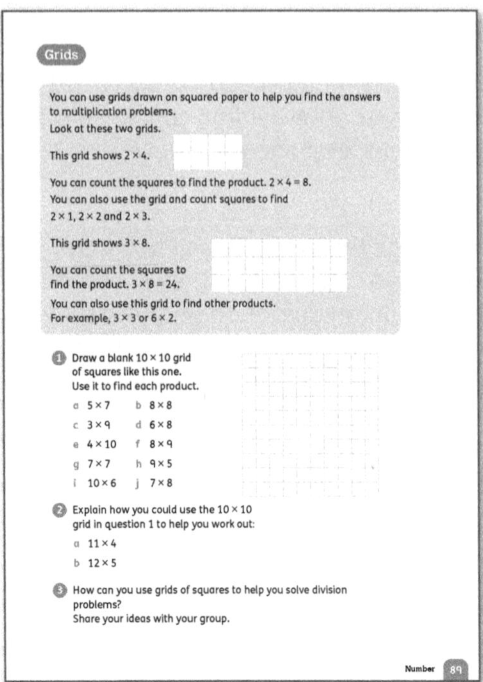

Materials

Squared paper.

Warm-up

As a mental warm-up, revise the 11 and 12 times tables facts, including divisions.

Focus

- Talk through the examples at the top of **Pupil Book 4 page 89** with the class.
- Let the children complete questions 1 and 2 in pairs. Using grids to multiply like this will help them develop their understanding of how to calculate the area of a rectangle in Unit 14. It will also help to reinforce the commutative property of multiplication.
- The children can explain their ideas in questions 2 and 3 to their group. Then have a class feedback session so the groups can share their ideas and ask each other questions.

Answers for Pupil Book 4 page 89

1 a 35   b 64   c 27   d 48
   e 40   f 72   g 49   h 45
   i 60   j 56

2 a Possible answer: Find 10×4 and 1×4 and then add the two products.
   b Possible answer: Find 10×5 and 2×5 and then add the two products.

3 Possible answer: Start to draw a grid for the number you are dividing with the number of rows that match the number you are dividing by. Count how many columns you have once you have the number of squares to show the number you are dividing. This is your answer.

Practise your facts

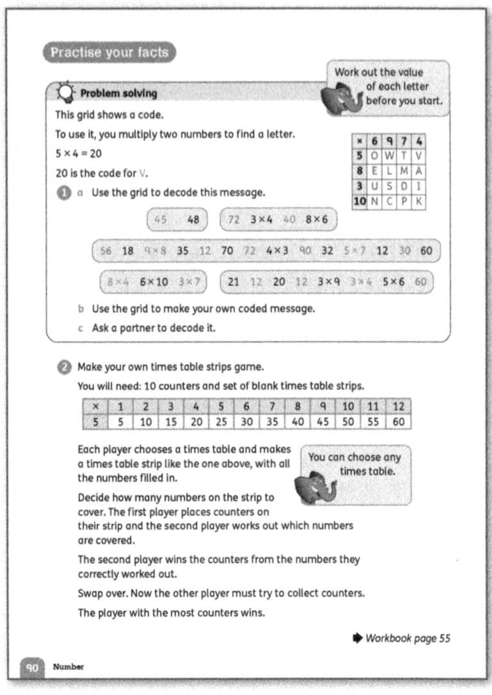

Materials

Counters; copies of blank table strips (see **Pupil Book 4 page 90** question 2) or squared paper to draw them on.

Warm-up

As a mental warm-up, give the children several products and let them say how they could get this number by multiplying two factors. For example, you say *33* and the children say '3 × 11' or '11 × 3'.

Focus

- <u>Problem solving</u>: Let the children work in pairs to complete question 1 on **Pupil Book 4 page 90.**
- They can continue working in pairs for question 2. Give them blank table strips to fill in, or squared paper to draw the strips on.
- There are no new concepts involved. The children need to use their multiplication and division facts to answer the questions.

Follow-up

Use **Workbook 4 page 55** to informally assess how well the children can recall multiplication table facts to 12 × 12.

Answers for Pupil Book 4 page 90

1 We like multiplication and division.
2 Individual answers

Answers for Workbook 4 page 55

1 The completed strips, with the numbers already in place given in brackets:
  a 5, 10, 15, 20, (25), 30, 35, 40, 45, (50), 55, 60
  b (6), (12), (18), 24, 30, 36, 42, 48, 54, 60, 66, 72
  c 7, 14, 21, 28, (35), 42, 49, 56, 63, (70), 77, 84
  d 8, 16, 24, 32, 40, (48), (56), 64, 72, 80, 88, 96
  e 9, 18, (27), (36), 45, 54, 63, 72, 81, 90, 99, 108
  f (10), (20), 30, 40, 50, 60, 70, 80, 90, 100, 110, 120
  g (11), (22), (33), 44, 55, 66, 77, 88, 99, 110, 121, 132
  h 12, 24, 36, 48, 60, 72, 84, 96, (108), (120), (132), (144)

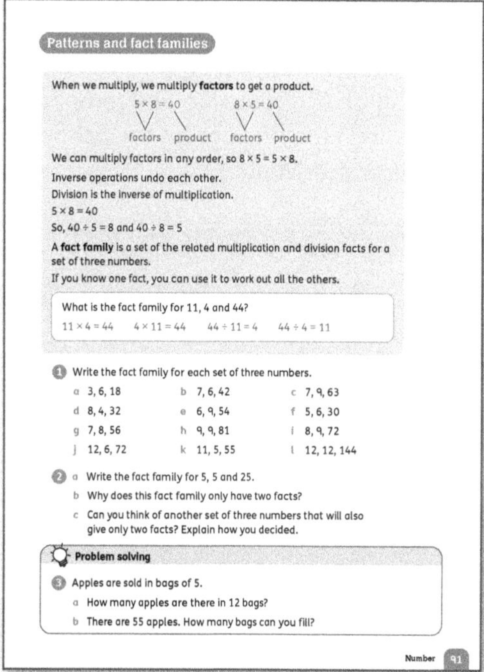

Warm-up

- As a mental warm-up, practise division facts around the class. You start by asking a child a question such as: *What is 36 ÷ 6?* (6) That child then makes up a division fact for another child to solve.
- You can do this in rows, by pointing to the child you want to answer or by letting children ask each other, as long as different children are asked.

Focus

- Turn to **Pupil Book 4 page 91** to revise the term *factor*. The children have learnt about factors previously, so they should know what a factor is.
- Revise the term *inverse operation* as well if necessary.
- Ask the children to explain in their own words how *fact families* work.
- Give them a few examples with two factors and a product and ask them to give the fact families. Include some examples where the product is not in the last position, for example 4, 48, 12.
- The children work individually to write the fact families in question 1. Let them compare answers and correct their own work.
- Do question 2 with the class. Let the children write the facts in part a before discussing the questions in parts b and c.
- <u>Problem solving</u>: Let the children discuss question 3 and how they would solve each part of the problem before sharing their ideas and solutions with the class.

Challenge

Ask the children to find the missing factor in a fact family. Give them one factor and a product and let them work out the missing factor before writing the fact families.

Interesting mistakes

The children may know their multiplication and division facts well, but they may not be able to work out how to use them to solve problems when they aren't expressed as simple calculations in the style of $a \times b$ or $x \div y$. Remind the children that drawing diagrams to show the problem (whatever type of diagram they choose) can help them to work out what to do.

Answers for Pupil Book 4 page 91

1 a $3 \times 6 = 18$; $6 \times 3 = 18$; $18 \div 6 = 3$; $18 \div 3 = 6$
  b $6 \times 7 = 42$; $7 \times 6 = 42$; $42 \div 6 = 7$; $42 \div 7 = 6$
  c $9 \times 7 = 63$; $7 \times 9 = 63$; $63 \div 9 = 7$; $63 \div 7 = 9$
  d $8 \times 4 = 32$; $4 \times 8 = 32$; $32 \div 4 = 8$; $32 \div 8 = 4$
  e $6 \times 9 = 54$; $9 \times 6 = 54$; $54 \div 6 = 9$; $54 \div 9 = 6$
  f $5 \times 6 = 30$; $6 \times 5 = 30$; $30 \div 5 = 6$; $30 \div 6 = 5$
  g $8 \times 7 = 56$; $7 \times 8 = 56$; $56 \div 7 = 8$; $56 \div 8 = 7$
  h $9 \times 9 = 81$; $81 \div 9 = 9$
  i $8 \times 9 = 72$; $9 \times 8 = 72$; $72 \div 8 = 9$; $72 \div 9 = 8$
  j $12 \times 6 = 72$; $6 \times 12 = 72$; $72 \div 6 = 12$; $72 \div 12 = 6$
  k $5 \times 11 = 55$; $11 \times 5 = 55$; $55 \div 5 = 11$; $55 \div 11 = 5$
  l $12 \times 12 = 144$; $144 \div 12 = 12$

2 a $5 \times 5 = 25$; $25 \div 5 = 5$
  b because 5 is multiplied by itself to give 25
  c Possible answer: 7, 7 and 49. I can choose any number multiplied by itself and their product.

3 a 60 apples   b 11 bags

Division facts

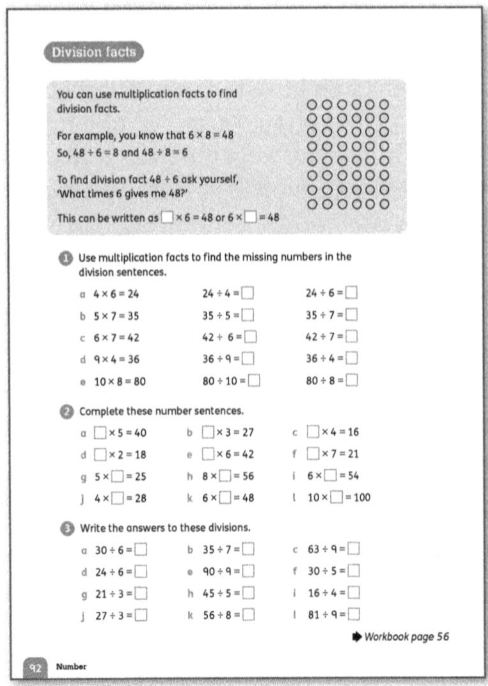

Warm-up

Repeat the division activity from the 'Warm-up' in the previous lesson.

Focus

This lesson consolidates inverse operations. The concept is not a new one as the children have dealt with it through the previous lessons.

- Read through the example on **Pupil Book 4 page 92** with the class.
- If necessary, do a few more examples to check that the children understand before they complete questions 1–3 on their own. They can compare and check answers.

Follow-up

Use **Workbook 4 page 56** as a class test to see how well the children can recall multiplication facts and use them to divide mentally.

Answers for Pupil Book 4 page 92

1 a 6; 4   b 7; 5   c 7; 6
  d 4; 9   e 8; 10
2 a 8   b 9   c 4   d 9
  e 7   f 3   g 5   h 7
  i 9   j 7   k 8   l 10
3 a 5   b 5   c 7   d 4
  e 10   f 6   g 7   h 9
  i 4   j 9   k 7   l 9

Answers for Workbook 4 page 56

1 a 45   b 9   c 3
  d 5   e 63   f 7
  g 7   h 3   i 6
  j 21   k 7   l 5
  m 9   n 48   o 18
  p 64   q 54   r 4
2 a (from top to bottom) 12, 8, 6, 4, 3
  b (from top to bottom) 15, 10, 6, 5, 3
  c (from top to bottom) 20, 10, 8, 5, 4
  d (from top to bottom) 24, 12, 8, 6, 4

Multiply and divide by 1 and 0

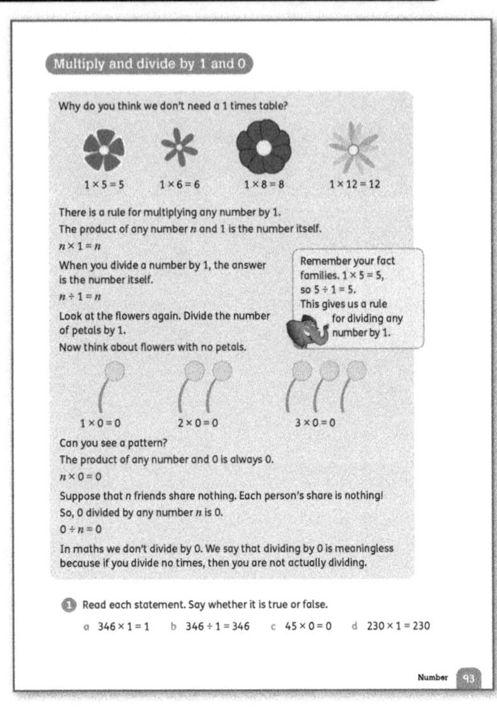

Materials

Calculators.

Warm-up

As a mental warm-up, give the children a multiplication or division number sentence (such as 56 ÷ 7 = 8) and ask them to complete the fact family.

Focus

- Ask: *Why don't we have a 1 times table?* Let the children share their ideas about this. They have already learnt that $2 \times 1 = 2$, $3 \times 1 = 3$ and so on. Do not expect them to give a generalised rule at this stage.
- Turn to **Pupil Book 4 page 93**. Use the pictures to talk in general terms about $1 \times$ any number. Explain that we can use n to stand for a number. So, $1 \times n$ means 1 times any number. The children can use calculators to check that this rule applies to numbers greater than 12.
- Ask: *What happens if you have 6 things and you divide them among 1 person only?* Elicit that the person would get all 6. Show this as $6 \div 1 = 6$. Let the children use their calculators to see what happens when they divide different numbers by 1.
- Again, explain that we can use n to stand for any number and that $n \div 1 = n$.
- Use the pictures to show that no petals (0) on a number of flowers will remain 0.
- Let the children use their calculators to multiply different numbers (including decimals and negative numbers) by 0.
- Tell the children that you have no crayons and that you are going to divide these among the class. Ask: *How many crayons will you each get?* (0 crayons)
- Explain that if we have nothing and we divide it among any number of people, they will get nothing. So, $0 \div n = 0$.
- Use the calculator to show some examples. Then, ask the children to type in any number and then enter ÷ 0. The calculator will give an error message.
- Explain that there is no such thing as dividing by 0. Let the children suggest everyday situations to show this. Start by giving them an example situation such as: *If I have 6 shirts and I divide them among no shelves I won't actually do anything with the shirts.*
- Read the statements in question 1 aloud and let the children indicate whether they think each statement is true or false. Ask for reasons and explanations.

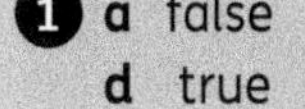

Answers for Pupil Book 4 page 93

1   a   false   b   true   c   true
    d   true

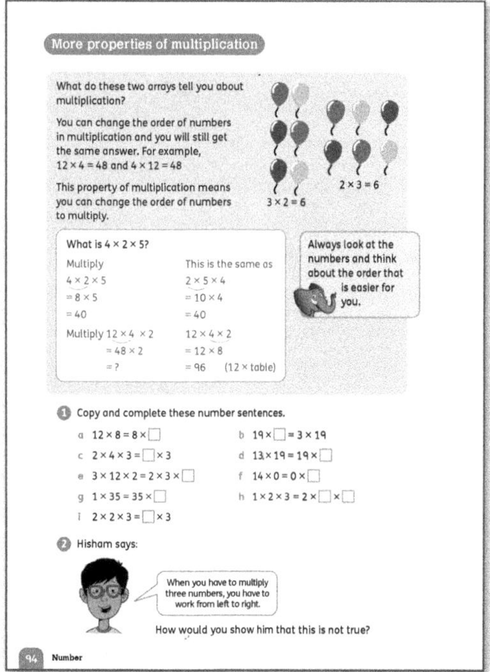

Warm-up

As a mental warm-up, give the children a multiplication or division number sentence and ask them to complete the fact family. Include 0 or 1 in some of the number sentences.

Focus

The children already know that multiplication is commutative and that they can change the order when they multiply two numbers. They are now going to apply this principle to help them multiply three numbers.

- Let the children work through the example on **Pupil Book 4 page 94** in pairs or small groups. Ask different children to tell the class what they learnt.
- The children can complete question 1 individually. Once they have done this, you could ask them to calculate the product for each number sentence.
- For question 2, the children think about what they have learnt and find a way of demonstrating it to others. They can discuss this in pairs.

Answers for Pupil Book 4 page 94

1   a   12   b   3   c   8   d   13   e   12
    f   any number   g   1   h   1, 3   i   4

2   Possible answer: Multiply them in a different order to show you get the same answer.

Use properties of multiplication and division

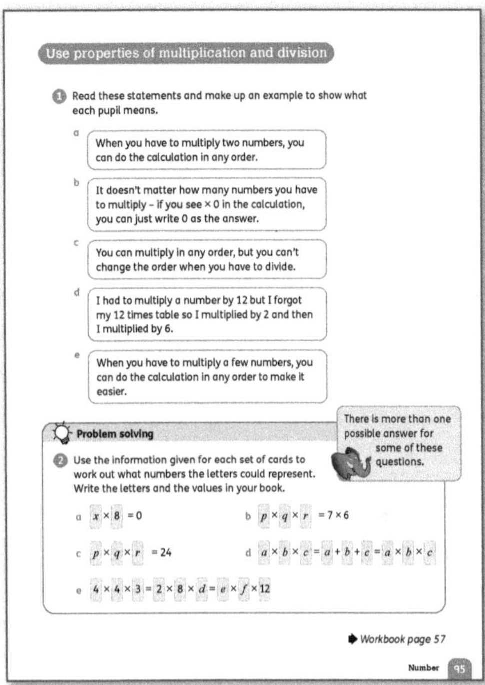

Warm-up

- As a mental warm-up, ask each child to write down one multiplication and one division that can be done mentally.
- Let the children stand up in turn and read out their calculations.
- Choose another child to give the answers. Then let that child read out their calculations.

Focus

This lesson involves children reasoning about the properties of multiplication and division.

- Write on the board:

  $4 \times 12 = 48$

  $5 \times 12 = 60$

  Ask the children how you could use these multiplication facts to work out

  9×12.

  Take suggestions and confirm that:

  $9 \times 12 = 4 \times 12 + 5 \times 12$

  $= 48 + 60 = 108$

  This example uses the distributive property of multiplication, but the children do not need to know this term.

- Repeat with more examples.
- Turn to **Pupil Book 4 page 95**. The children can work through the questions in pairs, talking about what they are doing as they work, so that they can make sense of concepts in their own way.
- For question 1, the children need to make up their own calculations to show what each statement means.

- Problem solving: For question 2, the children use the properties of multiplication to work out the unknown numbers (represented by letters) in the equations. For some of the letters, there is more than one possible answer.
- When children have completed the questions, have a class discussion about what they have done and discovered.

Follow-up

Ask the children to complete the questions on **Workbook 4 page 57**. This will consolidate what they have learnt and show you whether they are able to recall facts and apply properties of multiplication and division.

Interesting mistakes

The children may not immediately see how the distributive property can help simplify multiplication. If they cannot remember a fact like 12×8, they may not immediately see how to use factors to do an 'easier' calculation. For example, they could do:

$10 \times 8 + 2 \times 8 = 80 + 16$ to get 96.

Using 10×10 squared grids can help them to see this visually because they won't be able to make a 12×8 array. If they colour ten rows of 8 and then two columns of 8, they will have 12 eights, but 10 of them will be in one direction, and 2 in the other, so $10 \times 8 + 2 \times 8 = 96$.

Answers for Pupil Book 4 page 95

1 a–e Individual answers

2 a $x = 0$

b Possible answer: $p = 7, q = 3, r = 2$

c Possible answers: $p = 1, q = 2, r = 12$; $p = 2, q = 4, r = 3; p = 2, q = 2, r = 6$

d Possible answers: $a = 0, b = 0, c = 0$; $a = 1, b = 2, c = 3$

e $d = 3$ Possible answers: $e = 2, f = 2; e = 1, f = 4$

Answers for Workbook 4 page 57

1 Individual answers

2 a $8 \times 6; 2 \times 4; 4 \times 12; 48$

b $12 \times 5; 2 \times 30; 10 \times 6; 60$

c $30 \times 6; 10 \times 18; 60 \times 3; 180$

d $18 \times 3; 9 \times 6; 27 \times 2; 54$

e $4 \times 7; 1 \times 28; 28$

f $11 \times 0; 2 \times 0; 22 \times 0; 0$

3 (The numbers in each multiplication can be in any order.)

a $2 \times 3 \times 4$ b $3 \times 4 \times 6$ c $2 \times 5 \times 6$

d $2 \times 5 \times 7$ e $2 \times 3 \times 5$ f $4 \times 5 \times 6$

Ask questions like these to check how well the children can recall facts and apply the properties of division and multiplication.

- *What is 9 × 3? (27)*
- *What is three eights? (24)*
- *Double 9. (18)*
- *Is 40 a multiple of 5? How do you know? (Yes. Multiples of 5 end in 0 or 5.)*
- *What is the product of 8 and 9? (72)*
- *Share 36 between 4 people. (9)*
- *How many lengths of 8 cm can you cut from 56 cm of string? (7)*
- *Is 36 divisible by 8? How do you know? (No. 8 × 4 = 32 and 8 × 5 = 40)*
- *The answer to a division is 5. What could the division be? (for example: 30 ÷ 6)*
- *You start on 21 and make three jumps of 7 along a number line. What numbers do you land on? (28, 35, 42)*

- *How many groups of 5 do you need to make 40? (8)*
- *What is 63 divided by 9? (7)*
- *Write another multiplication sentence that has the same answer as 6 × 5. (for example: 3 × 10 or 3 × 5 + 3 × 5 or 2 × 5 + 4 × 5)*
- *Is 27 a multiple of 3? (Yes.) Is 10 a multiple of 3? (No.)*
- *A regular hexagon has sides of 12 cm. What is the perimeter of the shape? (72 cm)*
- *The sum of the sides of a regular octagon is 88 cm. What is the length of each side? (11 cm)*
- *What is 2 × 4 × 8? (64) Can you write this in another way using two factors? (8 × 8, 2 × 32, 4 × 16)*
- *Why is it silly to multiply the first five numbers in this calculation: 1 × 2 × 4 × 6 × 7 × 0? (You can see that there is a 0. Any number multiplied by 0 is 0, so the answer will be 0.)*
- *What is the missing number in the multiplication __ × 7 = 7? (1)*

UNIT 13 Negative numbers

Learning objectives

- Use negative numbers in context
- Count backwards through zero to include negative numbers
- Understand that temperature can be measured in degrees and that it can drop below 0 to give a negative temperature

Key words

negative number   temperature   minus sign
degrees Celsius   Fahrenheit   thermometer

Unit introduction

Teaching guidance

Present situations such as the ones to the right to encourage the class to think about numbers less than 0 and how to represent them. You can provide the diagrams or ask the children to draw their own.

1 This is a well. At the start of the day, the water level is 60 cm above ground level. The family use water to irrigate their vegetables and the water level drops to 40 cm below the ground level.

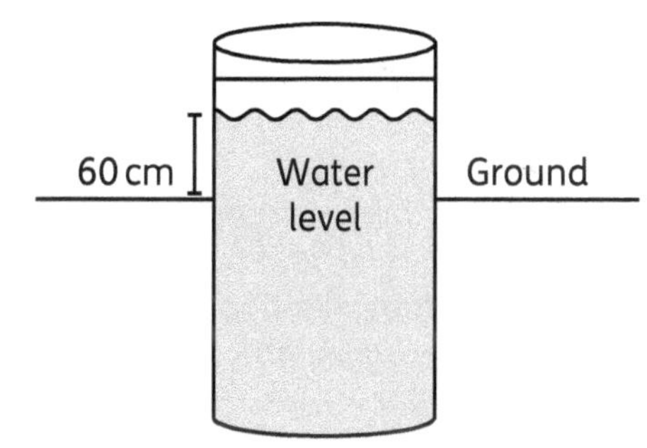

2 At the beach, a kite is flying 15 m above the level of the sea. One person is floating on the surface of the water. Another person is snorkelling 3 m below sea level. There are corals and starfish on the rocks 1 m below the snorkeller.

Discuss each situation and ask the children to think about how they could show each on a number line. Let them share their ideas, focusing on how they have represented above and below ground and above and below sea level (where the height of sea level is given as 0 m).

Numbers less than 0

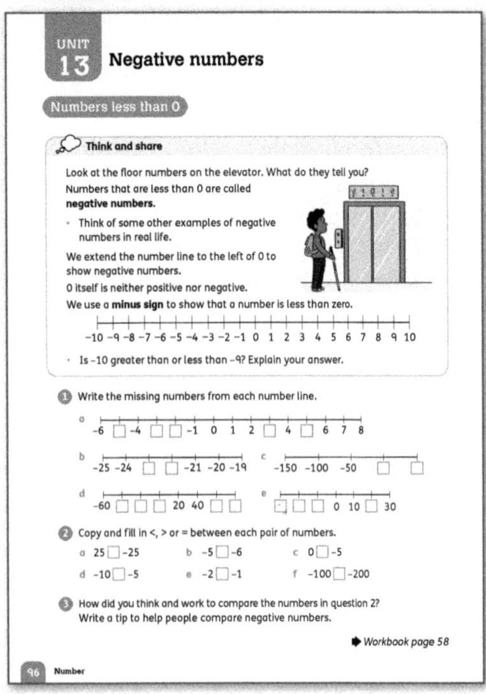

Materials
Thermometers with scales in degrees Celsius, including values below zero.

Warm-up
As a mental warm-up, revise counting forwards and backwards in given steps, starting at any number.

Focus
- Draw a picture of a tall skyscraper building. Explain that some buildings may have basement levels underground. Now label the floors above the ground 1, 2, 3, 4, 5 and so on. Label the ground floor 0.
- Ask: *If the ground floor is zero, what do we call the floor just below ground?*
- Elicit or explain the answer (−1). Then ask: *What about the next floor down?* (−2) and so on.
- <u>Think and share</u>: Turn to **Pupil Book 4 page 96**. Read through the information and then ask the children to think about where they might find *negative numbers* in real life.
- Take feedback. The children should realise that negative numbers are used in many different contexts. For example, children will be aware that in some regions, winter *temperatures* drop below freezing. Let them look at the scale on a *thermometer*. Point out the *minus sign* (also called the negative sign), which indicates the negative numbers.
- For children who have not experienced cold temperatures, talk about a kitchen freezer (where food is kept at low temperatures to keep it frozen).

Many modern freezers have a temperature display that shows negative numbers.
- Other contexts include: depth below sea level; underground parking levels; negative bank balances. There are many other examples, so encourage the children to suggest some or do their own research if possible.
- Draw a number line on the board. Model counting forwards and backwards from zero.
- Discuss negative numbers and how we represent them using the minus sign. Let the children then discuss whether −10 is greater or less than −9. Point out that as you count back 10, 9, 8, ... the numbers decrease in value (get smaller), and this continues as you count back through the negative numbers. So counting back from −9 to −10 shows that −10 is less than −9.
- Turn to the questions on **Pupil Book 4 page 96**.
- The children don't need to redraw the number lines in question 1, unless you think it would help them to do so.
- For question 2, encourage the children to draw rough number lines and position just the numbers they are comparing, to help them work out which is greater in each pair.
- The children can share the tips they write for question 3 in groups or with the class. Agree on some helpful guidelines for comparing negative numbers.

Follow-up
Use **Workbook 4 page 58** to informally assess the children's understanding of negative numbers and to consolidate counting back and forwards through zero.

Challenge
Let the children do their own research to find the heights above and below sea level of the highest and lowest points on each continent or use the table below. Ask the children to draw number lines to show the highest and lowest points for each continent and then let them work out the difference in height between these points.

Continent	Highest point	Lowest point
Africa	Kilimanjaro 5895 m	Lake Assal 150 m below sea level
Antarctica	Vinson Massif 4897 m	Ice 2538 m below sea level
Asia	Mt Everest 8848 m	Dead Sea 409 m below sea level
Australia	Kosciusko 2228 m	Lake Eyre 16 m below sea level
Europe	Mt Elbrus 5642 m	Caspian Sea 28 m below sea level
North America	Mt McKinley 6194 m	Death Valley 86 m below sea level
South America	Aconcagua 6960 m	Valdes Peninsula 40 m below sea level

Support

- Help the children to draw a number line to show negative numbers to −20 and have them refer to it as they need to.
- Write 'Less' with an arrow pointing left above the number line to remind the children that numbers decrease in value as we move to the left on any number line.
- You can also encourage the children to draw rough number lines to compare the position of two or more numbers. For example, to compare −4 and −10 they could draw:

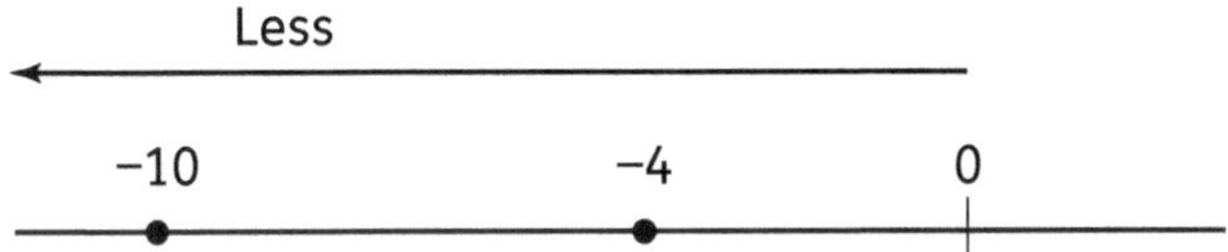

Interesting mistakes

The children may be able to continue counting once you have started a series (for example, −1, −2, −3, . . .), but need reminding that the 'bigger' numbers are actually 'smaller' when they have the minus sign in front of them. They may find the < and > signs difficult to apply as a result of this confusion (for example, if the children have to compare the numbers −3 and −5).

Keep reminding the children of the example of the elevator going down below ground level. You can ask: *Which is further down: the third floor below ground, or the fifth floor? (−5 is below −3, so −5 < −3.)*

Answers for Pupil Book 4 page 96

<u>Think and share:</u> Possible answers: thermometers, digital scales, sea levels

No, −10 is less than −9. The farther away a negative number is from zero when you count back to zero, the smaller it is.

1   a   −5, −3, −2. 3, 5   b   −23, −22
     c   0, 50   d   −40, −20, 0, 60, 80
     e   −30, −20, −10, 20
2   a >   b >   c >   d <   e <   f >
3   Individual answers. Possible tip: Draw a number line.

Answers for Workbook 4 page 58

1   a

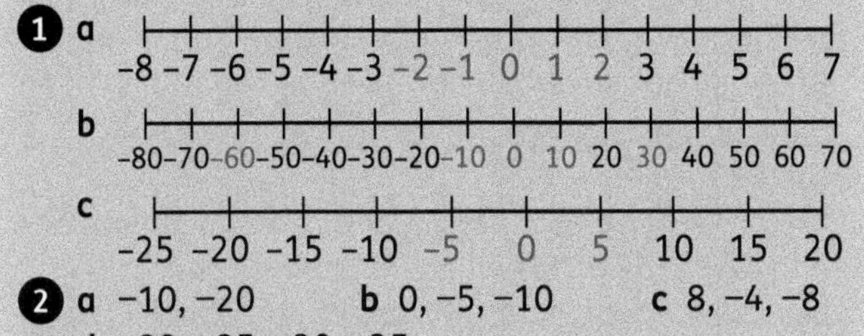

     b

     c

2   a   −10, −20   b   0, −5, −10   c   8, −4, −8
     d   −20, −25, −30, −35
     e   0, −16, −24   f   −80, −280, −380
3   (answers from left to right) 10; 14; 8, −2; 2; 0, −2, −12

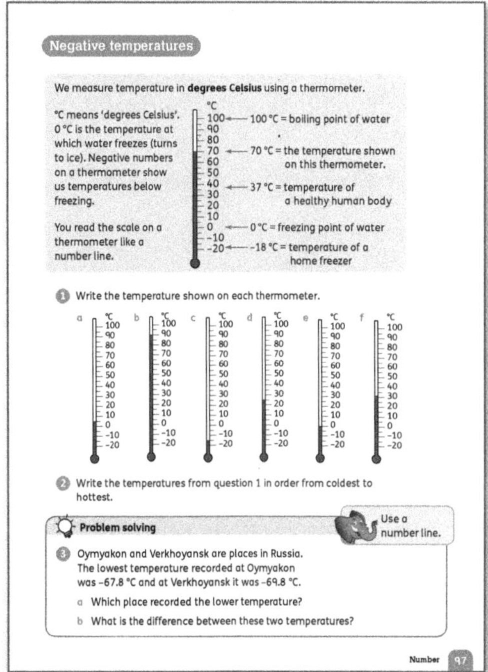

Materials

Thermometers; weather reports that include temperatures below zero (from news media or online sources); online interactive thermometer and thermometer worksheets (optional).

Warm-up

As a mental warm-up, revise counting forwards and backwards in given steps, starting at any number and extending below zero to include negative numbers.

Focus

- Discuss contexts in which we measure, record and report on temperatures, for example: radio, TV and newspaper weather forecasts; body temperatures when people are ill; choosing holiday destinations; cooking temperatures; swimming pool temperatures. Show examples of these where possible.
- Ask the children whether they know of any particular temperatures, such as the boiling and freezing points of water (100 °C and 0 °C) or normal body temperature (about 37 °C).
- Discuss units of temperature. We use *degrees Celsius* but countries such as the USA use the *Fahrenheit* scale. The children may also find the term Centigrade used in some sources. Explain that this is an old-fashioned name for Celsius, but technically the units are the same.
- Show the children a thermometer and demonstrate how we read the temperature. Make sure the children understand that we read the temperature shown at the end of the liquid in the tube.
- Let the children work in pairs to read through the explanation and complete question 1 and question 2 on **Pupil Book 4 page 97**.
- <u>Problem solving</u>: The children can continue working in pairs to solve question 3. They can draw a number line marked in tenths to help them.

Support

The children can use an online interactive thermometer to get an idea of how hot or cold temperatures are. There are also many worksheets available online that you can use to provide additional support and practice in reading and showing different temperatures.

Answers for Pupil Book 4 page 97

1 a 5 °C b 90 °C c −15 °C
  d 25 °C e 0 °C f 32 °C
2 −15 °C, 0 °C, 5 °C, 25 °C, 32 °C, 90 °C
3 a Verkhoyansk b 2 °C

Temperature changes

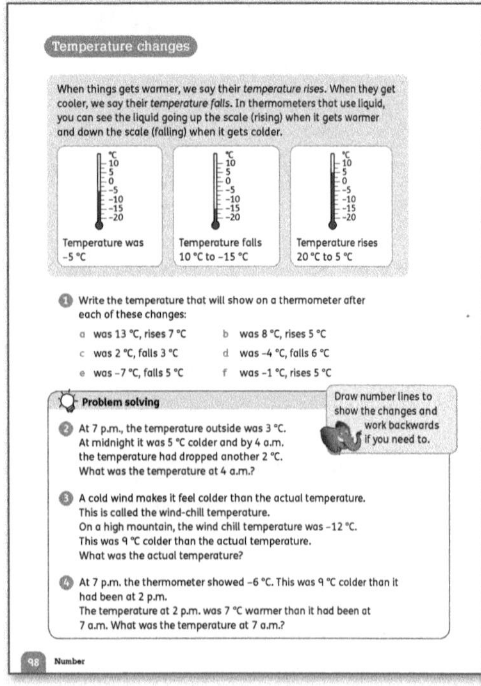

Warm-up

Practise mental addition or subtraction as a warm-up or use the 'Ordering numbers' activity from page 25.

Focus

- On **Pupil Book 4 page 98** the children start working with changes in temperature. Work through the examples carefully with the class.
- Go through the first few parts of question 1 with the class before asking the children to work out the rest. Encourage them to draw number lines if necessary.
- Problem solving: For questions 2–4, remind the children that they can use number lines and draw jumps to work out the answers.

Challenge

Let the children investigate minimum and maximum temperatures in different parts of the world. Explain that the term for the difference between the minimum and maximum temperature is the range of temperatures. Let them present their findings to the class.

Interesting mistakes

The children may recognise some situations involving temperature change as addition. For example, when they read 'at 8 a.m. it was 4 °C and at 2 p.m. it was 3 degrees warmer', they know that they need to add 3 degrees to find the temperature at 2 p.m. But when they read 'at 8 a.m. the temperature was 4 °C, which is 3 degrees cooler than at 2 p.m.', they may not recognise that they need to add 3 degrees to find the temperature at 2 p.m.

Using bar models (see page 22) or drawing number lines and jumps will help show what they need to do. For example:

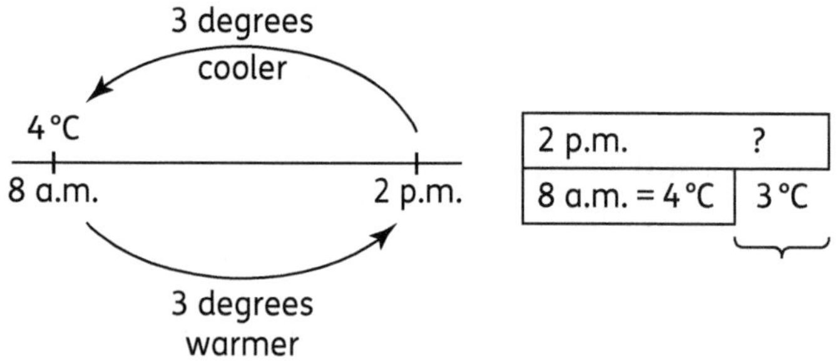

Answers for Pupil Book 4 page 98

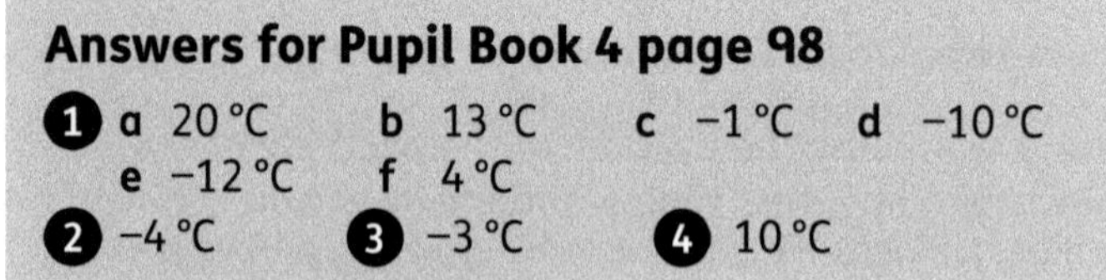

1 a 20 °C b 13 °C c −1 °C d −10 °C
  e −12 °C f 4 °C
2 −4 °C **3** −3 °C **4** 10 °C

End-of-unit check

Use some or all of these questions to assess how well the children have understood the concept of negative numbers and how well they can read, record and calculate temperatures.

- *What number is 1 less than zero? (−1) What number is 2 less than zero? (−2)*
- *How many whole numbers are there between −10 and 10? (21 if you include −10 and 10)*
- *Count backwards from −1 to −10. (−1, −2, −3, −4, −5, −6, −7, −8, −9, −10)*
- *Which number is less, −3 or 8? (−3)*
- *Which number is greater, −3 or −8? (−3)*
- *The temperature is 5 °C, then it drops by 10 degrees Celsius. What is the new temperature? (−5 °C)*
- *On the Moon, the temperature rises to 130 °C and drops as low as −150 °C. What is the difference between the highest and lowest temperatures? (280 °C)*
- *Micah's parents give him 5 points for good behaviour and take off 5 points when he behaves poorly. Is it possible for him to get −12 points one day? Explain your answer. (No. His total number of points must be a positive or negative multiple of 5.)*

Perimeter and area

Learning objectives

- Estimate, measure and calculate the perimeter of different shapes in centimetres and metres

- Find the area of shapes by counting squares

- Begin to apply formulae to find the perimeter and area of rectangles (including squares)

Key words

perimeter   rectangle   area   length   width   square   centimetre

Unit introduction

The children have worked with perimeters of simple shapes in earlier grades. Remind them that perimeter is the total distance round the outside of a 2D shape. The word *perimeter* comes from the ancient Greek words 'peri-', which means around, and 'meter', which means measure. So, perimeter means 'measure around'.

Materials

Squared paper; a selection of leaves of different shapes and sizes.

Teaching guidance

- Give each pair of children a sheet of squared paper, two leaves, a ruler and marker pen. Ask the children to draw a rectangular frame for each leaf to fit into. They will need to make sure that the leaf fits by placing it on the paper.

- Next, ask them to work out or measure the *perimeter* of each *rectangle*. Ask them to share how they worked out the perimeter, making sure that they give an answer with units.

- Then ask the children to put the leaves into the frames and tell them that they are going to estimate how many squares are covered by each leaf.

- Have a class discussion about how they will do this. You might decide to have a list of rules like the following, but allow the class to develop their own criteria first:

  ○ *Draw around the leaf and remove it so that you can see the squares to count them*

  ○ *Count all the whole squares. Put a dot in each one that you count so you don't count any squares more than once.*

  ○ *If more than half a square is covered, count it as one square, ticking them as you count them to make sure you don't count any more than once.*

  ○ *Combine half-covered squares to make wholes, colour or mark each pair and tally as you count them.*

  ○ *If less than half the square is covered, ignore it.*

- Give the children time to work out the answers and then take feedback about what was easy and what was difficult.

- Remind them that the number of squares covered by the leaf is the *area* of each leaf. Area is always measured in square units, but for this activity they can make statements such as: 'This long thin leaf has an area of 23 squares.'

- During this activity, check that the children can use a ruler to measure in centimetres or millimetres and that they understand the terms perimeter and area.

Perimeter

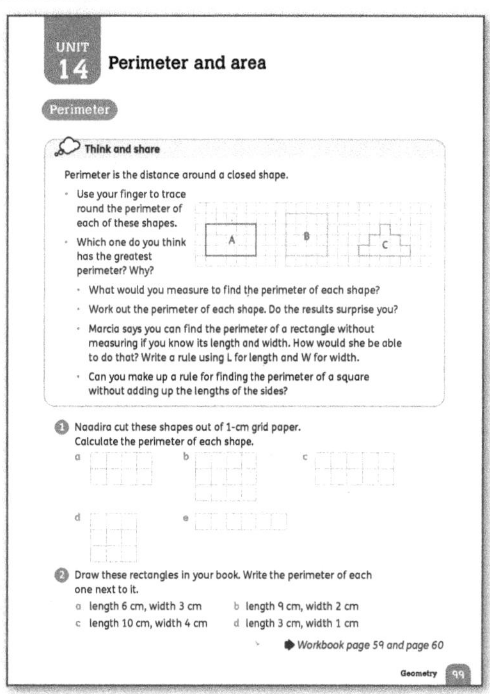

Materials

Printed sheets of rectilinear shapes drawn on a squared grid (all sides of the shapes should be on lines of the squared grid, as in **Pupil Book 4 page 99** 'Think and share').

Warm-up

Choose any mental addition activities from the 'Mental maths' activities on pages 27–28 as a warm-up.

Focus

- Start by putting the class into groups to do some practical activities on perimeter.

- Give the children the sheets of shapes drawn on a squared grid, and ask them to measure the sides and calculate the perimeter of each shape. They should record their work and show their calculations.

- Ask the children to work in pairs to measure and record the perimeter of various items in the classroom (for example, a sheet of paper, a floor tile, the top of your desk, the perimeter of a ruler).
- <u>Think and share</u>: Let the children work in small groups to carry out the activity and discuss the questions at the top of **Pupil Book 4 page 99**.

Spend time discussing the children's answers to the questions about how to use the *length* and *width* of a rectangle to find its perimeter. Use the following guidance to help shape the discussion.

- Ask: *What would you measure to find the perimeter of each shape?* The children may just say 'the lengths of the sides'. Ask questions to make them think about polygons and their properties, for example: *Do you need to measure the length of all the sides of a rectangle to find the perimeter?* (only two, the length and the width) *Why?* (*Opposite sides are the same length, so when you measure one, you know the other.*) *How many sides of a square do you need to measure to work out its perimeter?* (just one) *Why?* (*All four sides are the same length.*)
- Ask: *Work out the perimeter of each shape. Do the results surprise you?* The children may be surprised that all the shapes have the same perimeter. Ask questions such as: *How do the sides of the square compare with the sides of the rectangle?* (*The base is shorter, but the sides are higher, and this balances out.*) *What about the stepped shape?* (*The five horizontal 'step lengths' are the same length as the base, which is the same as the base of the rectangle, and the three vertical steps on each side are the same length as the width of the rectangle.*)
- Ask: *Marcia says you can find the perimeter of a rectangle without measuring if you know its length and breadth. How would she be able to do that?* Write a rule using L for length and W for width. The children may suggest '$L + W + L + W$'. If so, ask them how they can write this in a way that makes the calculations easier: ($L + L + W + W$ means that you can use doubling and $2L + 2W$ means that you can multiply by 2). Some children may get to $2(L + W)$ where you can add the lengths of the sides and then double to find the perimeter.
- Ask: *Can you make up a rule for finding the perimeter of a square without adding up the lengths of the sides?* Remind the children that multiplication is a short form of repeated addition, so $s + s + s + s$ can be written as $4 \times s$ or $4s$.

Turn to **Workbook 4 page 59**. The children should use a ruler to measure the length and width of each rectangle and then use their measurements to work out the perimeter. This activity revises measuring skills, but if the children measure slightly inaccurately, they should still be able to work out the perimeter.

Return to **Pupil Book 4 page 99**. Let the children work independently to complete question 1 and question 2.

Follow-up

Use **Workbook 4 page 60** as a problem-solving activity to consolidate thinking and reasoning about perimeter.

Answers for Pupil Book 4 page 99

<u>Think and share</u>: Possible answer: I would count how many squares on each side of the shape to find the perimeter.
Each shape has a perimeter of 16 cm. I am surprised because they all look different and shape C looks smaller than the other two shapes.
Marcia is correct. L + L + W + W = P or add L × 2 and W × 2
Perimeter of a square = Length of 1 side × 4

1. a 12 cm   b 14 cm   c 14 cm
   d 12 cm   e 14 cm
2. a 18 cm   b 22 cm   c 28 cm   d 8 cm

Answers for Workbook 4 page 59

1. a 5.6 cm, 3.8 cm   b 4.7 cm, 2.8 cm
   c 3.8 cm, 3 cm   d 4 cm, 2 cm
   e 5.8 cm, 4.8 cm   f 6.6 cm, 1 cm
2. a 18.8 cm   b 15 cm   c 13.6 cm
   d 12 cm   e 21.2 cm   f 15.2 cm

Answers for Workbook 4 page 60

1. Possible answers: rectangles with the following lengths and widths: 6 cm and 6 cm; 7 cm and 5 cm; 8 cm and 4 cm; 10 cm and 2 cm; 11 cm and 1 cm

Calculate perimeter

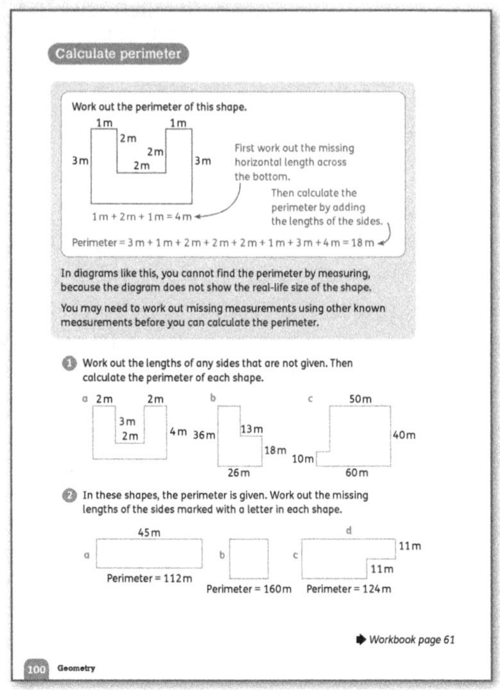

Warm-up

As a mental warm-up, use one of the 'Doubling' activities from page 29.

Focus

The focus in this lesson is on using properties of shapes to find missing lengths so that you have all the information you need to work out the perimeter.

- Work through the example on **Pupil Book 4 page 100** with the class.
- Refer the children to the diagram. Ask: *How long is the distance across the top?* (1 + 2 + 1 = 4 m) You may need to draw a dotted line on a copy of the diagram to show the children this:

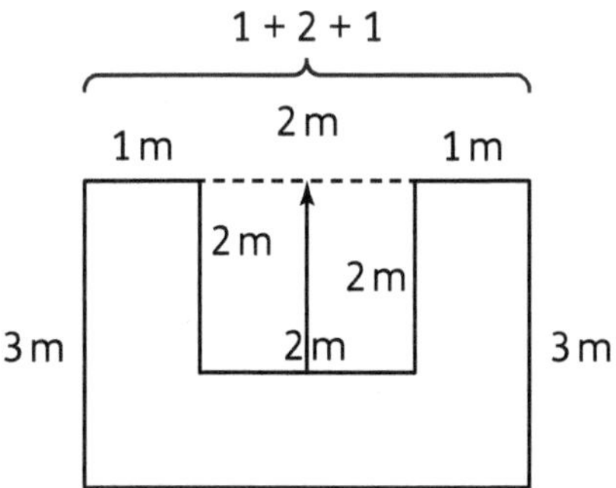

- Demonstrate how to start at the bottom left corner of the shape and go around it clockwise, writing each length in an addition calculation.
- The children could copy each shape in question 1 so they can write the missing lengths on their diagrams. Check that the lengths are correct before asking the children to calculate the perimeters.
- The children can work in pairs to discuss how to find the unknown lengths in question 2 before working these out on their own.

Follow-up

Use **Workbook 4 page 61** to assess the children's understanding of perimeter.

Challenge

Prepare a few perimeter word problems on cards and let the children select one or two of these to solve. Here are some challenging problems:

> *A rectangular field has a perimeter of 42 metres.*
>
> **a** *The field is twice as long as it is wide. Work out how long and how wide it is.*
> **b** *There is white line painted round the field. It is 1 m from the outside of the field. How long is the line?*

Answers: **a** 14 m long, 7 m wide   **b** 50 m

> *Marija has a rectangle of card that is 30 cm long and 15 cm wide. She cuts a rectangle out of the card to make an L shape. What is the perimeter of the L shape? Why?*

Answer: 90 cm, because whatever size rectangle she cuts out, the total length of all the sides will remain the same as the perimeter of the original rectangle.

> *Jose has an equilateral triangle with a perimeter of 54 cm and a rectangle with a perimeter of 120 cm. He puts them together like this:*
>
> *What is the perimeter of the shape he has made?*

Answer: 54 + 120 – 18 – 18 = 138 cm

Interesting mistakes

If the children don't really understand how the formula for perimeter works, they may need help to solve problems where they have to work backwards to find lengths when they are given the perimeter.

- You may need to show them some simpler problems. For example: *A square has a perimeter of 20 cm, what are the lengths of the sides?* Write out the formula 4 × side = 20, then rewrite this as $4 \times \square = 20$, reminding the children that they can use the inverse operation (that is, division) to work out the missing value.
- Next, say: *A rectangle has one side that is 5 cm long and a perimeter of 18 cm. What are the lengths of the other sides?*
- Draw a rough sketch and label one side 5 cm. Ask: *What do we know about the opposite side?* (It has to be 5 cm long.)
- Then ask: *If these two sides make up 10 cm of the perimeter, what do we know about the other two sides?* (They must make 8 cm of the perimeter, but they are the same length, so they must be 4 cm each.) You may also show this using the formula: $2 \times 5 + 2 \times \square = 18$.
- Do the first step, to show that we get $10 + 2 \times \square = 18$. Say: *10 plus something equals 18, what is the something?* (It must be 8.) We know that the something is 2 times a number which gives a product of 8, so what is the number? (4)

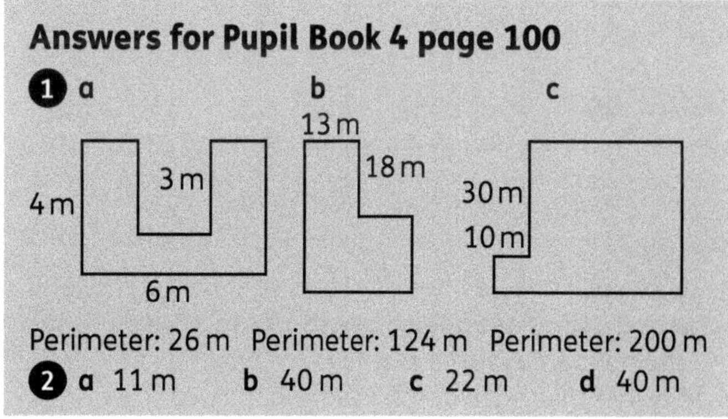

Answers for Pupil Book 4 page 100

1 a   b   c

Perimeter: 26 m   Perimeter: 124 m   Perimeter: 200 m

2 a   11 m   b   40 m   c   22 m   d   40 m

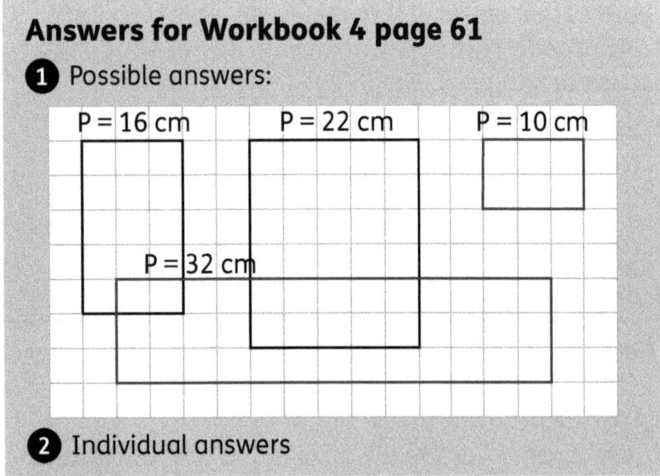

Area

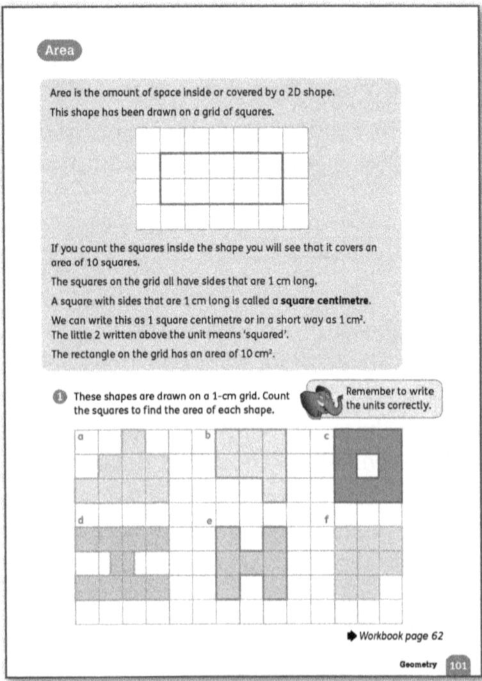

Materials

1-cm squared paper; scissors; tiling resources such as postage stamps, square cards, postcards, envelopes, sheets of paper, plastic shapes.

Warm-up

Revise multiplication and division facts to 12 × 12 as a mental warm-up.

Focus

- Give each child a sheet of 1-cm squared paper. (The size of the sheet doesn't matter as long as all the sheets are the same size.)
- Explain that each small square has sides of 1 cm and so they are called *square centimetres*. Show the children how to write the abbreviation cm².
- Together, work out how many square centimetres are on each sheet by counting (although some children may multiply to find the answer).

- Remind the children that finding out how many square units it takes to cover a flat surface is called finding the *area*.
- Use the sheets to work out how many square centimetres it will take to cover a desk and other areas in the classroom. Record the area of a desk in square centimetres on the board.
- Let the children draw a number of shapes that have straight sides (along the grid lines on their squared paper) and only right angles. Then ask them to count the squares to work out the area of each shape.
- Once you have revised the concept of area and the children know what a square centimetre is, work through the example on **Pupil Book 4 page 101** with the class. Remind the children that they worked out the number of squares in rectangular grids like these when they learnt multiplication facts.
- The children can do question 1 independently. Check that they include the units.

Follow-up

- Turn to **Workbook 4 page 62**. Let the children complete question 1 and question 2.
- Ask them to colour the shapes they have drawn using three different colours. Then ask them to cut out the shapes. The children then cut each shape into two parts and recombine these to make a different shape.
- Let them work out the area of the new shapes they make and discuss their results (the new shapes will have the same area as the original shapes).

Support

If any children have difficulty with the concept of area, it is useful to do some tiling activities to reinforce the ideas. Prepare some tiling activities for the children using available resources such as postage stamps, square cards, postcards, envelopes, sheets of paper, plastic shapes. Ask the children to cover the surface of a book cover, or their desk, and to say how many 'units' they needed to cover the area. Remind them that finding out how many units we need to cover something is called finding its area.

Interesting mistakes

- Some children may become confused and mix up the terms 'perimeter' and 'area', and they may sometimes add length and width and say it is the perimeter because they know that they multiply these to find area. With lots of practical activities, they will usually realise that perimeter is a measure of length, and area is a covering.
- The children may forget to use square units for area or leave out the units altogether. Remind them that we have to say what the area is covered or filled by (in this case, square centimetres).

Answers for Pupil Book 4 page 101

1
a 8 cm²	**b** 7 cm²	**c** 8 cm²
d 9 cm²	**e** 7 cm²	**f** 8 cm²

Answers for Workbook 4 page 62
1 Shape B. Individual answers
2 Individual answers

More area

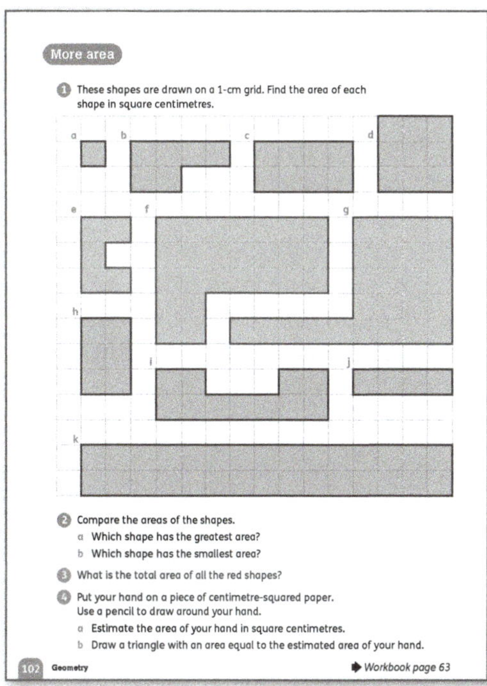

Materials
1-cm squared paper.

Warm-up
Continue to revise multiplication facts as a mental warm-up.

Focus
- There is no new teaching in this lesson.
- The children can work through questions 1–4 on **Pupil Book 4 page 102** independently to consolidate the work on area.

Follow-up
- Question 1 on **Workbook 4 page 63** provides more practice in calculating area on a squared grid. The children can complete this independently.
- When the children have completed question 2, working in pairs, talk about how they can work out the area of shapes when they cannot count the squares.
- Problem solving: The children can work independently on question 3, and share and discuss their results with the class.

Challenge
- Give the children a simple outline map of a fenced area of land like the one below.

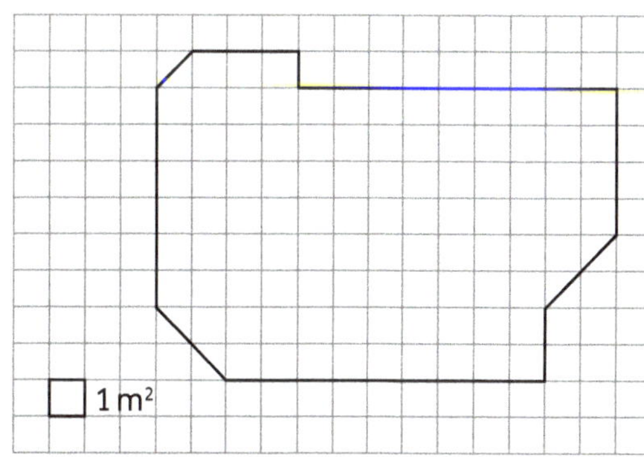

- Ask them to work out the total area of the land. Then give them the dimensions of some items that need to go on the map. For example:
  - a tool shed (area of 4 m²)
  - a round water reservoir (area of approximately $3\frac{1}{2}$ m²)
  - two vegetable patches (10 m² each)
  - a grazing area for goats ($16\frac{1}{2}$ m²).
- Let the children decide where they would place these on the map and then draw them using the areas given. Remind them that a map needs a key.
- When they have done this, let them calculate the area of land that is unoccupied.
- Problem solving: Referring to question 3 on **Workbook 4 page 63**, ask the children whether it is possible for Nisha to use the 18 m of rope to make a rectangle that has a perimeter of 18 m and an area of 18 m². (Do not say 'the same perimeter and area' as these are two different types of measurement and they cannot mathematically be the same.) (*A rectangle with length 6 m and width 3 m will have a perimeter of 18 m and an area of 18 m².*)

Support
Give the children a sheet of 1-cm squared paper and ask them to draw a different shape with the same area for each of the shapes in **Pupil Book 4 page 102** question 1. Let them compare and check each other's work.

Answers for Pupil Book 4 page 102
1 a 1 cm² b 6 cm² c 8 cm² d 9 cm²
  e 5 cm² f 25 cm² g 25 cm² h 6 cm²
  i 11 cm² j 4 cm² k 30 cm²
2 a k b a
3 55 cm²
4 a and b Individual answers

Answers for Workbook 4 page 63
1 a 6 cm² b 15 cm² c 13 cm²
  d 6 cm² e 7 cm²
2 a A = 15 cm²; B = 16 cm²; C = 20 cm²
  b Individual answers
3 The rectangle with the largest possible area is 4 m by 5 m and has an area of 20 m².

Use some or all of the following questions and activities to assess how well the children understand the concepts of perimeter and area, and whether or not they can apply what they have learnt.
- *What does perimeter mean?* (*the distance around a closed shape*)
- *What units can you use to measure the perimeter of small shapes?* (*for example: millimetres or centimetres*)
- Provide shapes with the side lengths labelled. Some side lengths that the children could work out should be missing. Ask the children to find the perimeter of each shape.
- Provide shapes drawn on squared paper or with the side lengths labelled. Ask the children to arrange the shapes in order, from the smallest to largest perimeter.

- Show a shape drawn on squared paper. *What is the area of this shape?*
- *Can two different shapes have the same area? Explain how.* (*Yes, because some multiples have different pairs of factors. For example, a 2 cm by 6 cm rectangle and a 3 cm by 4 cm rectangle.*)
- Provide shapes drawn on squared paper. Ask the children to arrange the shapes in order, from the largest to smallest area.
- *What is a square centimetre?* (*the area of a square with sides of 1 cm*) *How is a square centimetre different from a centimetre?* (*It is a unit for measuring area, and a centimetre is a unit for measuring length.*)
- *Draw three shapes with an area of 16 cm² and three shapes with a perimeter of 16 cm.* (*for example: for area: 1 cm by 16 cm rectangle, 2 cm by 8 cm rectangle, 4 cm by 4 cm rectangle; for perimeter: 1 cm by 7 cm rectangle, 2 cm by 6 cm rectangle, 4 cm by 4 cm rectangle*)
- *Which set of shapes was easier to draw? Why?*

UNIT 15 Fractions

Learning objectives

- Understand that fractions are equal parts of a whole and that as the whole is divided into more parts, the parts become smaller

- Recognise and show families of equivalent fractions using diagrams

- Know that two proper fractions can be equivalent in value

- Recognise and write decimal equivalents to $\frac{1}{4}$, $\frac{1}{2}$ and $\frac{3}{4}$

- Use knowledge of equivalence to compare and order fractions

- Reason about the location of mixed numbers on a number line

- Convert mixed numbers to improper fractions and vice versa

- Add and subtract proper fractions, improper fractions and mixed numbers with the same denominators

- Solve a range of problems involving fractions

Key words

fraction   half/halves   quarter   third   eighth
fifth   tenth   numerator   denominator
equivalent fraction   proper fraction
improper fraction   mixed number

Unit introduction

Teaching guidance

Prepare a display of different ways of representing $\frac{3}{4}$ such as the ones below. Include one diagram that does not represent $\frac{3}{4}$ (diagram H in the example below).

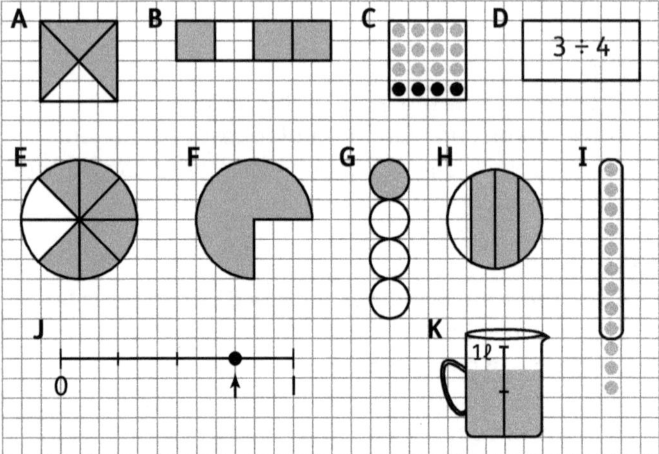

- Point to diagram A and ask: *What fraction of this square is shaded?* ($\frac{3}{4}$). Then ask the children to say how they know that the diagram shows $\frac{3}{4}$. (*There are four equal parts and three of them are shaded.*)
- Next, ask them to look at the other diagrams with a partner and to decide whether they show $\frac{3}{4}$ or not.
- Give the children time for discussion and then ask them to say which diagrams do not show $\frac{3}{4}$, giving reasons for their answers.
- The only diagram that doesn't show $\frac{3}{4}$ is H because the circle is not divided into four equal parts.

- The children may say that some of the other diagrams don't show $\frac{3}{4}$. Ask questions or do some practical work to help them to see that all the diagrams represent $\frac{3}{4}$. For example, to show them that $3 \div 4$ does represent $\frac{3}{4}$, refer to diagram A. Say: *If we had three squares like this and we divided them equally among four people, what would each person get?* (*three* quarters *of a square*)
- Similarly, they may decide that E is not $\frac{3}{4}$ because the circle is divided into *eighths*.
- Ask the children to show $\frac{1}{2}$ of the circle. They should be able to do this easily.
- Explain that *half* is equivalent to $\frac{4}{8}$. Then ask them to show $\frac{1}{4}$ of the circle. Ask: *How many eighths is that?* ($\frac{2}{8}$) Point out that if $\frac{1}{4}$ is $\frac{2}{8}$ then $\frac{3}{4}$ must be $\frac{6}{8}$.
- Talk through any other representations asking the others in the class whether they agree or not and why.
- It is important that the children realise that a fraction is not just a part of a shape or object and that fractions are numbers that can be represented on number lines.

Revisit fractions

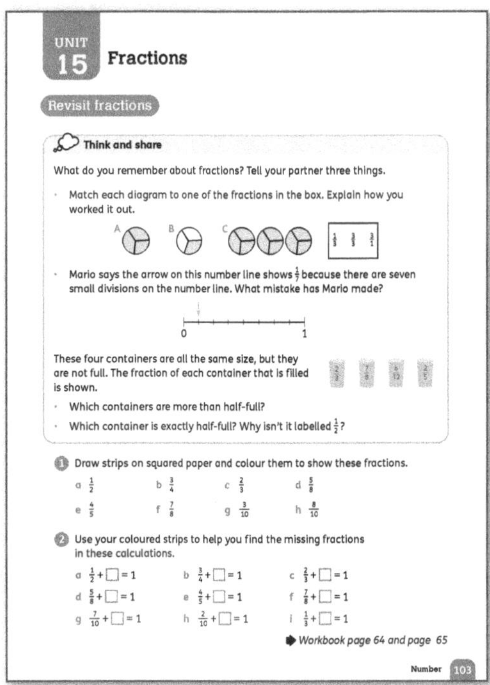

Materials

Squared paper; blank fraction walls (see 'Focus' for 'Equivalent fractions' on page 139).

Warm-up

Use any suitable 'Calculation skills' activities from pages 28–29 as a warm-up.

Focus

- <u>Think and share</u>: Ask the children to work in pairs to read through and discuss the questions about *fractions* on **Pupil Book 4 page 103**. The children's answers and explanations will help you assess their understanding of fractions.

Check the following:
- The children can recognise fraction notation such as $\frac{1}{3}$ as one *third*, $\frac{3}{3}$ as three thirds and $\frac{3}{1}$ as three ones, reminding them that 1 is a whole.
- They can read and make sense of number lines showing fractional divisions. It is important that the children can locate fractions on number lines before they draw fractional jumps to add or subtract fractions.
- They can compare fractions and remember that fractions can be expressed in different but equivalent ways. If necessary, show them a strip of 12 squares like the one below with $\frac{6}{12}$ shaded. Fold it to show them that $\frac{6}{12}$ is half the strip, so $\frac{6}{12}$ is just a different way of writing $\frac{1}{2}$.

- Turn to **Workbook 4 page 64** and ask the children to complete the drawings. Use the quarters shown in the diagrams to reinforce the idea that a quarter is not a fixed quantity: the quarters in the diagrams are all different sizes, but they are all quarters.
- For question 1 on **Pupil Book 4 page 103**, hand out squared paper to the children. Explain that they are going to draw strips and shade parts of them to show different fractions (like the strip showing twelfths, above). Tell them to look at the fractions they have to show (*halves*, quarters, thirds, eighths, *fifths* and *tenths*). Discuss suitable strip lengths for each fraction. For example, any even number of blocks can be used to show halves, but for eighths it is easier if they make the strip 8 (or 16) blocks long.
- For question 2, the children can use the strips as they need them to write the complementary fractions to make one whole.
- Introduce or revise the terms *numerator* and *denominator*. Discuss the fact that when the numerator is 1, the greater the denominator, the smaller the fraction, so $\frac{1}{12}$ is a smaller part of the whole than $\frac{1}{3}$. Let the children use their shaded strips to confirm this.

Follow-up

Use **Workbook 4 page 65** to consolidate fraction revision and to familiarise the children with the use of the terms numerator and denominator.

Challenge

Ask the children to cut out a number of rectangles. They don't need to be the same size. Challenge them to fold the shapes to show all the fractions from halves to twelfths.

Support

- Give the children a blank fraction wall and ask them to paste it on to card. Let them count and then label each part of each strip. For example, there are two parts, so each part is $\frac{1}{2}$.

- Then let them cut out the strips and cut each strip into its fractional parts.
- The children can then keep these in a small ziplock bag and use them for reference when they need to work with fractions.

Interesting mistakes

Children frequently find working with fractions difficult. Many of these difficulties seem to arise from an over-reliance on the 'part of a whole' model and that this is always presented through shaded diagrams.

It is important for the children to see fractions as numbers that can be ordered on a number line and compared. They also need to be aware that the 'whole' can be a quantity or amount, and so, for example, 12 out of 24 marks can be expressed as a fraction.

Answers for Pupil Book 4 page 103

<u>Think and share:</u> Possible answers: For A, there are three equal parts and all are shaded so $A = \frac{3}{3}$. For B, there are three equal parts but only one is shaded so $B = \frac{1}{3}$. For C I can see three whole shapes and I can use one to describe a whole so $C = \frac{3}{1}$.

Mario counted the divisions, not the parts and there are 8 parts so the number lines shows eighths.

The containers that are more than half-full are labelled $\frac{2}{2}$ and $\frac{7}{8}$.

The container labelled $\frac{6}{12}$ is half-full.

$\frac{6}{12}$ is an equivalent fraction to $\frac{1}{2}$.

1 **a–h** Individual answers

2 a $\frac{1}{2}$ b $\frac{1}{4}$ c $\frac{1}{3}$ d $\frac{3}{8}$
  e $\frac{1}{5}$ f $\frac{1}{8}$ g $\frac{3}{10}$ h $\frac{8}{10}$
  i $\frac{2}{3}$

Answers for Workbook 4 page 64

1 Possible answers:

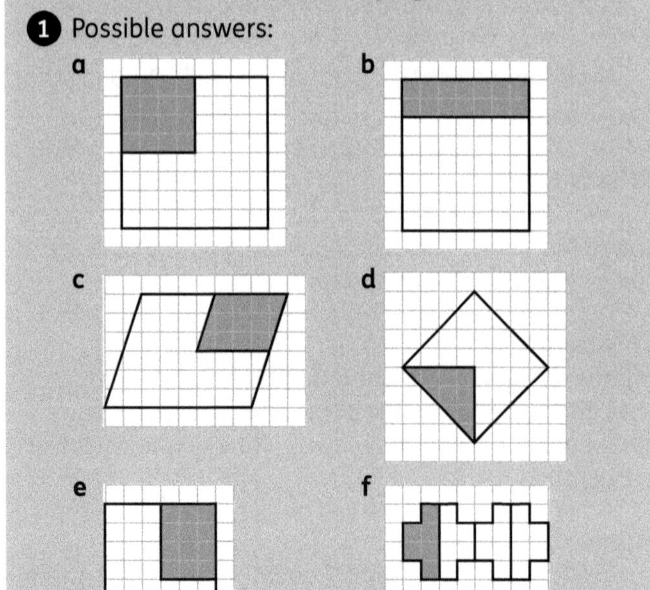

a   b   c   d   e   f

Answers for Workbook 4 page 65

1 a $\frac{1}{5}$ b $1\frac{3}{4}$ c $1\frac{2}{4}$

2 a $\frac{3}{5}$ b $\frac{3}{4}$ c $\frac{1}{2}$

3 6

4 8

5 a any 7 circles coloured
  b any 6 triangles coloured
  c any 6 squares coloured

Compare and order fractions

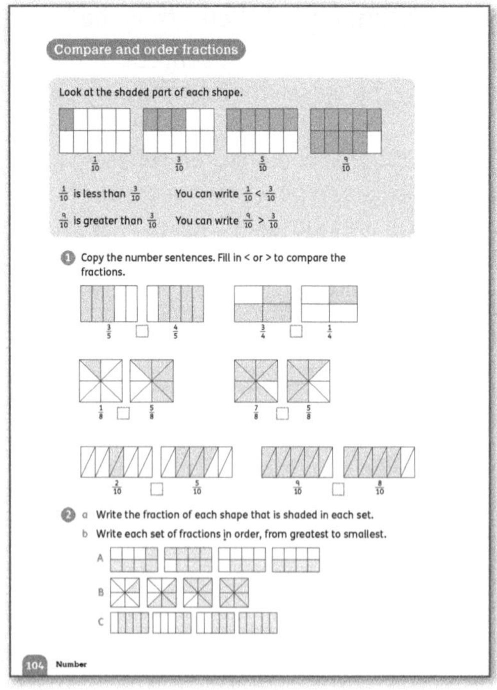

Warm-up

Use any suitable 'Calculation skills' activities from pages 28–29 as a warm-up.

Focus

- Work through the explanation at the top of **Pupil Book 4 page 104** with the class to revise ordering and comparing fractions with the same denominators as well as the use of the < and > signs.
- Let the children work independently through question 1 and question 2 and check that they are able to confidently compare and order simple fractions before moving on.

Follow-up

- Give the children a strip showing 4 eighths and ask them how many quarters this is.
- Show them, by folding the strip into four equal sections (quarters), that $\frac{4}{8}$ is equivalent to $\frac{2}{4}$.
- Repeat this for halves and then do the same with other *equivalent fraction* families.

- Then start to compare fractions that are not equivalent. For example, ask the children to compare $\frac{5}{8}$ with $\frac{1}{2}$. Remind them that $\frac{1}{2}$ is equivalent to $\frac{4}{8}$, so $\frac{5}{8} > \frac{1}{2}$.

Interesting mistakes

Children don't always realise that common fractions can be compared by converting them to the same denominator. This is an important concept that they will use later to calculate with fractions. Plenty of practice with strips of fractions, and the fraction wall used in the next lesson, will help to reinforce this.

Answers for Pupil Book 4 page 104

1. $\frac{3}{5} < \frac{4}{5}$; $\frac{3}{4} > \frac{1}{4}$; $\frac{1}{8} < \frac{5}{8}$; $>$; $\frac{7}{8} < \frac{5}{8}$; $\frac{2}{10} > \frac{5}{10}$; $\frac{9}{10} > \frac{8}{10}$

2. a $A = \frac{5}{8}, \frac{7}{8}, \frac{3}{8}, \frac{1}{2}$; $B = \frac{1}{8}, \frac{1}{2}, \frac{3}{4}, 1$; $C = \frac{4}{5}, \frac{2}{5}, \frac{3}{5}, 1$

   b $A = \frac{7}{8}, \frac{5}{8}, \frac{3}{8} \frac{1}{2}$; $B = 1, \frac{3}{4}, \frac{1}{2}, \frac{1}{8}$; $C = 1, \frac{4}{5}, \frac{3}{5}, \frac{2}{5}$

Equivalent fractions

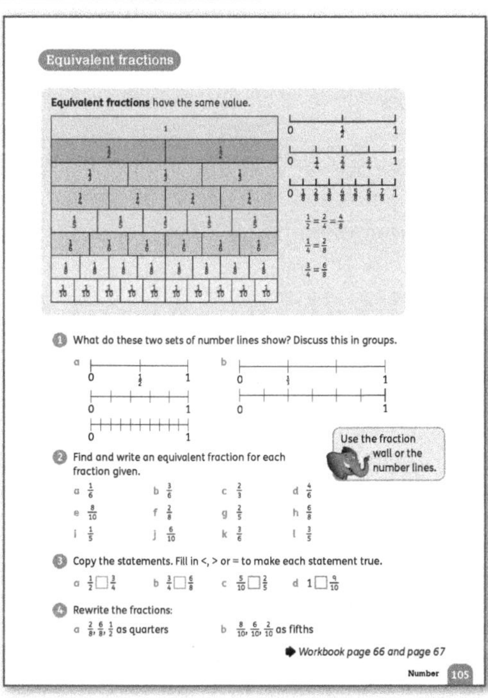

Materials

A large blank fraction wall (for classroom display, see 'Focus'); a copy of the fraction wall for each child (optional); equal lengths of string, paper strips or ribbon (for example, 30 cm long).

Warm-up

Use one of the 'Doubling' activities on page 29 as a mental warm-up.

Focus

- Display a large blank fraction wall, like this one:

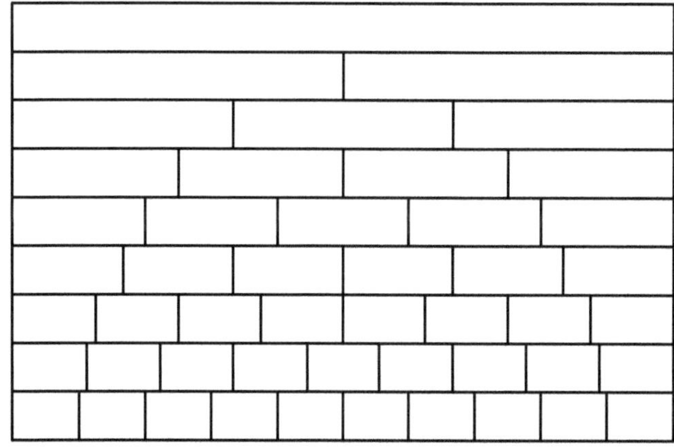

- Explain that the top row is one whole row. Each other row is divided into equal parts. Ask the children to determine the fractions represented by each part on each row and write these on the diagram. If you wish, the children can fill in the fractions on their own fraction walls (give a copy to each child).
- Write a fraction on the board, for example, $\frac{1}{4}$. Ask the children to use the fraction wall to find a fraction that is the same. Remind the children that the term for these fractions is *equivalent fractions*. Repeat until there are several pairs of equivalent fractions on the board.
- Ask the children what they notice about the numerators and denominators in each pair.
- Use their answers to work towards the idea that equivalent fractions can be found by multiplying or dividing the numerator and denominator by the same number.
- Display a fraction. Go round the class asking the children to say an equivalent fraction and record their answers. How many can they give? Ask: *Are there any more? How do you know?*
- Use **Workbook 4 page 66** to check that the children can compare and write equivalent fractions shown on diagrams. Check their answers before moving on.
- Turn to **Pupil Book 4 page 105**. Explain that the children can refer to the fraction wall to help them find equivalent fractions. Then focus on the number lines, and ask the children how these are similar to the fraction wall and how they are different.
- Let the children look at the number lines in question 1 and discuss what they represent. Have a class discussion in which the children share their ideas.
- The children can work through questions 2–4 independently.

Follow-up

Use **Workbook 4 page 67** to informally assess how well the children have understood the concepts in this lesson and whether or not they can use number lines to represent fractions.

Challenge

Let the children make posters comparing fractions. Starting with strips of equal length, give them specific non-unit fractions, such as $\frac{2}{3}$ and $\frac{3}{4}$ to fold, cut, mount and label (together with a whole length) in a similar way. Discuss which fraction is the longest and which is the shortest.

Support

- Let the children work in groups to make an equivalent fractions poster. Provide equal lengths of string, paper strips or ribbon (for example, 30 cm long).
- The children fold and cut the string, ribbon or strips to show halves, thirds, quarters, fifths, sixths and so on.
- On backing paper, they mount a whole length labelled '1, one whole', and underneath mount each fraction length appropriately labelled in words and symbols.

Answers for Pupil Book 4 page 105

1. Possible answers: $\frac{1}{2}$ is equivalent to $\frac{5}{10}$, $\frac{1}{2}$ is greater than $\frac{2}{5}$ but less than $\frac{3}{5}$.
2. Possible answers:
   a $\frac{2}{12}$ b $\frac{1}{2}, \frac{2}{4}, \frac{4}{8}$ c $\frac{3}{6}, \frac{6}{12}$
   d $\frac{2}{3}, \frac{8}{12}$ e $\frac{4}{5}$ f $\frac{1}{4}, \frac{3}{12}$
   g $\frac{4}{10}$ h $\frac{3}{4}, \frac{9}{12}$ i $\frac{2}{10}$
   j $\frac{3}{5}$ k $\frac{1}{2}, \frac{2}{4}, \frac{4}{8}$ l $\frac{6}{10}$
3. a $\frac{1}{2} < \frac{3}{4}$ b $\frac{3}{4} = \frac{6}{8}$ c $\frac{5}{10} > \frac{2}{5}$
   d $1 > \frac{9}{10}$
4. a $\frac{1}{2}, \frac{3}{4}, \frac{2}{4}$ b $\frac{4}{5}, \frac{3}{5}, \frac{1}{5}$

Answers for Workbook 4 page 66

1. a $\frac{1}{2} = \frac{4}{8}$ b $\frac{2}{5} = \frac{4}{10}$ c $\frac{2}{3} = \frac{4}{6}$
   d $\frac{1}{2} = \frac{3}{6}$ e $\frac{1}{4} = \frac{2}{8}$ f $\frac{1}{4} = \frac{3}{12}$
2. a $\frac{4}{6} = \frac{2}{3}$ b $\frac{6}{8} = \frac{3}{4}$ c $\frac{3}{9} = \frac{1}{3}$
3. a–c Individual answers

Answers for Workbook 4 page 67

1. a 1 b $\frac{1}{4}, \frac{1}{2}$ c $\frac{1}{8}, \frac{2}{8}, \frac{4}{8}, \frac{5}{8}, \frac{6}{8}$
2. a 2 b 2 c 3
   d 4, Individual answers
3. any 8 parts shaded to show $\frac{4}{5}$
   any 3 parts shaded to show $\frac{6}{8}$
   any 2 parts shaded to show $\frac{1}{4}$
   Possible equivalent fractions: $\frac{4}{5} = \frac{8}{10}$ $\frac{6}{8} = \frac{3}{4}$ $\frac{1}{4} = \frac{3}{12}$
4. a–c Individual answers

Fractions and equivalent decimals

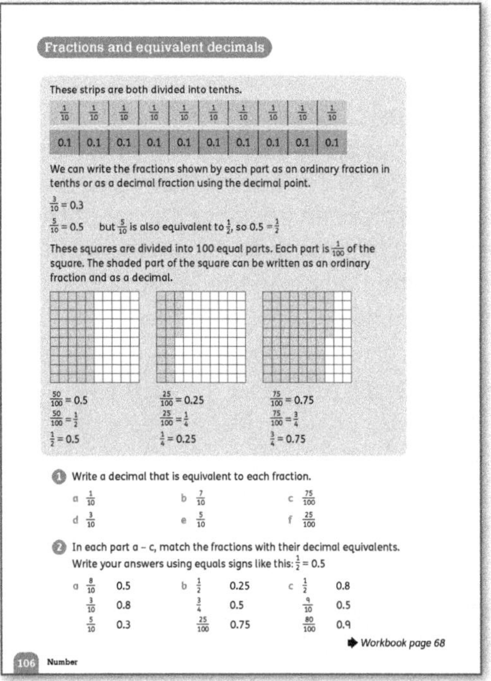

Materials

Blank 100 charts.

Warm-up

Revise counting in steps of 10 and 100 from any number or multiplying and dividing by 10 as a mental warm-up.

Focus

- Remind the children that they have already worked with decimal fractions and ask them what they remember. You could do this as a class discussion and record their ideas on the board.
- Make sure that the children realise that decimals are fractions and therefore they are similar in some ways to the fractions that the children are studying now.
- Work through the explanation and examples on **Pupil Book 4 page 106** with the class, or have the children work through this in small groups.
- Ask them to explain why we are able to find equivalent fractions for decimals. (*Decimals are simply a different way of writing tenths and hundredths, so we can write them as proper fractions and then find the smallest equivalent fraction.*)
- Let the children complete question 1 and question 2 on their own, and then compare and check each other's answers.

Follow-up

Use **Workbook 4 page 68** to check that the children are able to find and represent equivalent fractions and decimals.

Support

Provide more practice in finding equivalent fractions and decimals. Give the children blank 100 charts and some more fractions and decimals to colour on their charts.

Answers for Pupil Book 4 page 106

1   a   0.1   b   0.7   c   0.75   d   0.3
    e   0.5   f   0.25

2   a $\frac{8}{10} = 0.8$; $\frac{3}{10} = 0.3$; $\frac{5}{10} = 0.5$
    b $\frac{1}{2} = 0.5$; $\frac{3}{4} = 0.75$; $\frac{25}{100} = 0.25$
    c $\frac{1}{2} = 0.5$; $\frac{9}{10} = 0.9$; $\frac{8}{10} = 0.8$

Answers for Workbook 4 page 68

1   a   0.5   b   2.2   c   0.3

2   a $\frac{1}{4} = 0.25$ b $\frac{7}{10} = 1.7$ c $\frac{4}{10} = 2.4$

3   a   any 50 squares shaded
    b   any 75 squares shaded
    c   any 70 squares shaded
    d   any 35 squares shaded

4

Number line showing $\frac{35}{100}$, $\frac{1}{2}$, $\frac{7}{10}$, $\frac{3}{4}$ marked above the scale:

0   0.1   0.2   0.3   0.4   0.5   0.6   0.7   0.8   0.9   1

5   Decimals and fractions matched as follows:

0.73 and $\frac{73}{100}$ 0.8 and $\frac{8}{10}$

0.35 and $\frac{35}{100}$ 0.5 and $\frac{1}{2}$

0.99 and $\frac{99}{100}$ 0.03 and $\frac{3}{100}$

0.2 and $\frac{20}{100}$ and $\frac{4}{5}$

Mixed numbers

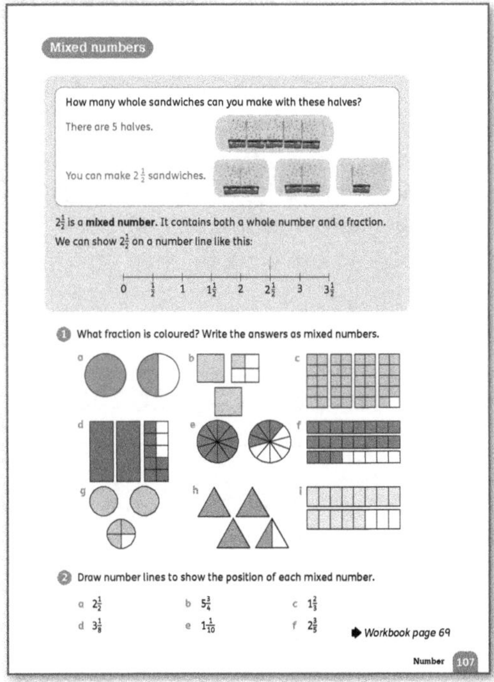

This is the first lesson in which the children formally deal with mixed numbers. However, the concept of a mixed number should not be unfamiliar to them as they may have come across amounts such as $1\frac{1}{2}$ litres, the age $7\frac{1}{2}$, $3\frac{1}{2}$ slices of bread and so on.

Materials

Pictures of items that hold quantities we can measure using mixed numbers, such as $2\frac{1}{2}$, $4\frac{1}{2}$, $1\frac{3}{4}$ (for example, litre bottles marked in quarter litres, egg boxes, yoghurt pots); materials that can be measured in fractional quantities (for example, sand or beans that can be poured).

Warm-up

Select any 'Place value and number sense' activity from pages 23–27 as a mental warm-up.

Focus

Give the class the following practical examples of mixed numbers from everyday life.

- Discuss what it means when you have 1 and a half cakes, or 2 and a quarter litres.
  Write these mixed numbers on the board and stress that each one is composed of a whole number and a part of the whole (the fraction) so it is called a *mixed number*.
- Give the children a set of containers (such as egg boxes or yoghurt pots) that hold the same amount and ask them to model amounts such as three and a half cups of sand.
- Draw a large 0–5 number line marked in halves, with only the whole numbers labelled. Point out positions on the number line for the children to identify as mixed numbers (for example, $1\frac{1}{2}$, $2\frac{1}{2}$).
- Do the same with number lines marked in quarters and fifths. Emphasise the need to count the sections between the whole numbers to work out what the fractional parts are. Make sure that the children realise that five sections between 0 and 1 represent fifths.
- Read through the example on **Pupil Book 4 page 107** to reinforce what you have taught, then ask the children to complete question 1 and question 2 on their own.

Follow-up

Use **Workbook 4 page 69** for additional practice of these concepts.

Challenge

Ask the children to write the mixed numbers on **Pupil Book 4 page 107** as decimal equivalents. They should be able to do this easily except for question 1 parts f and i and question 2 parts c and d but allow them to explore those by considering fractions as division and using a calculator to find the decimal equivalent of $\frac{5}{8}$ ($5 \div 8 = 0.625$), $\frac{2}{3}$ (0.666. . .) and $\frac{1}{8}$ (0.125).

Answers for Pupil Book 4 page 107

1 a $1\frac{1}{2}$ b $2\frac{1}{4}$ c $3\frac{9}{10}$

d $2\frac{7}{10}$ e $1\frac{2}{5}$ f $2\frac{1}{3}$

g $2\frac{3}{4}$ h $4\frac{1}{2}$ i $1\frac{5}{8}$

2 Possible answers:

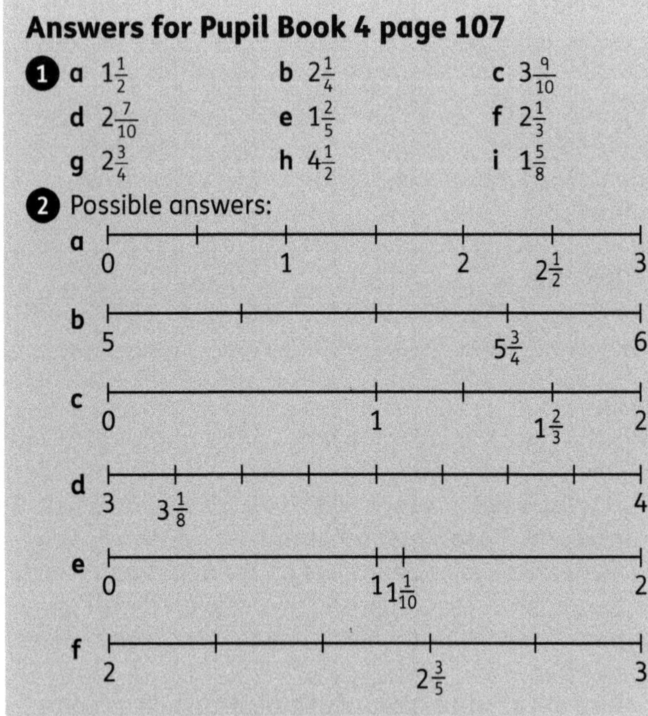

a 0 — 1 — 2 — $2\frac{1}{2}$ — 3

b 5 — $5\frac{3}{4}$ — 6

c 0 — 1 — $1\frac{2}{3}$ — 2

d 3 — $3\frac{1}{8}$ — 4

e 0 — $11\frac{1}{10}$ — 2

f 2 — $2\frac{3}{5}$ — 3

Answers for Workbook 4 page 69

1 a $2\frac{1}{2}$ b $3\frac{1}{4}$ c $2\frac{5}{8}$

d $3\frac{3}{4}$ e $1\frac{1}{4}$ f $1\frac{3}{4}$

2 $1\frac{1}{4}$ $1\frac{3}{4}$ $2\frac{1}{2}$ $2\frac{5}{8}$ $3\frac{1}{4}$ $3\frac{3}{4}$

0 — 1 — 2 — 3 — 4

3 $5\frac{1}{4}, 6\frac{2}{5}, 6\frac{3}{4}, 7\frac{1}{2}, 8\frac{3}{4}, 9\frac{3}{10}, 9\frac{9}{10},$

$5\frac{1}{4}$ $6\frac{2}{5}$ $6\frac{3}{4}$ $7\frac{1}{2}$ $8\frac{3}{4}$ $9\frac{3}{10}$ $9\frac{9}{10}$

5 — 6 — 7 — 8 — 9 — 10

4 Possible answer: I looked at the whole number part first, then I compared the fraction parts.

Mixed numbers and improper fractions

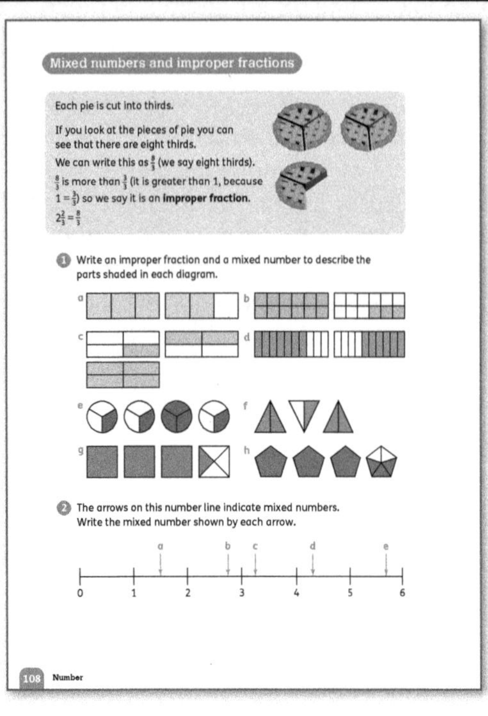

Materials

Sets of circles cut into halves, squares cut into quarters diagonally, rectangles cut into thirds, and other sets of shapes cut into different fractional parts, including one whole shape in each set.

Warm-up

Select any 'Place value and number sense' activity from pages 23–27 as a mental warm-up.

Focus

- At this level, it is useful to introduce *improper fractions* using concrete examples. For example, say:
  - *I cut some pizzas in half, keep a half for myself and now I have 7 halves. How many pizzas is that?* ($3\frac{1}{2}$)
  - *I cut circles of card into quarters. I give each group 9 quarters. How many circles do they have?* ($2\frac{1}{4}$)
- This helps children to understand that a numerator can be greater than a denominator and that improper fractions can be rearranged to make equivalent mixed numbers and vice versa.
- Divide the class into groups. Give each group a set of shapes cut into fractional parts and one whole shape for comparison. For example, give one group circles cut into halves. Give another group squares cut into quarters (diagonally), another group can have rectangles divided into thirds and so on.
- Try to provide quite a number of fractional parts for each group. Ask the children to arrange the parts to make whole shapes and arrange the leftover parts.
- Ask them to say how many they have as a *mixed number*. For example, they might say 'I have one and a half' or 'I have three and two thirds'. Then ask them to split up the parts again.
- Explain how to write each amount as an improper fraction. Write, for example, *I have 3 halves, I can write this as* $\frac{3}{2}$.
- Let the children make different groups using the parts they have and write them as mixed numbers and the equivalent improper fraction.
- Next, the children can work on their own or in pairs to complete question 1 and question 2 on **Pupil Book 4 page 108**.

Answers for Pupil Book 4 page 108

1 a $\frac{5}{3} = 1\frac{2}{3}$ b $\frac{15}{12} = 1\frac{1}{4}$ c $\frac{7}{4} = 1\frac{3}{4}$

d $\frac{13}{10} = 1\frac{3}{10}$ e $\frac{6}{3} = 2$ f $\frac{5}{2} = 2\frac{1}{2}$

g $\frac{13}{4} = 3\frac{1}{4}$ h $\frac{18}{5} = 3\frac{3}{5}$

2 a $1\frac{1}{2}$ b $2\frac{3}{4}$ c $3\frac{1}{4}$

d $4\frac{1}{3}$ e $5\frac{2}{3}$

Write improper fractions as mixed numbers

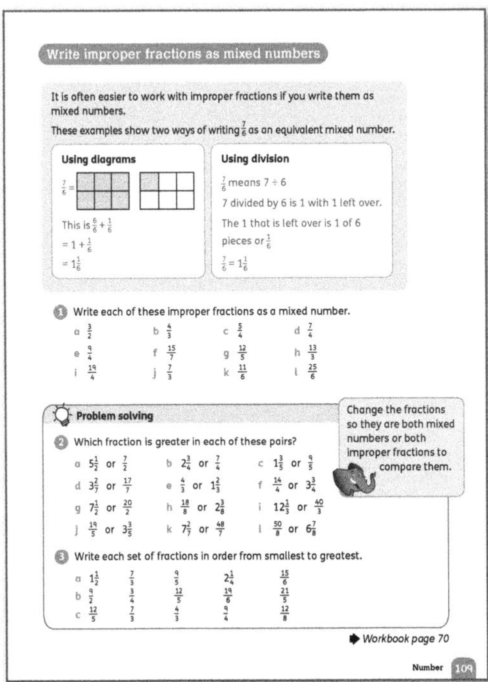

Write improper fractions as mixed numbers

It is often easier to work with improper fractions if you write them as mixed numbers.

These examples show two ways of writing $\frac{7}{6}$ as an equivalent mixed number.

Using diagrams

$\frac{7}{6} =$

This is $\frac{6}{6} + \frac{1}{6}$

$= 1 + \frac{1}{6}$

$= 1\frac{1}{6}$

Using division

$\frac{7}{6}$ means $7 \div 6$

7 divided by 6 is 1 with 1 left over.

The 1 that is left over is 1 of 6 pieces or $\frac{1}{6}$

$\frac{7}{6} = 1\frac{1}{6}$

1 Write each of these improper fractions as a mixed number.

a $\frac{3}{2}$ b $\frac{4}{3}$ c $\frac{5}{4}$ d $\frac{7}{4}$

e $\frac{9}{4}$ f $\frac{15}{7}$ g $\frac{12}{5}$ h $\frac{13}{3}$

i $\frac{19}{4}$ j $\frac{7}{3}$ k $\frac{11}{6}$ l $\frac{25}{6}$

Problem solving

2 Which fraction is greater in each of these pairs?

a $5\frac{1}{2}$ or $\frac{7}{2}$ b $2\frac{3}{4}$ or $\frac{7}{4}$ c $1\frac{3}{5}$ or $\frac{9}{5}$

d $3\frac{2}{7}$ or $\frac{17}{7}$ e $\frac{4}{3}$ or $1\frac{2}{3}$ f $\frac{14}{4}$ or $3\frac{3}{4}$

g $7\frac{1}{2}$ or $\frac{20}{2}$ h $\frac{18}{8}$ or $2\frac{3}{8}$ i $12\frac{1}{3}$ or $\frac{40}{3}$

j $\frac{19}{5}$ or $3\frac{3}{5}$ k $7\frac{2}{7}$ or $\frac{48}{7}$ l $\frac{50}{8}$ or $6\frac{7}{8}$

Change the fractions so they are both mixed numbers or both improper fractions to compare them.

3 Write each set of fractions in order from smallest to greatest.

a $1\frac{1}{2}$ $\frac{7}{3}$ $\frac{9}{5}$ $2\frac{1}{4}$ $\frac{15}{6}$

b $\frac{9}{2}$ $\frac{3}{4}$ $\frac{12}{5}$ $\frac{19}{6}$ $\frac{21}{5}$

c $\frac{12}{5}$ $\frac{7}{3}$ $\frac{4}{3}$ $\frac{9}{4}$ $\frac{12}{8}$

Workbook page 70

Number 109

Materials

Sets of shapes from the previous lesson; a large number line or fraction wall; fraction walls from the 'Equivalent fractions' lesson (see page 139); a worksheet with pictures of real items divided into fractional parts (see 'Support' below).

Warm-up

Revise division facts based on tables to 12×12 as a mental warm-up.

Focus

- Use the shapes cut into fractional parts from the previous lesson.
- Ask the children to work out how many pieces make 1 whole (for example, 3 thirds make 1 whole rectangle).
- Then ask them to take some other pieces, express the result as an improper fraction and say how many wholes and parts this is.
- Discuss the relationship between improper fractions and mixed numbers and let the children share the methods they used to convert from an improper fraction to a mixed number.
- Ask the children to complete **Workbook 4 page 70** next. This will help them to consolidate the connection between mixed numbers and equivalent improper fractions.
- Let the children compare and check each other's work.
- Write down two mixed numbers between 1 and 2 on the board or a sheet of paper, for example $1\frac{3}{5}$ or $1\frac{1}{4}$.
- Ask the children which one they think is greater. Ask why they don't need to think about the whole number parts to decide this (they are equal).

- Then ask the children to consider such things as whether $\frac{1}{4}$ is greater than $\frac{1}{5}$, how many fifths there are in $1\frac{3}{5}$ and how many quarters make $1\frac{1}{4}$.
- Ask the children to position each number on a number line or fraction wall to see whether their estimations were correct.
- Display two different mixed numbers, for example, $2\frac{1}{3}$ and $4\frac{3}{4}$. Ask the children how they can immediately tell that $4\frac{3}{4}$ is greater than $2\frac{1}{3}$. (*The whole number part, 4, is greater than 2, so the mixed number has to be greater too.*)

Turn to **Pupil Book 4 page 109**. Work through the two examples with the class.

- Some children will grasp the division principle here and use it to convert from improper fractions to mixed numbers. Other children may find this too abstract and need to use diagrams and jottings to help. Both methods are fine at this stage.
- The children can use diagrams or jotting or division to complete question 1.
- <u>Problem solving</u>: Let the children work together to decide how they will compare the mixed number and improper fractions in question 2.
- Encourage the children to sketch number lines and use them to order the fractions in question 3. Remind them that they may need to convert from one type to another to make this easier. They can use their fraction walls from the 'Equivalent fractions' lesson lesson, or the one on **Pupil Book 4 page 105**. Make sure they understand that they should give the answers using the fractions in the question, even if they convert them to compare their size.

Challenge

Ask the children to explore how to convert mixed numbers into improper fractions using diagrams and mathematical operations. Let them show their ideas and explain how they did the conversions.

Support

Prepare a worksheet with pictures of real items divided into fractional parts and let the children write the equivalent mixed number and improper fraction. For example, you could show pizzas, cakes, pies, chocolate bars or any other items in a table and the children write the answers alongside. Alternatively, you could give the mixed number and improper fraction and ask the children to draw the images.

Items	Mixed number	Improper fraction
	$2\frac{3}{6}$	$\frac{15}{6}$

Interesting mistakes

- Children sometimes reject the notion of an improper fraction because they have learnt that the numerator is always less than the denominator.

- You can address this misconception using physical and concrete examples, including parts of fraction strips. For example, say: *What fraction is this?* $(\frac{1}{5})$ *I have 7 parts like this, how many fifths is that?* (*7 fifths*) *How can we write that?* $(\frac{7}{5})$
- The children may automatically assume that a mixed number is greater than an improper fraction because they have been taught that a fraction is part of a whole (rather than understanding that fractions are numbers that lie between wholes on a number line).
- Use number lines to show both the mixed number and the fractional parts so children can see how they relate to each other. For example, if any children think that $2\frac{3}{8}$ is greater than $\frac{21}{8}$, show them on a number line why this is not the case. Encourage them to use numbers lines to compare if they are unsure.

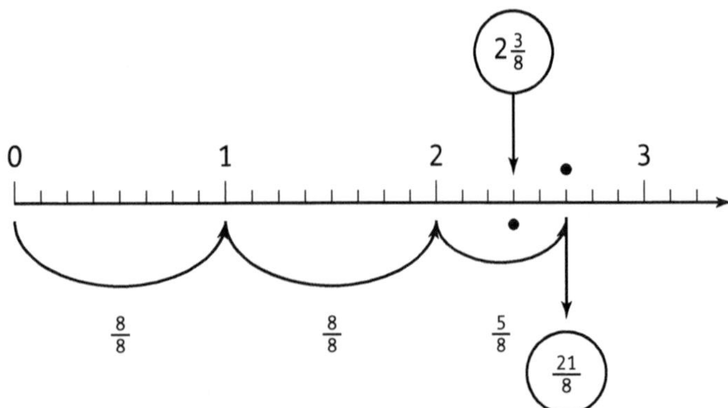

Answers for Pupil Book 4 page 109

1 a $1\frac{1}{2}$ b $1\frac{1}{3}$ c $1\frac{1}{4}$ d $1\frac{3}{4}$

   e $2\frac{1}{4}$ f $2\frac{1}{7}$ g $2\frac{2}{5}$ h $4\frac{1}{3}$

   i $4\frac{3}{4}$ j $2\frac{1}{3}$ k $1\frac{5}{6}$ l $4\frac{1}{6}$

2 a $5\frac{1}{2}$ b $2\frac{3}{4}$ c $\frac{9}{5}$ d $3\frac{2}{7}$

   e $1\frac{2}{3}$ f $3\frac{3}{4}$ g $\frac{20}{2}$ h $2\frac{3}{8}$

   i $12\frac{1}{3}$ j $\frac{19}{5}$ k $7\frac{2}{7}$ l $6\frac{7}{8}$

3 a $1\frac{1}{2}, \frac{9}{5}, 2\frac{1}{4}, \frac{7}{3}, \frac{15}{6}$

   b $\frac{3}{4}, \frac{12}{5}, \frac{19}{6}, \frac{21}{5}, \frac{9}{2}$

   c $\frac{4}{3}, \frac{12}{8}, \frac{9}{4}, \frac{7}{3}, \frac{12}{5}$

Answers for Workbook 4 page 70

1 The following should be shaded:

   a 3 whole shapes and 3 parts

   b 4 whole shapes and 3 parts

   c 3 whole shapes and 5 parts

   d 4 whole shapes and 1 part

   e 3 whole shapes and 7 parts

   f 2 whole shapes and 4 parts

   g 2 whole shapes and 4 parts

   h 3 whole shapes and 2 parts

2 a $\frac{13}{4}$ b $\frac{19}{4}$ c $\frac{29}{8}$ d $\frac{41}{10}$

   e $\frac{31}{8}$ f $\frac{14}{5}$ g $\frac{28}{10}$ h $\frac{11}{3}$

3 a 7   b 29   c $3\frac{1}{4}$ d $3\frac{4}{5}$

Add and subtract fractions

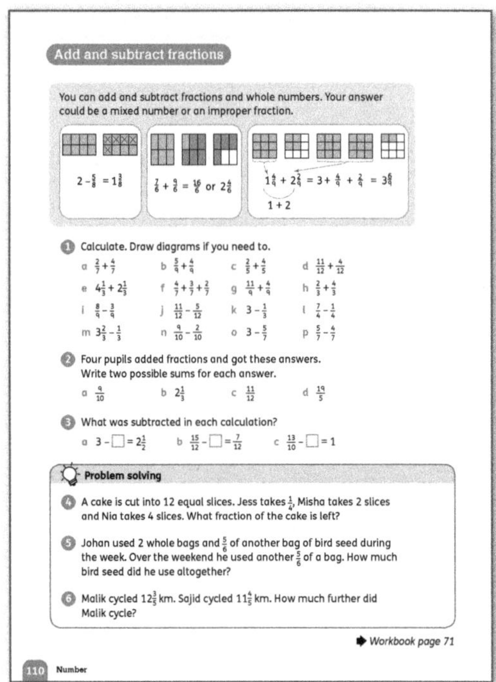

Add and subtract fractions

You can add and subtract fractions and whole numbers. Your answer could be a mixed number or an improper fraction.

$2 - \frac{5}{8} = 1\frac{3}{8}$

$\frac{7}{6} + \frac{9}{6} = \frac{16}{6}$ or $2\frac{4}{6}$

$1\frac{4}{9} + 2\frac{2}{9} = 3 + \frac{4}{9} + \frac{2}{9} = 3\frac{6}{9}$
$1 + 2$

1 Calculate. Draw diagrams if you need to.

   a $\frac{2}{7} + \frac{4}{7}$ b $\frac{5}{9} + \frac{4}{9}$ c $\frac{2}{5} + \frac{4}{5}$ d $\frac{11}{12} + \frac{4}{12}$

   e $4\frac{1}{3} + 2\frac{1}{3}$ f $\frac{4}{7} + \frac{3}{7} + \frac{2}{7}$ g $\frac{11}{9} + \frac{4}{9}$ h $\frac{2}{3} + \frac{4}{3}$

   i $\frac{8}{9} - \frac{3}{9}$ j $\frac{11}{12} - \frac{5}{12}$ k $3 - \frac{1}{3}$ l $\frac{7}{4} - \frac{1}{4}$

   m $3\frac{2}{3} - \frac{1}{3}$ n $\frac{9}{10} - \frac{2}{10}$ o $3 - \frac{5}{7}$ p $\frac{5}{7} - \frac{4}{7}$

2 Four pupils added fractions and got these answers. Write two possible sums for each answer.

   a $\frac{9}{10}$ b $2\frac{1}{3}$ c $\frac{11}{12}$ d $\frac{19}{5}$

3 What was subtracted in each calculation?

   a $3 - \square = 2\frac{1}{2}$ b $\frac{15}{12} - \square = \frac{7}{12}$ c $\frac{13}{10} - \square = 1$

Problem solving

4 A cake is cut into 12 equal slices. Jess takes $\frac{1}{4}$, Misha takes 2 slices and Nia takes 4 slices. What fraction of the cake is left?

5 Johan used 2 whole bags and $\frac{5}{6}$ of another bag of bird seed during the week. Over the weekend he used another $\frac{5}{6}$ of a bag. How much bird seed did he use altogether?

6 Malik cycled $12\frac{3}{5}$ km. Sajid cycled $11\frac{4}{5}$ km. How much further did Malik cycle?

➡ *Workbook page 71*

110   Number

Materials

Large number lines (for display); squared paper (for drawing fraction diagrams); a sheet showing some whole shapes and some divided into fractional parts (see 'Support' below).

Warm-up

Choose any 'Calculation skills' activity involving addition and subtraction from pages 28–29 as a warm-up.

Focus

The children have already learnt how to add and subtract fractions with the same denominators (within the range of 0 to 1). It is useful to revise this before moving on to adding and subtracting across whole numbers.

- Use number lines to revise the idea that you can 'jump' forwards or backwards a fractional amount. This example uses tenths because the children should be very comfortable adding and subtracting within 10.

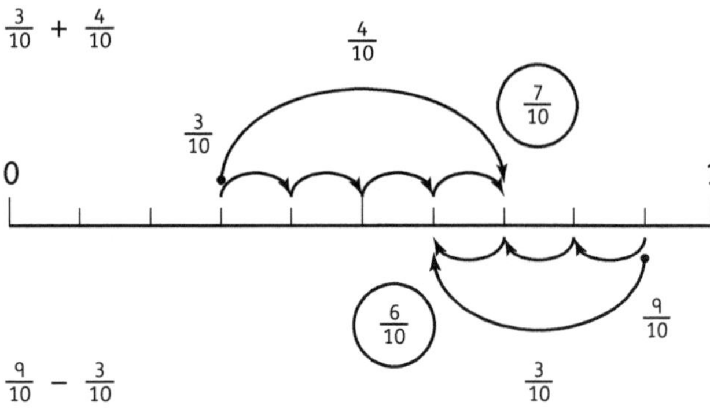

- Once you have revised this, move on to adding and subtracting fractions that bridge the whole number.

- Include a range of examples such as: $\frac{4}{5} + \frac{3}{5}$; $1\frac{2}{5} + 1\frac{4}{5}$; $3 - \frac{2}{5}$; $2\frac{2}{5} - \frac{4}{5}$. These examples are shown here:

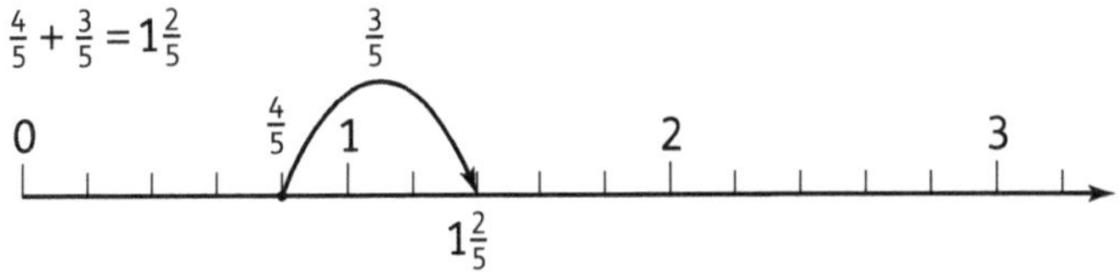

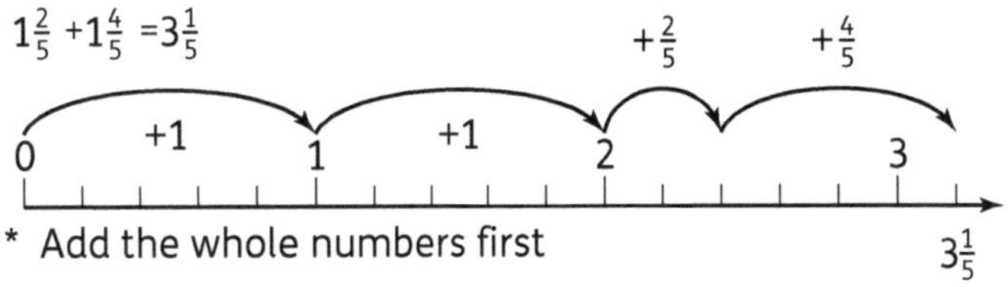

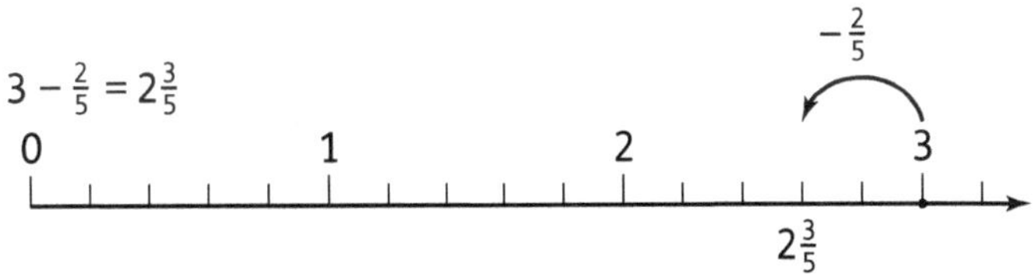

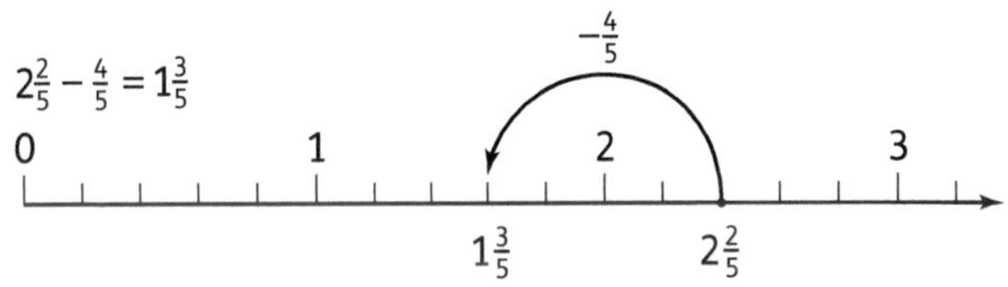

- After showing a few examples, ask the children to think about each one and how they might work it out without using the number line. Share their ideas and methods with the class.
- Turn to **Pupil Book 4 page 110**. Ask the children to work through the examples in groups, explaining what is happening and linking back to what they did using the number lines.
- Give the children a sheet of squared paper to use to draw diagrams for the calculations in question 1, if necessary. Spend some time talking about how to decide whether to give the answer as an improper fraction or a mixed number (they can choose at this stage).
- Let the children compare and check each other's answers to question 2. If their partner doesn't agree, they should try to convince them.
- Discuss with the class how they can work out question 3, reminding them that addition and subtraction are inverse operations, if necessary, and that this applies to adding and subtracting all numbers, including fractions.
- Problem solving: Read through questions 4–6 with the class and let the children share how they would tackle each one before asking them to solve the problems independently.

Follow-up

Use **Workbook 4 page 71** as additional practice and to consolidate adding and subtracting fractions.

Support

Prepare a sheet of shapes with some whole shapes and some divided into fractional parts. Ask the children to cut these out and show them how to use them to model adding and subtracting fractions.

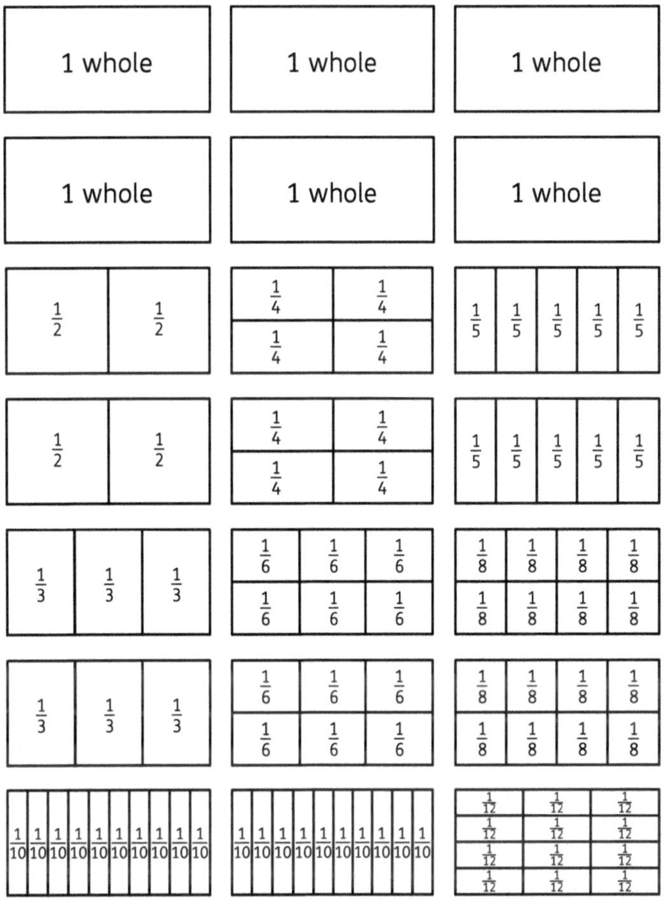

Answers for Pupil Book 4 page 110

1 a $\frac{6}{7}$ b $\frac{9}{9}$ c $\frac{6}{5}$ or $1\frac{1}{5}$
d $\frac{15}{12}$ or $1\frac{3}{12}$ e $6\frac{2}{3}$ f $\frac{9}{7}$ or $1\frac{2}{7}$
g $\frac{15}{9}$ or $1\frac{6}{9}$ h $\frac{6}{6}$ i $\frac{5}{9}$
j $\frac{6}{12}$ or $1\frac{1}{2}$ k $\frac{8}{3}$ or $2\frac{2}{3}$ l $\frac{6}{4}$ or $1\frac{1}{2}$
m $\frac{10}{3}$ or $3\frac{1}{3}$ n $\frac{7}{10}$ o $\frac{16}{7}$ or $2\frac{2}{7}$
p $\frac{1}{7}$

2 Individual answers

3 a $\frac{1}{2}$ b $\frac{8}{12}$ c $\frac{3}{10}$

4 $\frac{3}{12}$

5 $3\frac{4}{6}$ bags

6 $\frac{4}{5}$ km

Answers for Workbook 4 page 71

1 First diagram: $(\frac{8}{5})$, $\frac{6}{5}$ or $1\frac{1}{5}$ $(1\frac{1}{5})$, $\frac{8}{5}$ or $1\frac{3}{5}$
$(\frac{3}{5})$, $\frac{11}{5}$ or $2\frac{1}{5}$ (1), $\frac{9}{5}$ or $1\frac{4}{5}$ (2), $\frac{4}{5}$
Second diagram: $(\frac{7}{12})$, $\frac{10}{12}$ (1), $\frac{5}{12}$ $(\frac{1}{4})$, $\frac{14}{12}$ or $1\frac{2}{12}$
$(\frac{5}{12})$, 1 $(1\frac{2}{12})$, $\frac{5}{12}$ $(\frac{6}{12})$, $\frac{11}{12}$ $(\frac{7}{12})$, $\frac{10}{12}$

2 a $\frac{12}{20}$ or $\frac{3}{5}$ b $\frac{9}{16}$ c 3

Calculate with fractions

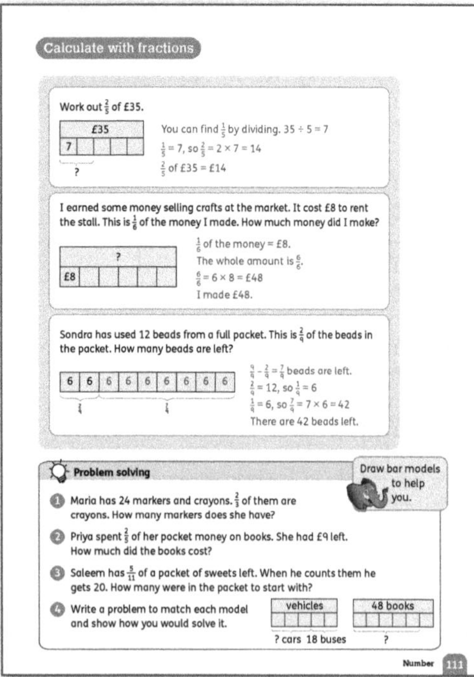

Fractions can act as operators in a mathematical situation. For example, $\frac{1}{5}$ of 100 means $\frac{1}{5} \times \frac{100}{1}$ $= \frac{100}{5} = 20$.

The children don't need to develop formal algorithms like the one above, but they do need to understand what to do to calculate a fraction of an amount or quantity.

The aim of this lesson is to demonstrate how the children can use bar models (see page 22) to represent problems with fractions and what they need to work out to solve a problem.

Materials
Squared paper (for drawing bar models).

Warm-up
Choose any 'Calculation skills' activity from pages 27–29 as a warm-up.

Focus
- Ask the children to work through the examples on **Pupil Book 4 page 111** in groups.
- Have a class discussion after they have done this. Ask questions such as: *What did you learn from this? How can this help you? Did anything surprise you? Do you have any questions about this?*
- Problem solving: Let the children work on their own or in pairs to solve the problems in questions 1–4 on **Pupil Book 4 page 111**. Give them squared paper to draw their bar models on.
- Spend some time afterwards talking about how they solved each problem. Identify any interesting mistakes and discuss what the class can learn from them.

End-of-unit check

Ask some or all of these questions to check that the children understand and can apply what they have learnt about fractions.

- *How many equal parts have you divided this shape into?*
- *What is each part called?*
- *What does the denominator (the number under the line in a fraction) tell you?* (the number of parts that the whole has been divided into)
- *What does the numerator (the number above the line in a fraction) tell you?* (the number of parts)
- *Give a fraction with a denominator of 10 that is equivalent to $\frac{3}{5}$.* ($\frac{6}{10}$) *How did you work this out?* (I multiplied both the numerator and denominator by 2.)
- *How many halves make one-and-a-half?* (3) *... two?* (4) *... three-and-a-half?* (7)
- *How many quarters make one-and-a-quarter?* (5) *... one-and-a-half?* (6) *... two?* (8)
- Show a picture of a mixed number. *What is this mixed number?*
- *How is a mixed number different from a proper fraction?* (It is greater than one whole.)
- *Shade $3\frac{1}{5}$ of these circles.*
- Show some mixed numbers and a number line. *Where do these mixed numbers go on this number line?*
- *What mixed number is shown by this arrow on the number line? How do you know?*
- *How would you change $\frac{11}{5}$ to a mixed number?* ($\frac{11}{5} = \frac{10}{5} + \frac{1}{5} = 2\frac{1}{5}$)
- *How would you explain to someone how to add and subtract proper fractions with the same denominator?* (Add or subtract the numerators and leave the denominator the same.)
- *What is $\frac{4}{5} + 2\frac{3}{5}$?* ($3\frac{2}{5}$)
- *Subtract $\frac{6}{10}$ from $1\frac{3}{10}$.* ($\frac{7}{10}$)
- *I cut $3\frac{1}{4}$ metres from a 5 m roll of fabric. How much do I have left?* ($1\frac{3}{4}$ m)
- *What is (give a fraction) of (give an amount)?*

Mixed practice 2

You can use Mixed practice 2 on **Pupil Book 4 pages 112–113** to assess the children's confidence in the concepts from Units 8–15.

1 a If you fold the shape along that line, the two parts will not fit exactly on top of each other.

b Any rectangle with one line of symmetry drawn. For example:

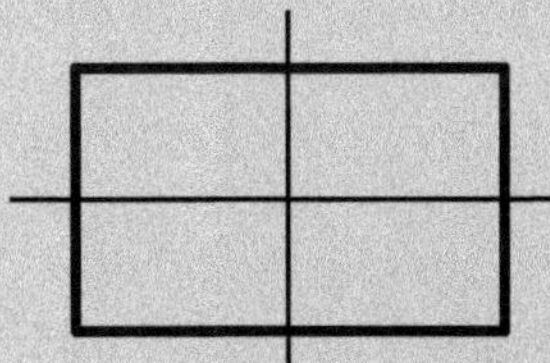

2 a A   **b** the shape in diagram C

3 a $\frac{1}{12}$ **b** heating water

c Yes, $\frac{2}{12} = \frac{1}{6}$ **d** $\frac{1}{6}$ of 240 = 40 units

e $\frac{3}{12} = 105$, so $\frac{1}{12} = 35$ units. $\frac{12}{12} = 35 \times 12 = 420$ units

4 a pentagon

b No, it is not regular because, although the sides are all the same length, the angles are not the same size.

c *b* and *e*   **d** acute

e 2 (*c* and *d*)   **f** $12\frac{1}{2}$ cm

5 a 158   **b** 1542   **c** 4412   **d** 377

6 a 4   **b** 80

c Possible answers:

Chart A:
- The vertical axis should show the number of vehicles not the number of groups of tallies.
- 'Vehicle types' should be the horizontal axis label, not the chart title.

Chart B:
- There are no numbers on the vertical axis, so we don't know what the scale is.
- The names of the types of vehicle should be along the horizontal axis instead of the number of each type of vehicle.
- 'Type of car' should say 'Type of vehicle' and it should be the horizontal axis label.
- There is no title.

Chart C:
- The scale does not need to go up to 50 when the maximum number of vehicles of one type is 35.
- There is no label for the frequency axis.
- Instead of labelling the bars T, T, B, C, it would be clearer to label them bicycle, truck, bus and car.

UNIT 16 Position and movement

Learning objectives

- Understand that position can be described using coordinates

- Read and plot points in the first quadrant of a coordinate grid

- Use given coordinates to draw polygons in the first quadrant

- Describe movement in a straight line as a translation of any given unit up/down and left/right

- Translate points and shapes in the first quadrant

- Reason and solve problems related to position, movement and symmetry properties (reflect shapes) on the coordinate grid

Key words

coordinates point coordinate grid horizontal vertical axis/axes *x*-axis *y*-axis vertex/ vertices translate translation

Unit introduction

Materials

A large coordinate grid labelled with letters and numbers and with different shapes positioned at points on the grid (like the one shown in the 'Teaching guidance' below); sticky notes or small pieces of paper; sticky tape or adhesive tack.

At Level 3, the children worked with alphanumeric grids (letters and numbers) to give the position of items. Now they will work with pairs of numbers to give the *coordinates* of different *points* on a *coordinate grid*.

Note that the points are marked at the intersections of lines of the grid and not within the squares of the grid.

Teaching guidance

- Start with a memory game using a grid labelled with letters and numbers.

- Write the letters along the *horizontal axis* so that the children get used to reading the horizontal axis before the *vertical* axis.

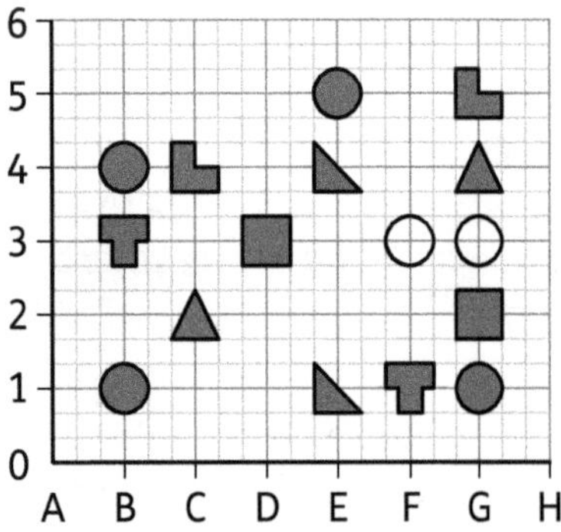

- Show the whole grid to the class to let them see what it looks like.
- Then cover the items using sticky notes or pieces of paper held on by sticky tape or adhesive tack.
- Play a game by asking different children to say the positions of two shapes on the grid. For example, they might say 'B1 and E1'. Show those two shapes. They don't match, so cover them again.
- When the children find two matching shapes, leave the shapes uncovered.
- If the children enjoy the activity, repeat it using a different grid or ask them to work in pairs to make a grid to play the game against another pair.

Describe positions on a grid

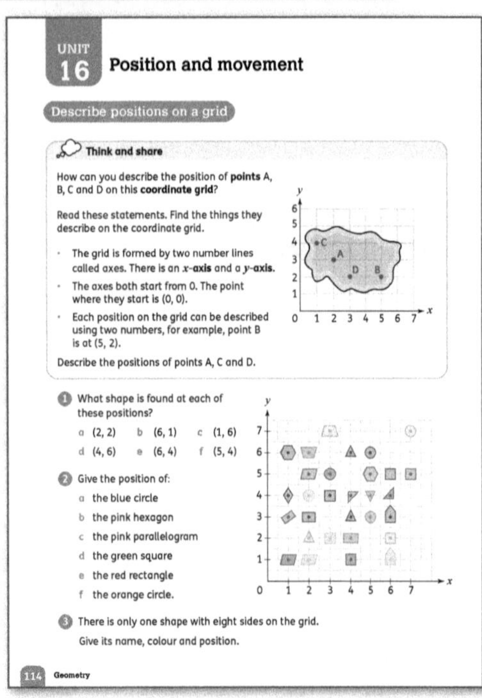

Warm-up
Use any of the 'Mental maths' multiplication activities on pages 27–29 as a warm-up.

Focus
- <u>Think and share</u>: Place the children in groups and ask them to talk about how they could give the position of each point marked on the coordinate grid on **Pupil Book 4 page 114**.
- Take feedback and use it to stress that this grid has two sets of numbered *axes* (as opposed to one lettered and one numbered). Introduce the terms *x-axis* and *y-axis*.
- Ask: *Which axis shows how far across the grid the point is?* (the horizontal or *x*-axis) *What does the y-axis show?* (*how far up the grid the point is*). Explain that we always give the number on the horizontal or *x*-axis before the number on the vertical or *y*-axis.
- Discuss why it is important to know which number to use by referring to the letters C and D. C is shown at (1, 4) and D is shown at (3, 2).
- Once you are satisfied that the children understand the grid and how it can be used to give positions, let them read through the statements and use the grid to make sense of each one.
- Note that the word coordinates is introduced formally on **Pupil Book 4 page 115**.
- For question 1, let the children work on their own or in pairs to identify the shapes at each position. They can refer to these by colour and shape, for example, 'red triangle' or they can draw them in their books.
- To complete questions 2 and 3, the children need to find the shapes and then give the coordinate pair that gives the position of each one. Explain that these are written in brackets so that we know the numbers are giving a position.

Challenge
Let the children map the classroom and use coordinates to give the position of their desks.

Interesting mistakes
The children may get the order of coordinates confused and forget which to give first. Remind them that the axes are labelled *x* and *y*: *x* comes before *y* in the alphabet. Also use the example that we walk across the floor before we can climb up or down stairs.

Answers for Pupil Book 4 page 114
<u>Think and share:</u> A (2, 3) C (1, 4) D (3, 2)

1 a green triangle   b blue pentagon
   c yellow hexagon   d red triangle
   e yellow triangle   f blue triangle

2 a (5, 3)   b (5, 5)   c (2, 5)
   d (6, 2)   e (2, 3)   f (3, 5)

3 blue octagon, (2, 4)

Use coordinates

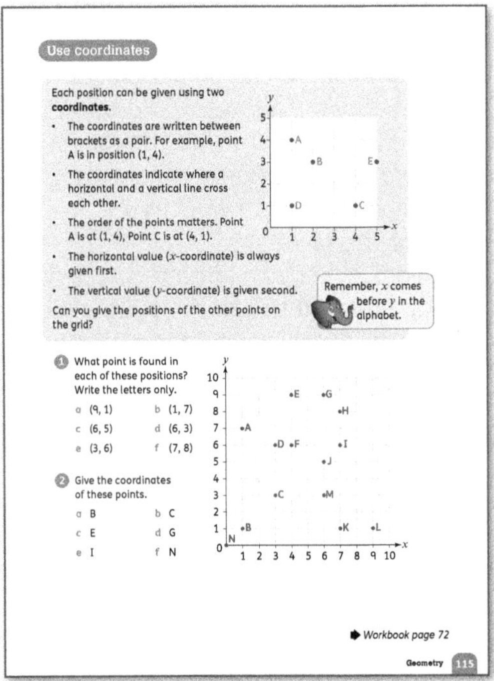

Materials
Chalk or string; a range of objects.

Warm-up
Select any of the 'Mental maths' multiplication activities from on pages 27–29 as a warm-up.

Focus
In this lesson, the children learn that the term *coordinates* describes the numbered pairs that give the position of points.

- Read through the information at the top of **Pupil Book 4 page 115** with the class, reminding them as you go that they have already worked with coordinates in the previous lesson.
- Point to points A and C as you read the coordinates of each. Then ask the children to tell you the coordinates of the other points on the grid. These are:
  Point B (2, 3)
  Point D (1, 1)
  Point E (5, 3).
- The children can work through question 1 and question 2 independently or in pairs. Check their answers as a class.

Follow-up
Let the children work on their own to complete the map of the marine park on **Workbook 4 page 72**. When they have completed this, let them compare with others to check their work. Ask different children to say where they have positioned the octopuses and to give reasons for choosing this position.

Support
- Make a large grid on the ground outside. You can either draw it using chalk or make the lines using string. Put objects on some of the intersections of gridlines. Arrange the children into pairs.
- One child chooses an object and says the coordinates for the position of that object. The other child walks along the x-axis from 0 and then walks 'up' in the y-axis direction to find the object in that position.
- If they identify it correctly, they may move the object to another position and then have their turn to select an object and say the coordinates for their partner. If they are incorrect, their partner has another turn.

Interesting mistakes
- It is important for the children to know that the x-axis is the horizontal line and that the y-axis is the vertical line, and that the coordinates are given in the order (x, y). Keep reminding them that the first number represents the distance along the x-axis, and the second represents the distance along the y-axis.
- The children may also sometimes identify the coordinate on the x- or y-axis, and forget about the second coordinate. If necessary, they can use two rulers (or any two straight-edged items, such as books or sheets of paper) to form a corner so that they can easily find the coordinates.

Answers for Pupil Book 4 page 115
1. a L   b A   c J
   d M   e D   f H
2. a (1, 1)   b (3, 3)   c (4, 9)
   d (6, 9)   e (7, 6)   f (0, 0)

Answers for Workbook 4 page 72

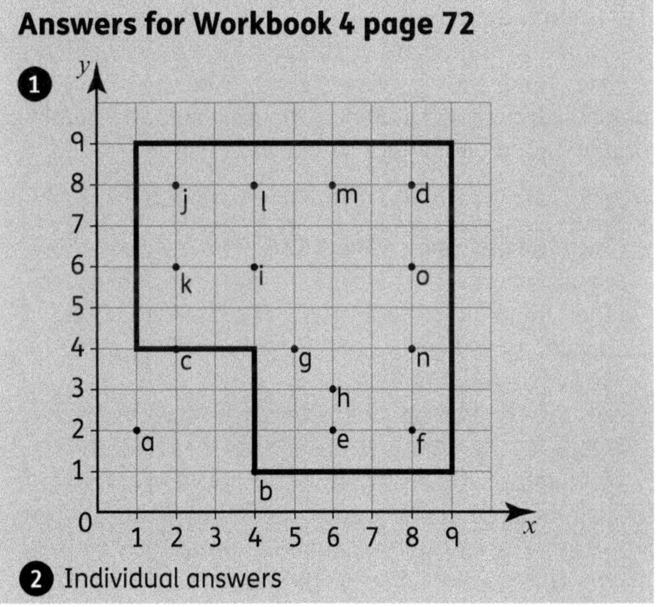

1.
2. Individual answers

Translations

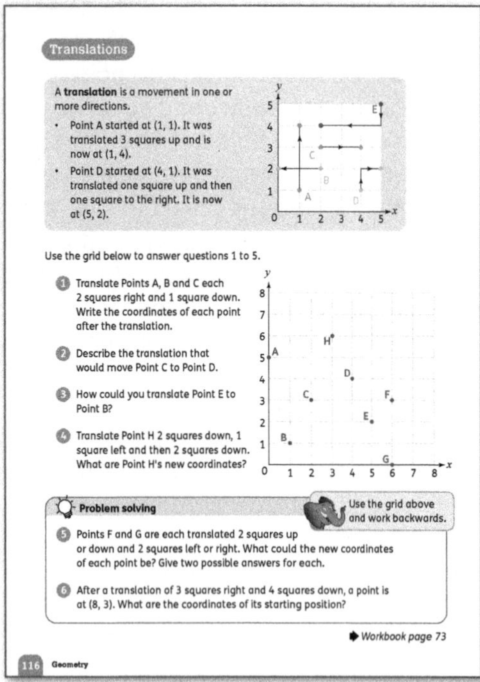

Translations

A **translation** is a movement in one or more directions.

- Point A started at (1, 1). It was translated 3 squares up and is now at (1, 4).
- Point D started at (4, 1). It was translated one square up and then one square to the right. It is now at (5, 2).

Use the grid below to answer questions 1 to 5.

1. Translate Points A, B and C each 2 squares right and 1 square down. Write the coordinates of each point after the translation.

2. Describe the translation that would move Point C to Point D.

3. How could you translate Point E to Point B?

4. Translate Point H 2 squares down, 1 square left and then 2 squares down. What are Point H's new coordinates?

Problem solving

Use the grid above and work backwards.

5. Points F and G are each translated 2 squares up or down and 2 squares left or right. What could the new coordinates of each point be? Give two possible answers for each.

6. After a translation of 3 squares right and 4 squares down, a point is at (8, 3). What are the coordinates of its starting position?

♦ Workbook page 73

116 Geometry

Materials

Blank coordinate grids for the children to work on (or squared paper for drawing their own grids); a very small item to use as a counter (a seed, a small bead or a little piece of eraser); tracing paper (optional).

Warm-up

Ask the children what it means if you translate a sentence from one language to another. (You change the sentences into the new language. The new sentence is a translation of the original sentence.)

Tell them that in maths, we use the term *translate* to describe moving points or shapes to a different position. *Translation* doesn't change the shape or size of the object being moved, it simply moves it side to side or up or down (or a combination of both).

Focus

- Turn to **Pupil Book 4 page 116**. Read through the explanation with the class and let them trace with their fingers the movement of Points A and D as described. Make sure that they can tell the difference between left and right.
- Let the children describe the other translations.
- Ask a few questions to check that the children understand. For example: *How could you translate Point A to get it in the same position as Point C?* (one square right, one square up) *How would you get the translated Point E back to its original position?* (three squares right, one square up)
- Before completing questions 1–4, ask the children to give the coordinates of each of the points A to H on the grid.
- Then let the children complete questions 1–4 on their own, but encourage them to talk to each other and to ask questions if they are not sure what to do.

- Problem solving: Discuss question 5 with the class. Ask them how many different translations are possible for these movements. (There are four: up and right, up and left, down and right, down and left.) Note that Point G can be translated four ways too, but that moving it down would move it off the coordinate grid shown and into the fourth quadrant, with negative y-values. Some children may work this out, but they only need to work in the first quadrant of the grid at this level.
- For question 6, ask the children to say how they would undo different translations. For example, if the movement was up, they would move down to undo it and work out where the point started.

Follow-up

- Once the children can confidently describe translations and explain how points were translated, turn to **Workbook 4 page 73**. Remind them that translation is a movement and does not change the shape or size of the shape.
- Discuss how we can translate a whole shape. The *vertices* (corners) of the shape are points with coordinates. We have to translate each *vertex* the same number of squares in the same direction. If necessary, do one or two of the translations with the class. Allow those who need support to make a small rectangle and to use it to model the translations.
- Problem solving: Once the children have done the translations in questions 1 and 2, let them work together to solve question 3. Note that there are many possible solutions.

Support

If any children need more time to visualise the translations, let them use a very small item as a counter (a seed, a small bead or a little piece of eraser) to physically model the movements on the grid. For **Workbook 4 page 73** question 1, they can trace and cut out the shape and use that in the same way to model the movements.

Interesting mistakes

When the children are given a translation up or down first, they may get confused between the moves and the coordinates (for example, by moving one up and changing the x-coordinate because it comes first in the pair). Modelling the translation with a physical object or actually drawing the moves and plotting the new points will help them to avoid this confusion.

Answers for Pupil Book 4 page 116

1. A = (2, 4); B = (3, 0); C = (4, 2)
2. 2 squares right and 1 square up
3. 4 squares left one square down
4. (2, 2)
5. Possible answers: F = (8, 1), (8, 5), (4, 1), (4, 5); G = (4, 2), (8, 2)
6. (5, 7)

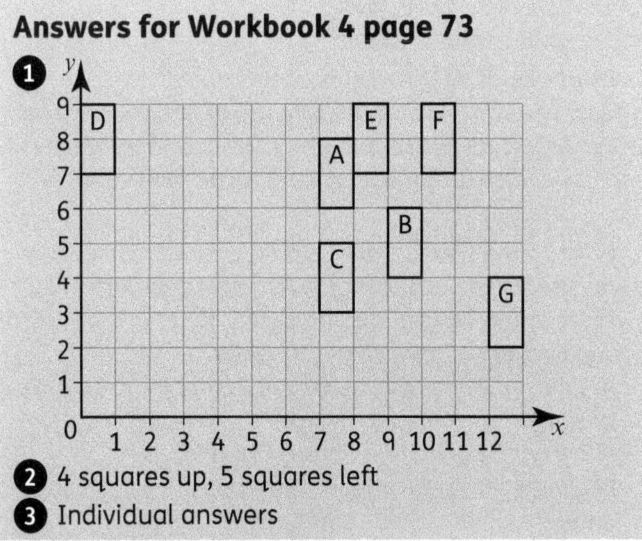

Answers for Workbook 4 page 73

1 (coordinate grid showing shapes D, E, F, A, B, C, G plotted)

2 4 squares up, 5 squares left

3 Individual answers

Shapes on the grid

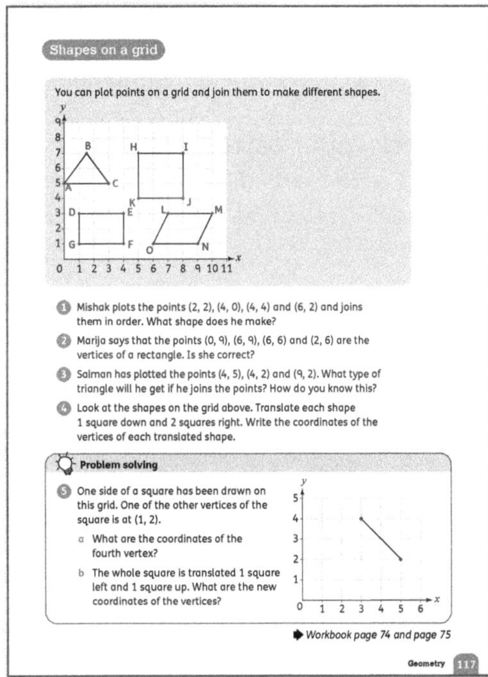

Materials
Blank coordinate grids or squared paper; small mirrors (for Workbook questions).

Warm-up
Look at the coordinate grid on **Pupil Book 4 page 117** with the class. Let the children name each shape as accurately as possible and ask them to give the coordinates of the vertices of each shape. Highlight how the points are generally plotted in order and then joined up to make the shapes.

Focus
- Use **Workbook 4 page 74** to consolidate plotting points and drawing shapes on the grid. Let the children check and correct each other's work.

- Turn back to **Pupil Book 4 page 117**.
- Work through question 1 with the class. Let the children plot the points on blank coordinate grids and join them if they need to. Alternatively, they could draw their own coordinate grids on squared paper.
- Do question 2 orally with the class. Ask the children to explain their answers.
- Then work through question 3 orally with the class. Accept all sensible reasons.
- The children can do question 4 on their own. They could trace the shapes and physically move them on the grid to work out where the shapes end up.
- <u>Problem solving</u>: Let the children work in pairs to solve question 5.

Follow-up
- Use **Workbook 4 page 75** as an extension activity. The children need to apply what they have learnt in this lesson and also what they know about symmetry and reflection of shapes.
- <u>Problem solving</u>: For question 2, the children may find it helpful to turn their page round so that the mirror line is either horizontal or vertical.

Challenge
Give the children a sheet of four blank grids on centimetre squared paper with 0–6 on each axis. They will also need two sets of digit cards 1–6. They pick a card from each set and use the two numbers as x- and y-coordinates. Repeat this four times for each grid. If the children get the same coordinates, they can either switch the order or roll again. They should plot the points and join them to make a polygon. Let them name each polygon and work out its area in square units.

Answers for Pupil Book 4 page 117

1 square

2 no

3 right-angled triangle; Possible answer: the line joining (4, 5) and (4, 2) has the same x-coordinate, so it is a vertical line; the line joining (4, 2) and (9, 2) has the same y-coordinate, so it is a horizontal line; horizontal and vertical lines meet at a right angle.

4 triangle: A (2, 4), B (3.5, 6), C (5, 4); rectangle: D (3, 2), E (6, 2), F (6, 0), G (3, 0); square: H (7, 6), I (10, 6); J (10, 3), K (7, 3); parallelogram: L (9, 2), M (12, 2), N (11, 0), O (8, 0)

5 a (3, 0)   b (2, 5), (4, 3), (0, 3), (2, 1)

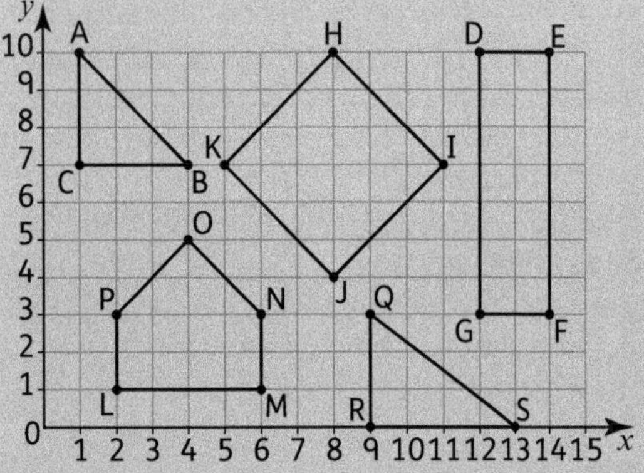

1   **a**   right-angled triangle   **b**   rectangle
   c   square   **d**   pentagon
   e   right-angled triangle   **f**   Individual answers

Answers for Workbook 4 page 75

1   Any four points joined to make a quadrilateral, and reflected in the mirror line. For example:

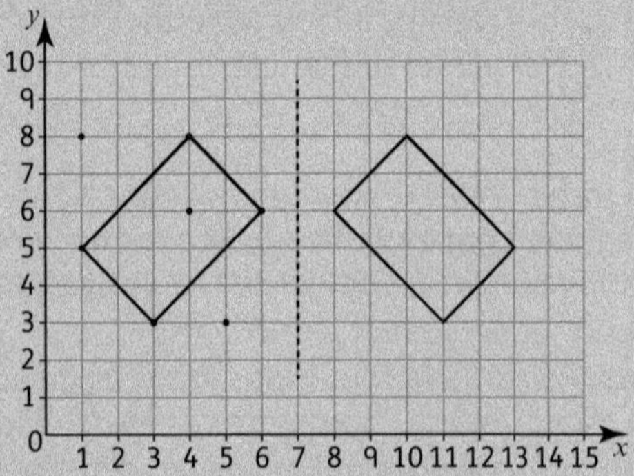

2 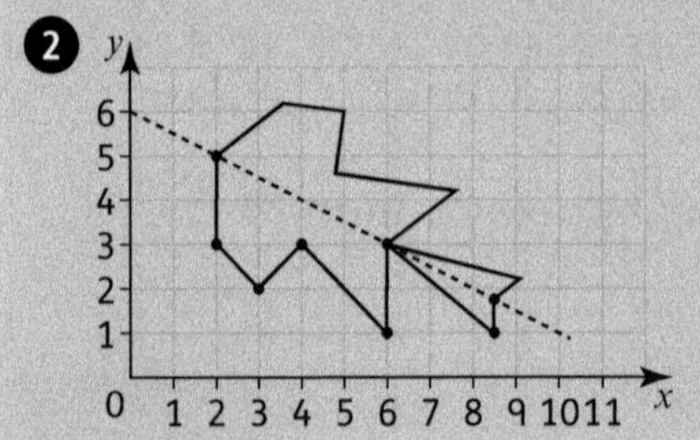

Use some or all of these questions and activities to assess the children's understanding of the concepts in this unit. The most useful way to check understanding is to ask the children to complete an activity on a coordinate grid.

- Point to the horizontal axis on a grid. *What is the horizontal line in a pair of axes called?* (x-axis)
- Point to the vertical axis on a grid. *What is the vertical line in a pair of axes called?* (y-axis)
- *At which number do the axes intersect (cross)? (0) Find this point on your pair of axes.*
- *Show where point (4, 5) is on your grid.*
- Show a coordinate grid with shapes drawn at some different points. *What is at (2, 3)?*
- Give the children a set of coordinates. *Plot the points on your grid.*
- Show some labelled points on a coordinate grid. *Translate Point A 2 squares right and 3 squares up.*
- *Plot these points and join them in order: (1, 1), (1, 4), (6, 1). What shape have you drawn? (a right-angled triangle)*
- *Starting at point (6, 4), plot the vertices of a rectangle with area $3 \times 5 = 15$ cm². (for example: (6, 4), (11, 4), (11, 1), (6, 1) or (6, 4), (9, 4), (9, 9), (6, 9))*

UNIT 17 Multiplication

Learning objectives

- Use place value and known facts to multiply mentally
- Revise multiplication and division facts to 12×12
- Multiply any number by 10 and 100 and recognise that this makes the numbers 10 or 100 times greater
- Convert measurements by multiplying by 10 or 100

- Use the distributive property to multiply 2-digit numbers by 1-digit numbers
- Multiply 2-digit and 3-digit numbers by a 1-digit number using formal written layout
- Solve problems involving multiplication, including scaling problems

Key words

times   multiply   product   array
multiplication fact   division fact   partitioning
grid method   column method   estimate

Teaching guidance

- Start by discussing all the words that the children know to talk about multiplication and write them on the board. They might say: *times, multiply, product, by, ×, groups of, lots of, twice, double, three times as many, repeated addition, array, grid*.
- Ask different children to make up an example of each word in use. Write their examples next to the words and leave them on display in the classroom.
- Display the number 100 (or 120). Ask the class to find as many ways as they can to make 100 by multiplying. (It is likely the children will stick to whole numbers for this, but some children may explore using decimals and/or fractions. This will lead to interesting classroom discussion, so allow for it.)
- Give the children time to jot down as many ways as they can think of. Then go round the class and record their methods.
- When children give a commutative pair (for example, you have recorded 2 × 50 and another child gives 50 × 2), write these next to each other and circle them.
- Let the children choose another target number in groups and repeat this activity by themselves.
- Discuss why we can write more multiplications for some target numbers than for others. For example, 100 has lots of factors, so we can write a lot of multiplications. But if the target number is prime (such as 23) there will only be two factors, so there are only two ways of making that number by multiplying whole numbers (1 × 23 and 23 × 1).

Remember your facts

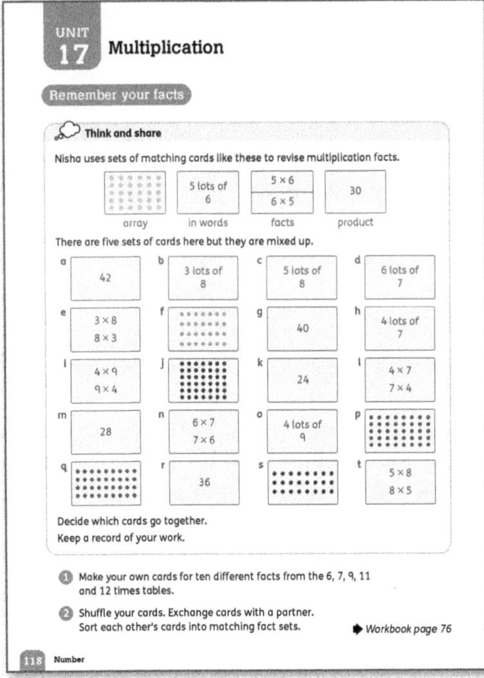

Materials

Card; scissors.

Warm-up

Revise *multiplication facts* and *division facts* up to 12 × 12.

Focus

The aim of this lesson is to revise the facts that the children already know so that they can use them to multiply larger numbers.

- Remind the children that there are different ways to represent the same multiplication. Make sure that they understand the terms *array* and *product*.
- Think and share: Turn to **Pupil Book 4 page 118**. Let the children work in pairs to discuss and then sort the cards in this activity.
- The children then work in pairs to complete question 1 and question 2. They each make their own set of cards, then exchange them with their partner and sort each other's sets.

Follow-up

Use **Workbook 4 page 76** for additional practice.

Support

Some children may still have difficulty recalling multiplication and division facts, particularly for the times tables that they have only learnt this year. Allow the children to use table grids as a memory aid and reference if necessary while practising recall of facts on an ongoing basis.

Answers for Pupil Book 4 page 118

Think and share: Set 1: a, d, j, n   Set 2: m, l, f, h
Set 3: r, q, o, I   Set 4: g, c, p, t   Set 5: k, e, s, b
1 and **2** Individual answers

Answers for Workbook 4 page 76

1 Possible answers:

1st row: 5 by 7 array	5 × 7 and 7 × 5	5 lots of 7	35 (provided as an example)
2nd row: 5 by 9 array	6 × 9 and 9 × 6	6 lots of 9	54
3rd row: 3 by 12 array	3 × 12 and 12 × 3	3 lots of 12	36
4th row: 9 by 7 array	9 × 7 and 7 × 9	9 lots of 7	63
5th row: 5 by 8 array	6 × 8 and 8 × 6	6 lots of 8	48
6th row: 4 by 10 array	4 × 10 and 10 × 4	4 lots of 10	40
7th row: 3 by 8 array	3 × 8 and 8 × 3	3 lots of 8	24
8th row: 6 by 4 array	6 × 4 and 4 × 6	6 lots of 4	24

Multiply tens

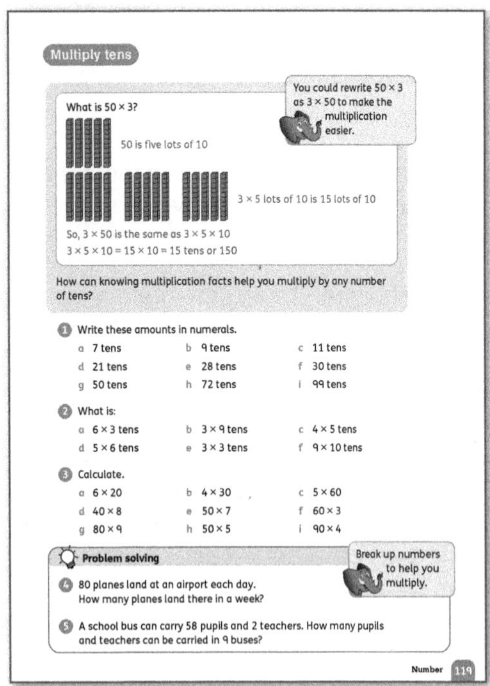

Materials
Sets of ten interlocking cubes or base-ten 10-rods
(or other apparatus showing tens – see pages 20–21).

Warm-up
- As a mental warm-up, display a number of
  multiples of 10, for example: 120, 80, 130, 230, 2000,
  1500, 20, 700.
- Point to a number and ask the children to express it
  as a number of tens. For example, 120 is 12 tens.
- Then say a number of tens and ask the children to
  say the number, for example, *26 tens* (*260*).

Focus
- Turn to **Pupil Book 4 page 119**. Read through the
  example at the top of the page with the class.
- Use interlocking cubes to do a number of practical
  examples with the class. Show them the group of
  tens (for example, 3 tens making 30) and ask: *What
  is 4 × 30?*
- Ask the class how knowing multiplication facts
  helped them calculate this. Accept all reasonable
  explanations. For example: 'I know that 4 × 3 is 12.
  4 × 30 is the same as 4 × 3 × 10, which is 120.'
  Explain that 120 is 10 times greater than 12.
- The children work independently to complete
  questions 1–3.
- <u>Problem solving</u>: The children should be able to do
  question 4 by writing 80 as 8 tens and applying what
  they have just learnt.
- The children can solve question 5 by adding 58 + 2
  to get 60 and then multiplying by 9. However, some
  children may work out 58 × 9 and 2 × 9 and add
  the products. If they do, discuss which method is
  most efficient.

Multiply larger numbers by 10

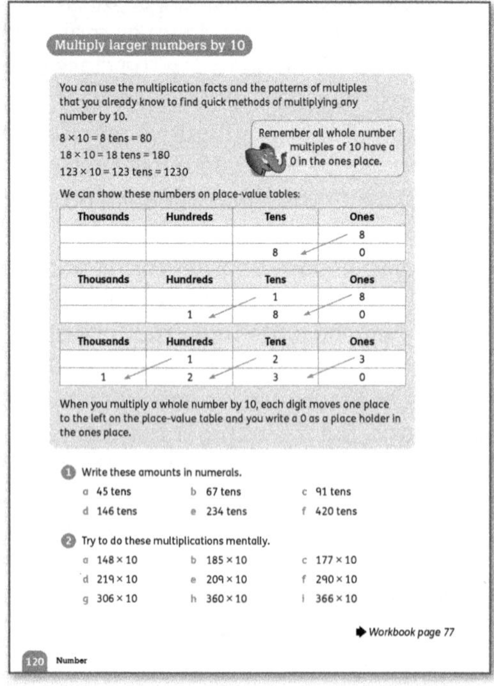

Materials
Place-value tables (see page 22) and counters.

Warm-up
- As a mental warm-up, revise adding and subtracting
  hundreds or thousands from any number. (The
  children will use this skill in the Workbook questions.)
- Ask questions such as: *Add 200 to 3456.* (*3656*)
  Subtract 2000 from 5436. (*3436*) *What is 500 more
  than 345?* (*845*) *What is 300 less than 4894?* (*4594*)

Focus
- Use place-value tables and the examples on
  Pupil Book 4 page 120 to revise and demonstrate
  multiplication by 10. Do a few more examples using
  other starting numbers if necessary. Ask questions
  such as: *What number is 10 times greater than 15?*
  (*150*) . . . *than 230?* (*2300*)
- Then, let the children complete question 1 and
  question 2 mentally if possible. If not, encourage
  them to use a place-value table and counters.

- Check the answers with the class. Examine any mistakes and let the children suggest what they did wrong and what they learnt from it.

Follow-up
Use **Workbook 4 page 77** for additional practice and to consolidate multiplying by 10 and multiples of 10.

Interesting mistakes
The rules for multiplying and dividing by 10 and multiples of 10 may confuse children who do not fully understand place value. Use manipulatives such as place-value cards, base-ten blocks and place-value tables to model calculations (see pages 21–22).

Answers for Pupil Book 4 page 120
1 a 450   b 670   c 910
    d 1460   e 2340   f 4200
2 a 1480   b 1850   c 1770
    d 2190   e 2090   f 2900
    g 3060   h 3600   i 3660

Answers for Workbook 4 page 77
1 a 50, 500   b 230, 2300
    c 770, 7700   d 390, 395, 3950
    e 1210, 210, 2100   f 4260, 260, 2600
    g 5090, 90, 900
2 5000 g sugar, 40 eggs, 6500 g flour, 2500 ml milk, 1750 ml vegetable oil, 120 g baking powder, 50 ml vanilla

Multiply by 100

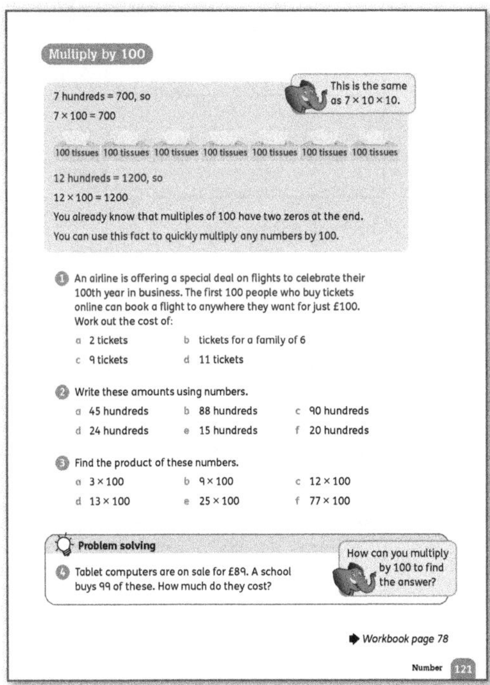

Materials
Place-value cards and place-value tables (see page 22).

Warm-up
As a mental warm-up, do some counting in steps of 100 with the class.

Focus
- Look at examples of items that are packed in hundreds on **Pupil Book 4 page 121**. Ask the children to suggest other things that come packed in hundreds (for example: cotton balls, ear buds, paperclips, freezer bags, stickers, thumbtacks, hairpins, boxes of chalk, paper napkins, tea bags, elastic bands). The children can find as many examples as possible in the classroom or at home.
- You could do a class display project of items that come packed in hundreds or in measurements of 100 (100 ml bottles, 100 g packets of spices and so on).
- Discuss how many items would be in 2, 3 or 5 packs. Let the children share their ideas for working this out. For example, *If there are 100 teabags in one box, I'd double it to find out how many there are in two boxes and get 200; or two boxes means 2 hundreds, so there would be 200.*
- Work through the concept of a number of hundreds using the example on **Pupil Book 4 page 121** and the children's own examples.
- Point out that when we have pictures like the ones in the Pupil Book, we can skip count in hundreds to find the total. However, if there are many items, (for example, 315 packets of beads with 100 beads in each packet), it makes sense to think about the number of hundreds to work out the product (total). For this example, you would think '315 hundreds is 315 × 100 which is 31 500'.
- It is important to use language such as *12 hundreds* so the children realise that 1200 is the same as 12 hundreds. Reinforce the idea using place-value cards and tables if necessary.
- Ask questions such as: *What number is 100 times greater than 34? (3400) … than 9? (900)*
- Let the children try as far as possible to use mental methods to complete questions 1–3 on **Pupil Book 4 page 121**.
- <u>Problem solving</u>: The children should be able to do question 4 by multiplying 89 by 100 and then subtracting 89 from 8900. Continue to encourage the children to try as far as possible to use mental methods.

Follow up
- Use the measurement conversion activities on **Workbook 4 page 78** to consolidate multiplying by 100 in different real-life contexts.
- <u>Problem solving</u>: In question 3, make sure the children understand that '10 stitches per centimetre' means '10 stitches in every centimetre'. When they have answered the question, discuss their strategies.

Challenge

Give the children a number of cards with products written on them. Include units of measure and context clues on some of the cards. Ask the children to make up a very simple and a very difficult word problem for each product.

Here are some examples to include:

£2600   4200 litres   300 kilometres further
600 kilograms   400 grams of sugar
500 millilitres of medicine

Let the children share the problems they made up with the class, showing the solution. Ask them to say what makes some problems easy to solve and some more difficult.

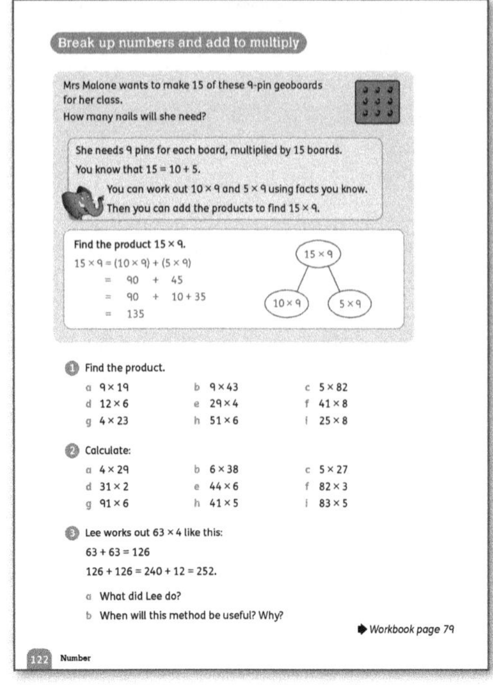

Answers for Pupil Book 4 page 121

1 a £200   b £600   c £900   d £1100
2 a 4500   b 8800   c 9000   d 2400
   e 1500   f 2000
3 a 300   b 900   c 1200   d 1300
   e 2500   f 7700
4 £8811

Answers for Workbook 4 page 78

1 a 400   b 3400   c 7300   d 9900
2 a 200   b 300   c 2700   d 2300
3 3200 stitches

Break up numbers and add to multiply

Materials

Calculators.

Warm-up

Select any suitable 'Place value and number sense' activity from pages 23–27 as a mental warm-up. The children will be *partitioning* numbers into tens and ones in this lesson.

Focus

- Turn to **Pupil Book 4 page 122** and read the problem about the geoboards at the top of the page.
- Ask the children to work out the product of 15 and 9 on their own, using jottings.
- Then give different children the chance to share their methods. Show these visually as they do this. For example, a child might say: 'I doubled 15 to get 30. Then I multiplied $3 \times 9 \times 10$ to get 270 and I halved that to get the answer'.

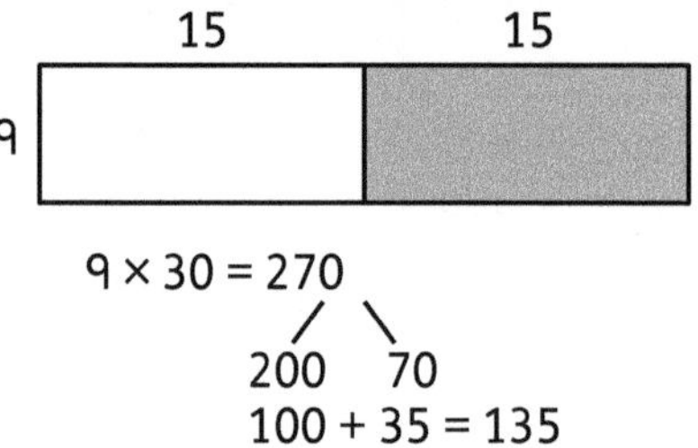

$$9 \times 30 = 270$$
$$200 \quad 70$$
$$100 + 35 = 135$$

- Other methods might include:
  - splitting 15 into fives

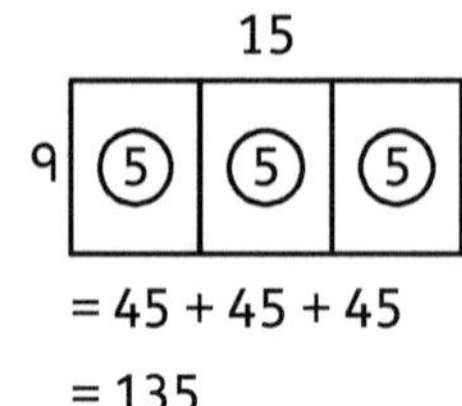

$$= 45 + 45 + 45$$
$$= 135$$

  - splitting 9 into threes

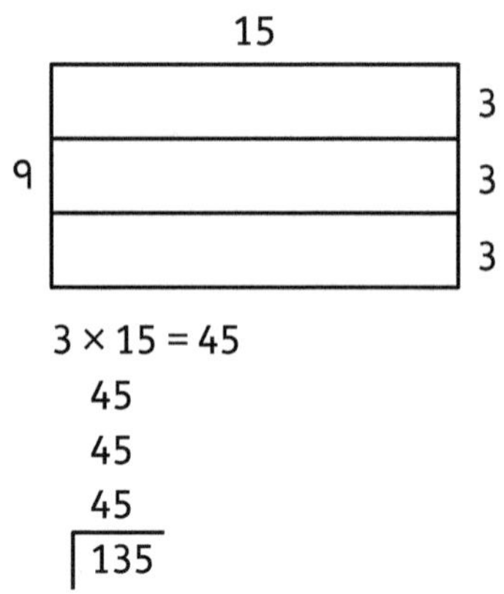

$$3 \times 15 = 45$$
$$45$$
$$45$$
$$45$$
$$\overline{135}$$

  - splitting 15 into 10 and 5

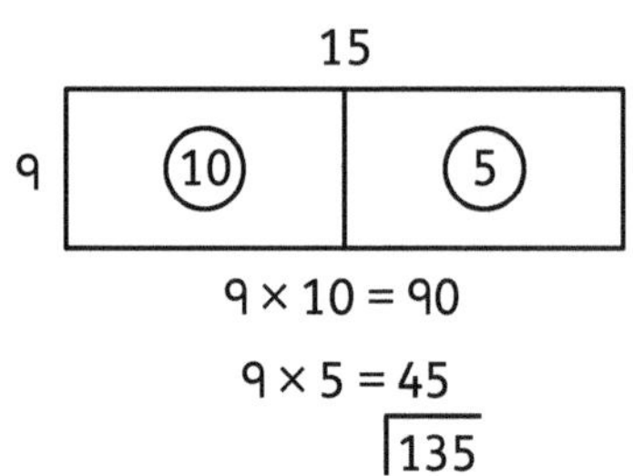

$$9 \times 10 = 90$$
$$9 \times 5 = 45$$
$$\overline{135}$$

◦ adding 1 to multiply by 10 then subtracting.

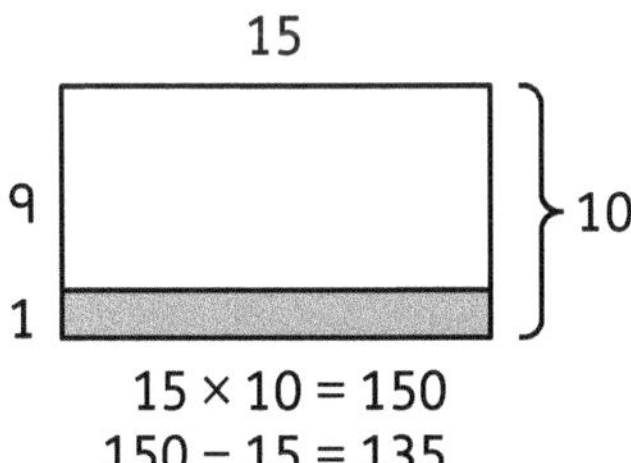

$$15 \times 10 = 150$$
$$150 - 15 = 135$$

- The important point of this discussion is that there are different ways of thinking about numbers and the way we think about them often affects how we calculate with them.
- Point out that one method of multiplying numbers is to split them into parts (partition them). The children used this method with small numbers in Unit 12 using the distributive property of multiplication.
- Explain that partitioning numbers using place value is a very useful strategy for multiplying by a single-digit number because then we can use times tables to work out the product in parts.
- Ask the children to turn to the example of 15 × 9 on **Pupil Book 4 page 122**. Let them work through this in pairs and ask them to say how this is similar to or different from how they worked.
- Ask the children to try partitioning to solve question 1 and to draw a part–whole diagram to show their working. Let them use a calculator to check their own answers.
- Let the children work independently to find the product in question 2. They don't need to draw a diagram, but can if they want to. They may also be able to do some of the calculations without partitioning the numbers.
- Discuss question 3 as a class. Ask the children how they would calculate the product.

Challenge

Use **Workbook 4 page 79** as a challenge to encourage the children to think about patterns and properties of numbers. Have a class or group discussion so that the children can share their ideas.

Answers for Pupil Book 4 page 122

1 a 171   b 387   c 410
   d 72   e 116   f 328
   g 92   h 306   i 200

2 a 116   b 228   c 135
   d 62   e 264   f 246
   g 546   h 205   i 415

3 a Possible answer: He worked out that double 63 was 126 by adding and because doubling twice is the same as multiplying by 4, he added 126 and 126 to find the answer. He broke up 126 and 126 to make them easier to add.
   b Individual answers

Answers for Workbook 4 page 79

1 a 66; 156; 216; 36; 96; 186
   b 174; 354; 474; 114; 234; 414

2 a 200; 500; 700; 100; 300; 600
   b Each digit moves one place to the left and you put a 0 in the ones position
   c 50; 110; 150; 30; 70; 130; when you double a number ending in 5, you get a number ending in 0

3 a 104; 224; 304; 64; 144; 264
   b It is a 4 because 4 × 6 = 24.

4 a (provided as an example)
   b 7 × 48; 280 + 56; 336
   c 9 × 80; 720 + 54; 774

Use written methods to multiply

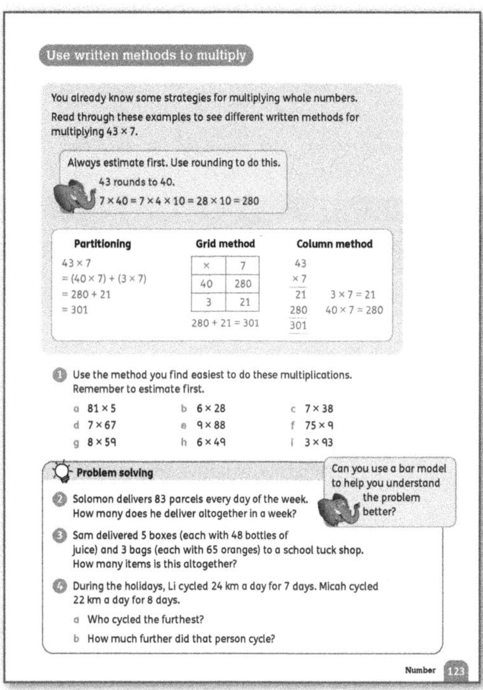

Warm-up

As a mental warm-up, write sets of three related numbers on the board. For example, write 45, 9 and 405. Let the children make all the possible multiplication and division calculations using the numbers. (45 × 9 = 405, 9 × 45 = 405, 405 ÷ 9 = 45, 405 ÷ 45 = 9)

Focus

- Spend some time working through the different methods of finding 43 × 7 shown on **Pupil Book 4 page 123**: partitioning, *grid method* and *column method*.
- Discuss the methods as you work and encourage the children to suggest any other methods they use.
- Let the children work on their own to complete question 1. Remind them to *estimate* first.
- <u>Problem solving</u>: The children should be able to solve the problems in questions 2–4 on their own. Check their answers and discuss the methods they used.

Challenge

Challenge the children to write their own worded multiplication problems. Write the criteria for the problems on the board. For example: it must be solved by multiplying a 2-digit number by a 1-digit number; it must have a product less than 400; it must result in a product that is a multiple of 5. Let the children write a word problem to fit each criterion.

Support

The children may find the more compact column methods difficult if they do not understand how they work. Remind the children to have a conversation in their heads about what they are doing and model this when you work through examples by telling the class what you are doing at each stage of the working.

Use written methods for 3-digit numbers

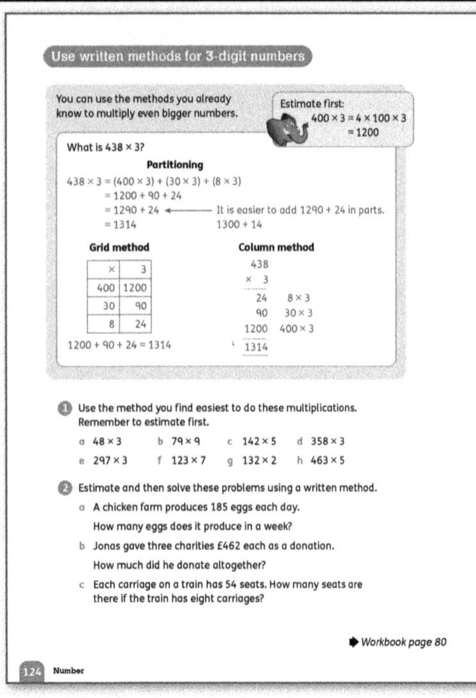

Materials

Calculators; squared paper.

Warm-up

Choose any suitable 'Rounding and estimating' activity from pages 26–27 as a mental warm-up.

Focus

- Before showing the children methods for multiplying 3-digit numbers by 1-digit numbers, ask the children how they would do this. Let them share their ideas and demonstrate their methods for the others, either as a whole-class session or in smaller groups.
- Stress that they already have a number of ways of working and that all are adaptable to multiply larger numbers.
- Work through the examples on **Pupil Book 4 page 124**, relating the methods to what they have already done as you go.
- Ask the children to complete question 1 and question 2 on their own. They can use a calculator to check their own answers.
- If they make a mistake, ask them to find it and circle it. Make time to talk about what they can learn from the mistakes with the class.
- Once the children can use formal written methods successfully, there is little value in them simply doing more of the same kind of calculation. Instead, move towards more challenging and problem-solving activities in which they apply the method to solve a problem.

Follow-up

Let the children discuss and then solve the questions on **Workbook 4 page 80**. These problems involve thinking and reasoning rather than rote completion of a set of calculations. Make time in class to talk about how children worked, particularly to find the missing digits in question 2.

Support

If any children find it difficult to keep numbers in the correct places, one below the other, when they use column methods, give them squared paper to work on. If necessary, they can label the places.

Challenge

Prepare a grid of nine products. Then provide some 3-digit factors and 1-digit factors and let the children find and write the multiplication that matches each product. They need to use estimation skills as well as number sense and reasoning to help them rule out options.

For example:

The grid contains the products of nine different 3-digit by 1-digit multiplications.

1164	1552	2457
3256	1365	2884
2060	3704	5621

Factors:

194	273	407	412	463	803
5	6	7	8	9	

Use the factors given above to work out what each multiplication could be. Write each multiplication above its product in the grid.

Interesting mistakes

The children may make mistakes in calculations due to mental arithmetic errors, such as incorrect times tables or errors in addition or subtraction. Encourage them to estimate first and then to check their work to decide whether their answer is reasonable or not.

> ### Answers for Pupil Book 4 page 124
> **1** a 144   b 711   c 710   d 1074
>   e 891   f 861   g 264   h 2315
> **2** Accept any reasonable estimates. The calculated answers are:
>   a 1295   b £1386   c 432

> ### Answers for Workbook 4 page 80
> **1** a (provided as an example)
>   b 1800 + 240 + 18 = 2058
>   c 2800 + 0 + 36 = 2836
> **2** a 805 × 3 (provided as an example)
>   b 803 × 5
>   c 850 × 3
> **3** a–c Individual answers

Multiply and solve

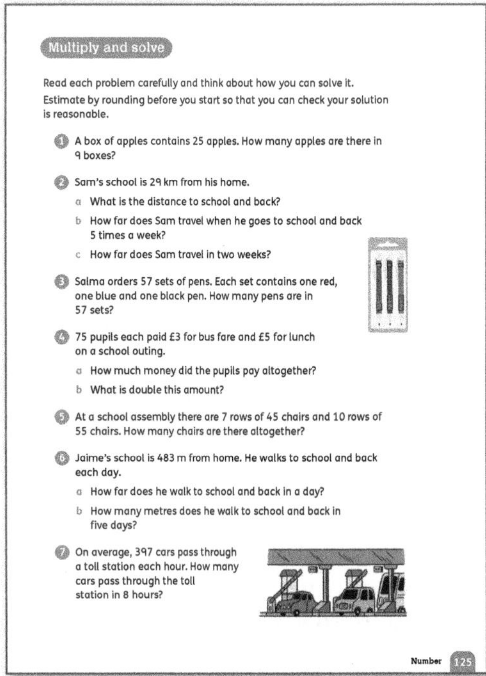

Warm-up

As the children will be estimating again in this lesson, use 'Which number?' on pages 26–27 as a mental warm-up.

Focus

There are no new concepts in this lesson. The children are expected to apply what they have learnt to solve problems.

- Turn to **Pupil Book 4 page 125**. Allow time for reading and discussing what is required in questions 1–7 before asking the children to solve these problems.

Support

For children who find word problems difficult, spend some time reading the problems and asking them to find and highlight words that tell them what they need to do and what they need to find out. Continue to encourage them to draw diagrams to represent problems.

Interesting mistakes

Children often just write the number in response to a word problem. Remind them that they are being asked a question about something and that their answer should contain units or items so that it is clear that they have understood the problem and given an answer related to it.

> ### Answers for Pupil Book 4 page 125
> **1** 225 apples
> **2** a 58 km   b 290 km   c 580 km
> **3** 171 pens
> **4** a £600   b £1200
> **5** 865 chairs
> **6** a 966 m   b 4830 m
> **7** 3176 cars

Think and solve

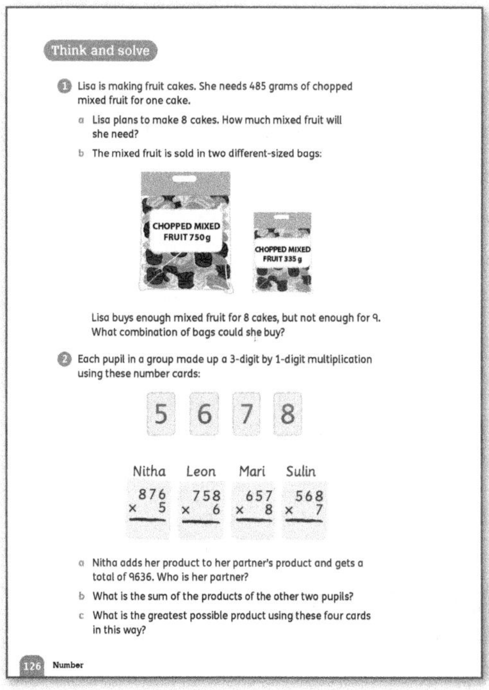

Warm-up

Use 'The 10 questions game' on page 25 as a warm-up.

Focus

- Question 1 and question 2 on **Pupil Book 4 page 126** allow the children to consolidate multiplication and to apply their problem-solving skills in more complex situations.
- Discuss each problem with the class, asking them to share their ideas for how they would solve it. Then

ask the children to work out the answers. Spend some time afterwards talking about what they found easy and what they found challenging. Discuss what they learnt by solving these problems.

Answers for Pupil Book 4 page 126

1 **a** 3880 g
  b Any answer(s) from these combinations:

Big bags	Small bags
0	13
0	12
1	10
2	8
3	5
3	6
4	3
4	4
5	1

2 **a** Mari   **b** 8524   **c** 6120

End-of-unit check

Use some or all of these questions and activities to assess how well the children can multiply and use multiplication to solve problems.

- *What is* (give a number) *multiplied by 10 (or 100)?*
- *Describe an easy way of multiplying by 50 (or 70 or 80). (Multiply by 5 (or 7 or 8) and then multiply by 10.)*
- Show the children some completed multiplications, some with mistakes. *Which of these calculations are correct/incorrect? What has the person done wrong in this calculation? How could you help them to correct it?*
- *Order these calculations, from greatest to smallest product: a 712 × 3, b 507 × 6, c 414 × 9. (c (3726), b (3042), a (2136))*
- *Make up a calculation d which will be fourth in the sequence. (any multiplication with a product less than 2136)*
- Give the children a multiplication word problem. *What clues in this word problem can help you decide what to do to solve it? (For example: Words such as product, altogether and lots of often mean multiply.)*
- *Write a word problem that can be solved by multiplying a 1-digit number by a 3-digit number.*

UNIT
18

UNIT 18 Work with line graphs

Learning objectives

- Interpret and present data using time graphs.
- Solve comparison, sum and difference problems using data presented on time graphs

Key words
line graph   axis/axes   horizontal   vertical   value

Unit introduction

The children have not previously worked with line graphs in maths although they may have seen and used these in science or social studies. These graphs are generally used to show numerical data that changes over a period of time. The data points are plotted and then joined with straight lines.

Teaching guidance

- Use a local weather graph, called a meteogram, to introduce the idea of a *line graph* and to show how line graphs are different from bar charts. You can find very good graphs online with accurate data for most places in the world. An example is shown at the top of the next page.
- Display the meteogram for the class and talk about what it shows.
- Note that this is a weather prediction, so it will be interesting to display the most recent version for your area and to then check its accuracy over the next two days.
- The grey bars show the rainfall expected at different times during the next 48 hours. The amount in millimetres is shown above the bars.
- The time is shown on the *horizontal axis*, using 24-hour notation. The line beneath the weather icons shows the predicted temperature at different times over the next 46 hours.

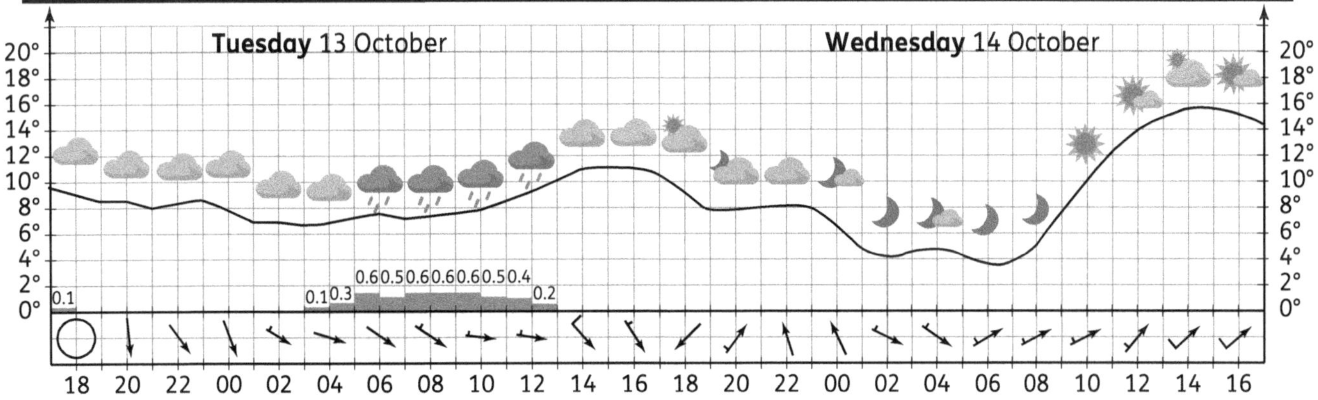

- Temperature is read from the *vertical* scale, which is marked in degrees.
- The slope and direction of the line gives information about how the data changes. When the line beneath the weather icons slopes up it means that the temperature is rising, when it slopes down, it means that the temperature is decreasing.
- In statistics, line graphs are normally used to show measurements such as temperature or height (continuous data) and, in these cases, all points on the line have a *value*.
- Where a line graph is used to show discrete data, such as the number of children absent each day for a week, only the points plotted for each day have a value. In statistical terms, such data would be better represented as a bar or bar line graph.
- If the children are interested, let them choose different places in the world and download the meteograms. They can then compare the data shown on these graphs in groups.

Line graphs

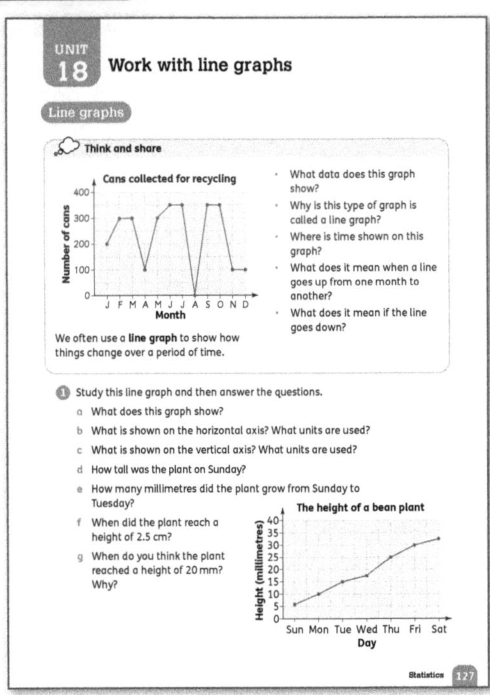

Warm-up

Use any mental 'Calculation skills' activities that involve addition and subtraction from pages 28–29 as a warm-up. The children will be working with the sum and difference of data.

Focus

- <u>Think and share</u>: Let the children look at the graph on **Pupil Book 4 page 127** and answer the questions in small groups. They should be able to do this confidently if you discussed line graphs as part of the unit introduction.
- Take feedback to make sure that the children answer correctly. Then point out the features that all line graphs need: title, axis titles, axis labels, time on the horizontal axis, clearly plotted points joined with straight lines. As you work through these, let the children find them on the graph.
- The children can answer the questions in question 1 orally in groups or they can write their answers and you can check them as a class.

Answers for Pupil Book 4 page 127

<u>Think and share:</u> Possible answers: How many cans were collected over 12 months. It is a line graph because it uses a line to show how things change over time. The time is shown on the horizontal axis. When the line goes up there is an increase, when the line goes down there is a decrease.

1 a The height of a bean plant on different days of the week

 b days   c height in mm   d 6 mm

 e 9 mm   f Thursday

 g on Wednesday; the line passes through 20 mm between Wednesday and Thursday

Draw and interpret line graphs

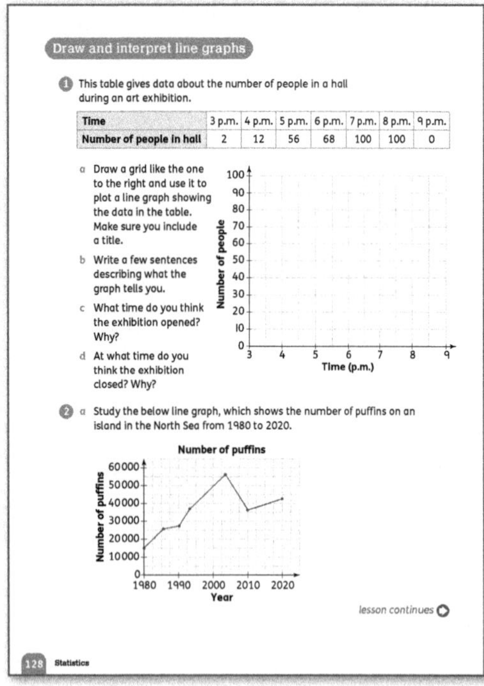

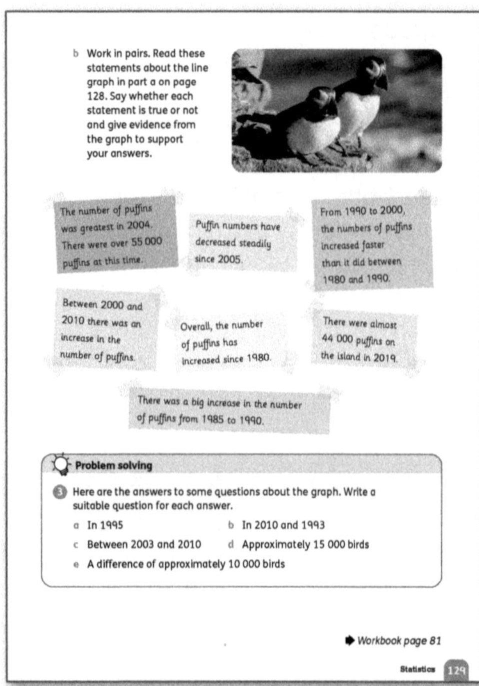

Before this lesson, if possible, record the temperature in the classroom each day for a week using a thermometer and record the data. Use this to draw a line graph to show the class how this is done. If this isn't possible, use the data from the table on **Pupil Book 4 page 128** to demonstrate how to do this.

Materials

A thermometer; graph paper or squared paper; simple line graphs (search online).

Warm-up

Use any mental 'Calculation skills' activities that involve addition and subtraction from pages 28–29 as a warm-up. The children will be working with the sum and difference of data.

Focus

- Turn to **Pupil Book 4 pages 128–129.** Work through question 1 step-by-step with the class. Start by looking at the data with the class and ask:
  - *What title could we use for this graph? (People at an art exhibition) Why?*
  - *What information is on the horizontal axis? (Time of day) What scale has been used? (Two squares represent 1 hour.)*
  - *What information is on the vertical axis? (Number of people) What scale has been used? (One square represents 10 people.)*
- Next, provide graph paper or squared paper for the children to draw and label the *axes* as discussed. Then talk about how to plot the points. Make sure that the children see how they need to move up from the time and across from the number of people to plot each point. Remind them that this is similar to plotting coordinates on the grid, but here the coordinates are time and number of people.
- Let the children complete the graph and compare their final versions to check that they are correct.
- The children then use their completed graphs to answer the questions in question 1 parts b–d.
- For question 2, the children will use thinking and reasoning skills to read and interpret a graph based on a real-world situation involving population change over time.
- The children can work in pairs to decide whether the statements in question 2b are true or not. They must find evidence from the graph to support their decisions. Once they have done this, read out each statement and invite the children to share their answers.
- <u>Problem solving</u>: The children can work on their own to do question 3 or they could discuss and write questions in their groups or pairs.

Follow-up

Use **Workbook 4 page 81** to check that the children can draw a line graph accurately and use data from a line graph to answer questions.

Challenge

Present the children with a set of different scenarios related to data and time. Let them draw rough sketches to show the general shape that a line graph would have for each scenario. Once they have done this, ask them to exchange graphs and write a different scenario to match each graph their partner has drawn. Here are some possible scenarios:

- The mass of a baby increases quickly in the first few weeks of its life, then it stays the same for a little while before increasing at a slower rate than before.
- The temperature in a classroom is very high at lunch time. It decreases steadily over the course of the afternoon.
- A tank filled with water has a hole in it and the water leaks out slowly over a period of a few hours. Once the water gets below the level of the hole, it stops leaking and stays at the same level.
- The temperature of water in a kettle rises steadily as it gets to boiling point. Once it has boiled, the kettle switches off and the water slowly starts to cool.
- The temperature of an ice block taken from the freezer rises steadily as the ice melts and turns to water. The water temperature remains at room temperature.

Let the children choose an endangered species from your country (or elsewhere) and do some research to find out how numbers of the species have changed over time. They can draw a graph to show the results of their research. Rhinoceros numbers are quite useful as conservation efforts in Africa led to an increase in the number of animals, but these gains are now threatened by poaching.

Support
To encourage the children to think about graphs and what they show us, ask the children to make up five questions about a simple line graph. (You can find many examples online.) Let them write the answers to their questions as well. Use their questions to assess their level of understanding of the graph and how it works.

Interesting mistakes
Some children may forget to label graphs or fail to see the importance of labelling their graphs. If so, present them with some unlabelled graphs and ask questions about them. Once they realise it's impossible to answer, they are likely to realise the importance of labelling.

Answers for Pupil Book 4 pages 128–129

1 a The number of people in a hall during an art exhibition

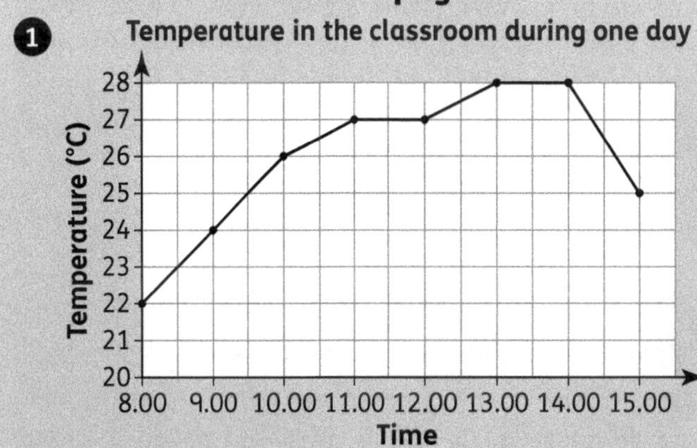

b Possible answers: The graph tells the number of people at the exhibition between 3 p.m. and 9 p.m. There was no on in the hall at 9 p.m.

c 3 p.m.; there were only 2 people at 3 p.m.

d 9 p.m.; because there were no people at 9 p.m.

---

2 (purple statement) True. The highest point on the graph is over 55 000 puffins in 2004.

(blue statement) False. From 2010 to 2020 the number of puffins increased.

(pink statement) True. Between 1990 and 2000 the graph is steeper than between 1980 and 1990.

(first green statement) False. There was only an increase between 2000 and the highest point in 2004.

(yellow statement) True. In 1980 there were 1500 puffins and in 2020 there were over 40 000, which is an increase.

(second green statement) False. The final point on the graph shows almost 44 000 puffins in 2020 (not 2019).

(orange statement) False. There was only a small increase from 1985 to 1990.

3 Possible questions:
   a In which year were there about 40 000 puffins?
   b In which two years was the number of puffins about the same? / In which two years were there about 37 000 puffins?
   c Between which years did the number of puffins decrease?
   d How many puffins were there in 1980?
   e What is the difference between the number of puffins in 1990 and in 1993?

Answers for Workbook 4 page 81

1 Temperature in the classroom during one day

2 a 26 °C
   b It stayed the same.
   c Because the temperature dropped.
   d 9.30 a.m. because when the graph line is at 25 °C the time is halfway between 9.00 and 10.00.

Use line graphs to answer questions

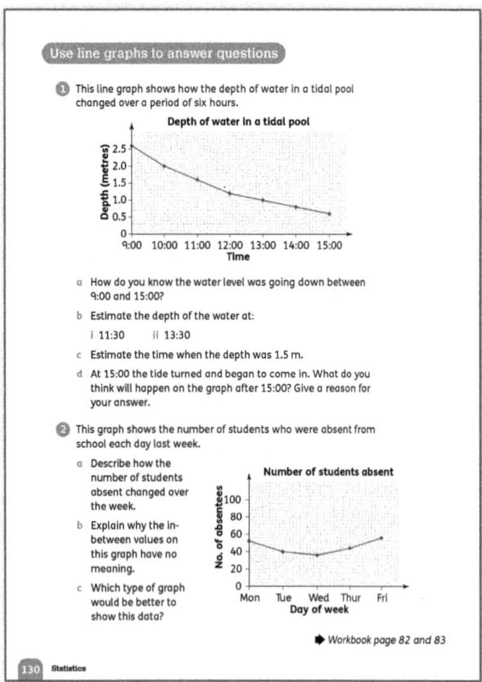

Warm-up
Revise counting in decimal steps, for example steps of 0.1 or 0.5, as a mental warm-up.

Focus
- Use **Pupil Book 4 page 130** to consolidate the work on reading and interpreting graphs.
- Ask the children to work independently to answer the questions in question 1 and question 2.

Follow-up
Use **Workbook 4 pages 82–83** to provide further practice.

Answers for Pupil Book 4 page 130
1 a The line slopes downwards, which shows the depth of the water was decreasing.
  b i 1.4 m   **ii** 0.9 m
  c 11:20
  d It will slope upwards as the depth of the water will increase as the tide comes in.
2 a Possible answer: More students were absent at the start and end of the week, and the fewest were absent on Wednesday.
  b Possible answer: There is nothing 'in between' Monday and Tuesday; you can only record the number of students absent on Monday and the number absent on Tuesday.
  c a bar chart

Answers for Workbook 4 page 82
1 a (1 month) 53 cm; (2 months) 57 cm; (3 months) 60 cm; (4 months) 62 cm; (5 months) 64 cm; (6 months) 66 cm
b and g

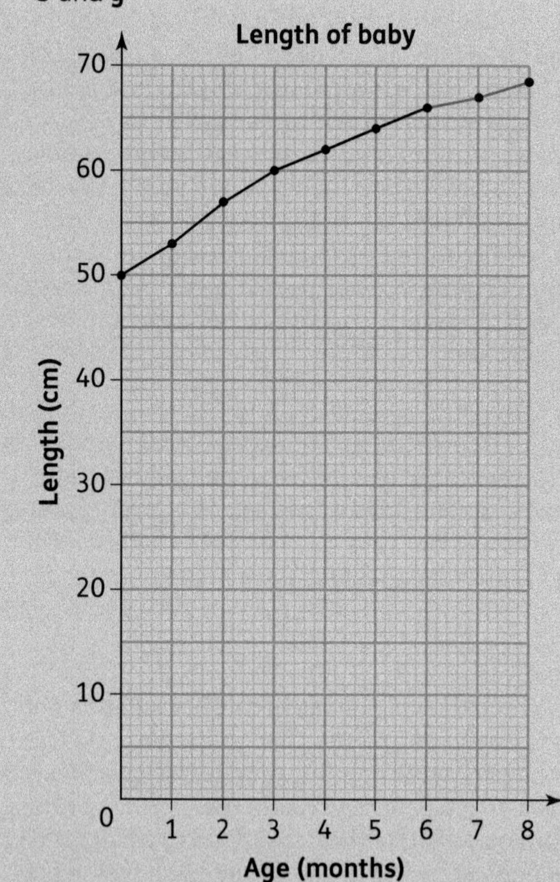

  c 4 cm   **d** 7 cm
  e Between month 1 and month 2
  f The baby grew 4 cm between month 4 and month 6.
  g 68.5 cm

Answers for Workbook 4 page 83
1 a

Week	Time (s)
1	80
2	72
3	68

b

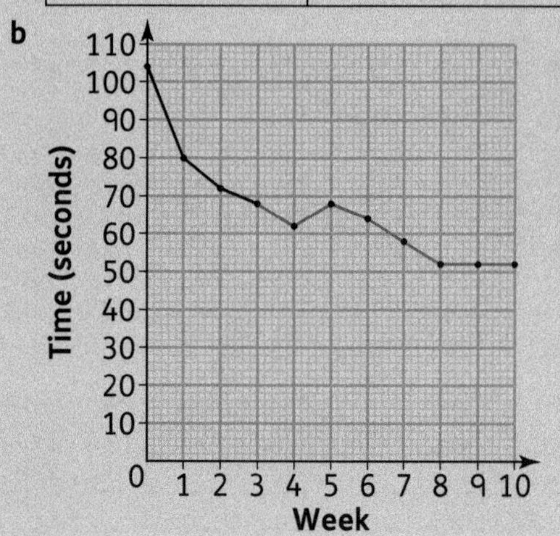

2 a Possible answer: Anjali made great progress in the first week, when she improved her time by 24 seconds, then her time improved steadily until week 4. In week 5 her time got a little bit worse, then her time improved steadily again until week 8. In weeks 8 to 10, her time remained the same.

b 8 weeks   **c** It was 28 seconds faster.

d Possible answer: No because there is a limit to how much more she can improve by. She has already reached a low time and that has remained the same for three weeks.

End-of-unit check

Use real graphs from this unit or elsewhere and ask the children some of these questions about them to check their understanding of the concepts in this unit.

- Show a line graph.
  - *What type of graph is this?*
  - *What is the title of this graph?*
  - *What scale is used on the vertical axis?*
  - *What does this graph show?*
- Show a line graph about class absences in a week. *How many children were absent on Tuesday?*
- Show two graphs with different scales. *How are these two graphs similar/different?*
- Show a line graph with a line sloping down to the right. *Why does the line on this graph slope down to the right? (As the values on the horizontal axis increase, the values on the vertical axis decrease.)*
- Show a graph with one or more errors, missing labels, and so on. *What is wrong with this graph?*

You can also give the children a table of data and ask them to use it to draw a line graph. An interesting graph is produced by graphing the height to which a ball bounces when dropped from a height.

The ball bounces a few times until it comes to a stop. Each time it bounces, it rises up less than the previous time. This set of data produces a rising and falling pattern. It includes decimal values, so the children will need to think carefully about the scale.

Ask: *Draw a graph of this data for a ball that is dropped from a height. Describe what happens.*

Height of ball (m)	2	0	0.9	0	0.5	0	0.2	0
Time (seconds)	0	0.6	1.1	1.5	1.8	2.2	2.4	2.6

UNIT 19 Patterns

Learning objectives

- Count on and back in given steps, and extend beyond zero to include negative numbers
- Use objects to represent numbers
- Recognise the term-to-term rule and extend linear and non-linear number sequences
- Recognise and extend the spatial pattern of square numbers

Key words

sequence   term-to-term rule   term
square number

Unit introduction

This unit offers the opportunity to explore patterns around us. Pattern is one of the key concepts in mathematics.

Patterns can help children to learn mathematics, and mathematics often results in patterns that they didn't know about or expect.

Patterns can be found anywhere. The key element of patterns, whether they are number patterns or shape patterns, is some kind of repetition.

Teaching guidance

- Let the children choose different types of patterns to research and present. They could consider patterns in nature, art, architecture, fabrics, tiles, flooring, engineering and structures. If they are more interested in numbers, they could look at different number patterns (Fibonacci numbers are a good example for this level.)
- The children can work on their own, in pairs or in small groups to find patterns in the area they have chosen and then make a poster or other display to show them patterns and describe how they work.

Investigate patterns and rules

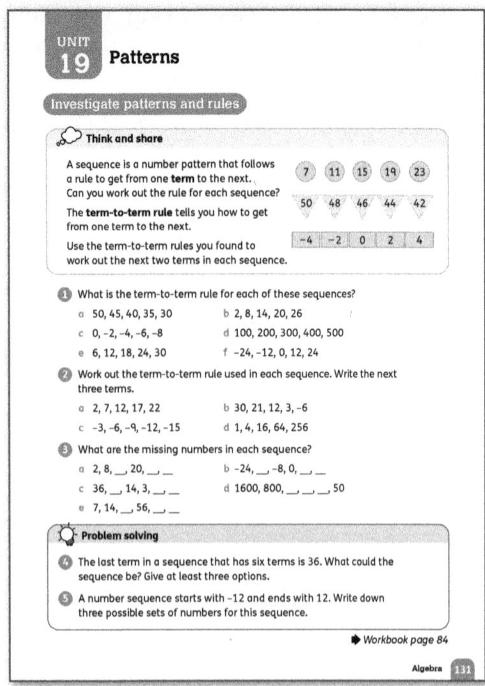

Warm-up

As a mental warm-up, give the children a starting number and ask them to count on or back from that number in given steps. Include some negative numbers.

Focus

The children already know what a number *sequence* is and they worked with sequences and rules in earlier levels.

- <u>Think and share</u>: Use the activity at the top of **Pupil Book 4 page 131** for revision of number sequences. Let the children work on the activity in groups. Then ask them to share their *term-to-term rules* and to give the next two *terms* in each pattern.
- Do question 1 orally with the class. Read each sequence as the children follow it in their books and give them time to think about the term-to-term rule before asking different children for their answers.
- The children can discuss the sequences in question 2 to identify the rules before applying them to find the next three terms.

- Let the children work independently to find the missing terms in question 3. Next, let them explain their thinking to a partner. This is a key step in the process because often their partner will have worked or thought in a different way, so discussion exposes the children to different approaches.
- <u>Problem solving</u>: Question 4 and question 5 are open-ended problems with a number of different solutions. Let the children start on their own and then encourage them to share their ideas with their groups.

Follow-up

Ask the children to complete **Workbook 4 page 84** and check their answers to make sure that they can identify and describe number sequences.

Answers for Pupil Book 4 page 131

<u>Think and share:</u> Possible answers: The rule for the first sequence is, add 3. The next two terms are 26 and 29.
The rule for the second sequence is, subtract 2. The next two terms are 40 and 38.
The rule for the third sequence is, add 2. The next two terms are 6 and 8.

1 a subtract 5   b add 6   c subtract 2
   d add 100   e add 6   f add 12

2 a add 5 27, 32, 37   b subtract 9 –15, –24, –33
   c subtract 3; –18, –21, –24
   d multiply by 4; 1024, 4096, 16 384

3 a 14 26 32   b –16 8 16   c 25 –8 –19
   d 400 200 100   e 28 112 224

4 Possible answers:
Add 6: 6, 12, 18, 24, 30, 36
Add 9: – 9, 0, 9, 18, 27, 36
Subtract 3: 51, 48, 45, 42, 39, 36

5 Possible answers:
Add 3: –12, –9, –6, –3, 0, 3, 6, 9, 12
Add 4: –12, –8, –4, 0, 4, 8, 12
Add 6: – 12, –6, 0, 6, 12

Answers for Workbook 4 page 84

1 a 154, 158, 162   b 135, 140, 145
   c 200, 240, 280   d 75, 78; 90, 93
   e 154, 151; 139, 136

2 a rule: add 5 (provided as an example); 145, 150, 155
   b rule: add 100; 820, 920, 1020
   c rule: subtract 100; 599, 499, 399
   d rule: subtract 10; 139, 129, 119

3 a 523, 623, 723, 823   b 315, 305, 295, 285
   c 180, 175, 170, 165   d 280, 330, 380, 430
   e 8, 4, 0, –4

Number patterns

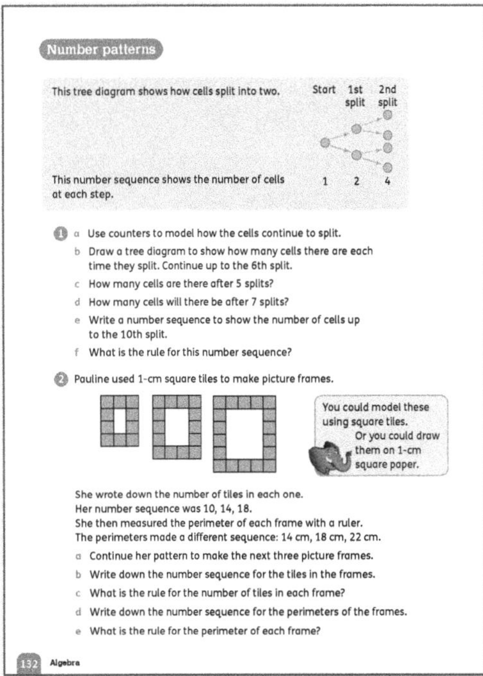

Materials

Counters; small tiles or cubes to model picture frames or 1-cm squared paper to draw them.

Warm-up

Select suitable 'Counting backwards and forwards' activities from page 23 as a warm-up.

Focus

- Turn to **Pupil Book 4 page 132**. Look at the diagram with the class. Discuss what it shows, making sure that the children realise it starts with one cell and that the first pair of branches (2) is the first split. (This is an example of a non-linear pattern. If you graphed this pattern it would not produce a straight line, rather it would make a curve that slopes up quickly (an exponential curve).)
- Hand out counters to the children and let them model the pattern in question 1. Remind them that the cell isn't split to start with, the first split is when it breaks into two.
- For question 2, the children can use small tiles or cubes to model the frames but they could also draw these on 1-cm squared paper.

Answers for Pupil Book 4 page 132

1 a Individual answers

  b A tree diagram showing the following:
   First split, 2 cells   Second split, 4 cells
   Third split, 8 cells   Fourth split, 16 cells
   Fifth split, 32 cells   Sixth split, 64 cells

  c 32   d 128

  e 2, 4, 8, 16, 32, 64, 128, 256, 512, 1024

  f Multiply by 2

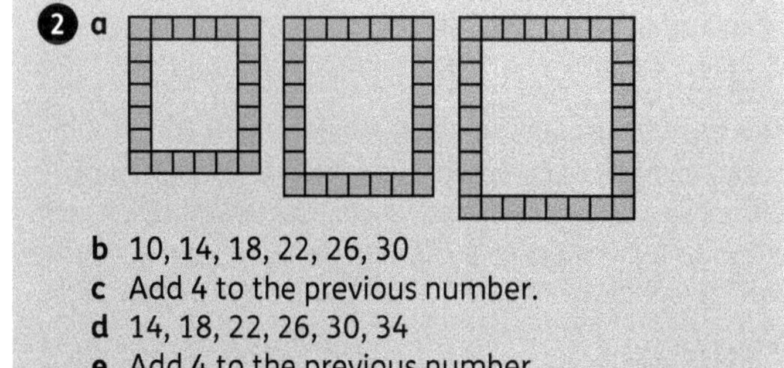

Square numbers

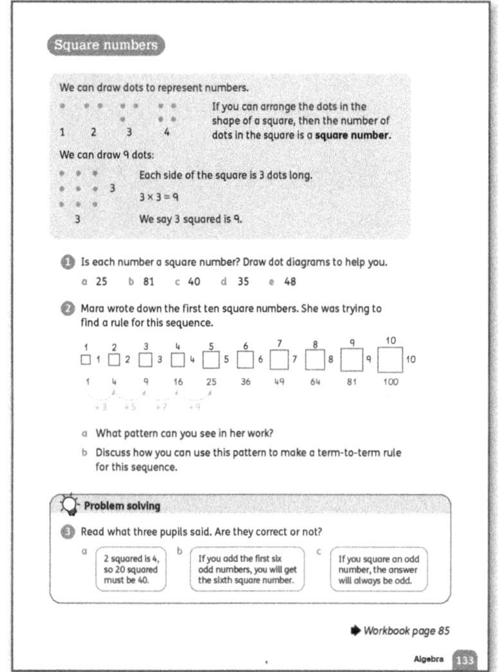

Materials

Counters; 12 × 12 multiplication grid (**Workbook 4 page 54**).

Warm-up

Select suitable 'Calculation skills' activities from pages 28–29 as a warm-up.

Focus

- Use counters to model *square numbers*. Give each group a handful of counters and let them arrange them to make squares.
- Note that some children might do arrays and others might arrange the counters around the perimeter of a square. Both versions will give the same square number when you consider area.
- Turn to **Pupil Book 4 page 133**. Look at the counters shown and ask the children which show square numbers. Make sure that they recognise that 1 is a square number. Remind them that when they worked with area, 1 cm × 1 cm gave them 1 square centimetre.
- Ask the children to predict which numbers in question 1 will be square numbers before they model them by drawing dot diagrams.
- Give the children time to consider and then discuss the pattern shown in question 2. The difference between terms is not constant, so this is also a non-linear pattern.

- Problem solving: Let the children do calculations to test their ideas, and then explain their thinking.

Follow-up
Use **Workbook 4 page 85** to extend the idea of representing square numbers spatially. The children can compare answers and check each other's work. Elicit that any number multiplied by itself produces a square number, pointing out that the sides of a square are always equal.

Challenge
Let the children investigate larger square numbers using graph paper and calculators. Ask them to summarise what they find out.

Support
Let the children use their 12×12 multiplication grid (from **Workbook 4 page 54**) to colour or circle all the square numbers. Then ask: *What pattern can you see?*

Answers for Pupil Book 4 page 133
1 a yes   b yes   c no   d no   e no
2 a Possible answers: You add the next odd number. The difference between the last two terms increases by 2 each time.
  b Possible answer: Add 2 to the difference between the last two terms, then add this amount to the last term to get the next term.
3 a Incorrect: when you multiply 20 by itself you get 400,
  b Correct: $1 + 3 + 5 + 7 + 9 + 11 = 36$ and $6 \times 6 = 36$
  c Correct: the product of two odd numbers is always odd.

Answers for Workbook 4 page 85
1 a 25; 5×5 square shaded
  b 9; 3×3 square shaded
  c 64; 8×8 square shaded
2 a 2   b 49   c 9

End-of-unit check
Use some or all of the following questions and activities to consolidate work on patterns and sequences and to assess how well the children have understood the ideas in this unit.
- *How do you decide whether something is a pattern? (There must be some elements that repeat.)*
- *Each car parked on a road has four wheels. Write a number sequence to show how many wheels there are on five cars. (4, 8, 12, 16, 20) What can you call this sequence? Why? (The 4 times table because the term-to-term rule is 'add 4'.)*
- Show the children a pattern of shapes.
  - *What rule was used to make this pattern?*
  - *Is there any other way of carrying on this pattern?*
  - *What comes next in this pattern?*
- *Use counters to make your own shape pattern. Write a number pattern to match your shape pattern.*
- *What is a square number? (a number that is made by multiplying a number by itself)*
- *Can you arrange 12 counters to make a square array? Explain why or why not. (No. You cannot make 12 with the same number of rows as columns because it is not a square number.)*
- *Write the first five numbers in the sequence of square numbers. (1, 4, 9, 16, 25)*

UNIT 20 Division

Learning objectives
- Recall multiplication and division facts for tables up to 12×12
- Use place value, known and derived facts to divide mentally
- Divide whole numbers by 10 and 100
- Solve scaling problems using multiplication and division
- Estimate and divide whole numbers, including division with a remainder

Key words
divide   equal groups   share   inverse   fact family
remainder   multiple   inverse operation

Unit introduction

Materials
A large 'Division' chart like the one shown at the top of the next page (for display); large sheets of paper and marker pens.

Teaching guidance
- Prepare and display a large chart like the one at the top of the next page. Give each group a large sheet of paper and marker pens to make their own.

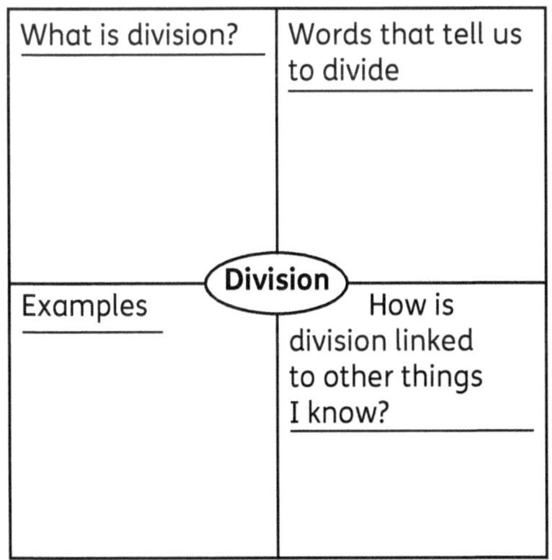

- Tell the children that they are going to learn about division in this unit.
- Ask them to work cooperatively in their groups to complete their chart.
- Next, take feedback and use the children's ideas to complete the large class chart. Leave this on display for the duration of this unit.
- The children can choose the definitions that they think are most useful for 'What is division?'
- Some words and phrases for 'Words that tell us to divide' include: *divide*, divided by, *equal groups*, half, how many in each . . . , one group, *share*, per, split equally, equal parts, goes into.
- Use the children's examples of divisions in the 'Examples' section.
- For 'How is division linked to other things I know?' the children will give different answers, but they should include these concepts: division is the *inverse* of multiplication; we can use grids and arrays to show division; fractions such as $\frac{1}{4}$ mean 1 divided by 4. These concepts have been covered in earlier units.

Divide using facts you know

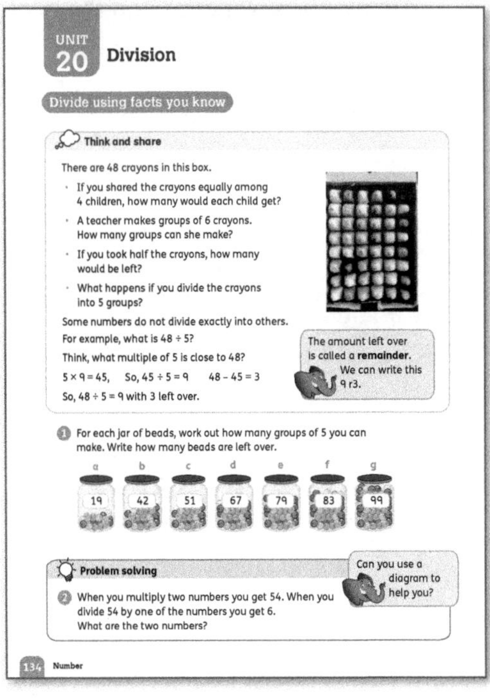

Materials

Objects for sharing and grouping; squared paper; sets of cards with numbers 10–70 (one set for each group); blank cards.

Warm-up

Select suitable activities involving multiplication and division facts from the 'Activity bank' on pages 23–30 or develop your own.

Focus

The children have already learnt that division is the inverse of multiplication and they have made *fact families* of multiplication and division facts using two factors and their product. In this lesson, they will develop the ideas to use known facts to solve divisions where there is a *remainder*.

- Ask the children to shuffle the 10–70 number cards and then deal out 12 from the pack. Choose a divisor (number to divide by), for example, 5 and let the children sort their cards according to the remainder when that number is divided by 5. For example:

No remainder	Remainder 1	Remainder 2	Remainder 3
25, 60	36, 21	42, 67	43, 23, 58

- Repeat this for different divisors and ask the children to talk about what they notice each time. It is important for them to realise that the remainder is always less than the divisor. If you divide by 5, you cannot have a remainder greater than 4.
- <u>Think and share</u>: Revise division as sharing, using the example on **Pupil Book 4 page 134** and concrete apparatus if necessary. Give the children time to answer the questions about the crayons before working through the example ($48 \div 5$) with the class.
- For question 1, ask the children to recall what they know about *multiples* of 5, counting up in fives from zero if you need to, to remind them. They should remember that all multiples of 5 have either a 5 or a 0 in the ones position. This can help them work out the division facts they need.
- <u>Problem solving</u>: Let children work on question 2 individually, and then discuss their strategies in pairs. They may not realise that one of the two numbers must be 6, and may use trial and error to find the two numbers.

Support

If any children have difficulty recalling multiplication facts, they will not be able to do the mental divisions. Ask them to make a set of fact cards with the multiplication fact on one side and the related division facts on the other. For example, they write $6 \times 7 = 42$ and $7 \times 6 = 42$ on one side and $42 \div 6 = 7$ and $42 \div 7 = 6$ on the other. Let them use these cards as they need to, and also develop games and activities to help develop fluency and rapid recall of facts.

Interesting mistakes

The children may make mental arithmetic errors when working out calculations. The mistakes might include incorrect times table facts or errors in repeated subtraction. Encourage the children to always check their calculations, and estimate to consider whether an answer is likely to be correct.

Answers for Pupil Book 4 page 134

Think and share: Each child would get 12 crayons. She can make 8 groups of 6 crayons.

If you took half the crayons, there would be 24 left. If you divide the crayons into 5 groups, you will have 9 in each group with 3 left over.

1 a 3 r4   b 8 r2   c 1 r1   d 13 r2
e 15 r4   f 16 r3   g 19 r4

2 6 and 9

More division

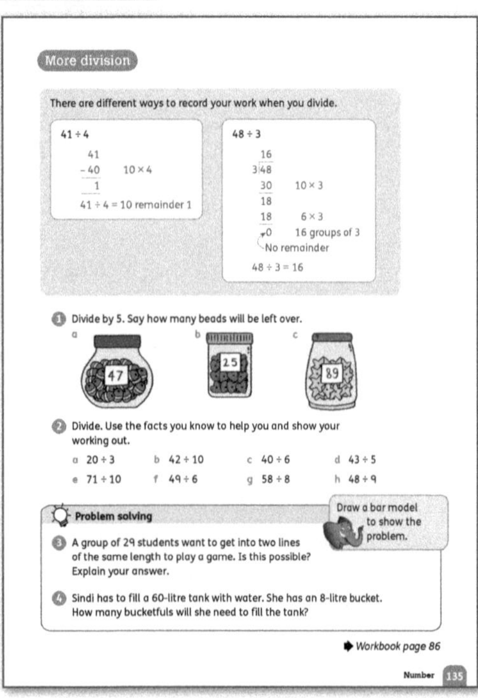

Materials
Counters.

Warm-up
Select suitable activities involving multiplication and division facts from the 'Activity bank' on pages 23–30 or develop your own.

Focus
- Work through the examples on **Pupil Book 4 page 135** with the class.
- Point out that division can be recorded using a standard format (sometimes called a division house), as shown in the right-hand example, as well as with a division sign.
- Spend some time talking about other ways the children can record their work when they do division. Some children might have a grasp of the short division algorithm.

- Let the children work independently to complete question 1 and question 2.
- Problem solving: Let the children work cooperatively to discuss and solve the problems in questions 3 and 4. Make sure that they record their work clearly.

Follow-up
Let the children complete **Workbook 4 page 86** and spend some time discussing what they did with the remainders in questions 1 and 2. Let them check each other's answers to question 3.

Challenge
Let the children explore division with remainders using a calculator. Explain that the calculator divides exactly and gives the remainder as a decimal. Let the children work out how they can use the calculator answer to find the remainder as a number. For example, if they divide 57 ÷ 4, they will get a result of 14.25. This tells them that there are 14 fours in 57. 14 × 4 = 56 and 57 – 56 = 1, so the answer is 14 r1.

Support
Let the children use counters to physically model the division if necessary. For example, give them 25 counters and ask them to share these between 4 people. Encourage them to use groups to do this rather than sharing one by one. Also allow them to use the 12 × 12 times table grid to find the facts they need if they have difficulty recalling them.

Interesting mistakes
- The children may not be able to make sense of written divisions (such as 49 ÷ 6). Help them by modelling how to read this (*forty-nine divided by six*) and how to reframe it as a multiplication so that they can use known facts (*How many sixes make forty-nine?*)
- The children may not always find an efficient recording method for division. At this stage, we are not trying to get them to use a formal algorithm and we are not dealing with long division, so as long as they can show their workings and the related calculations, any method they use is acceptable.

Answers for Pupil Book 4 page 135

1 a 2 left over   b 0 left over   c 4 left over
2 a 6 r2   b 4 r2   c 6 r 4
d 8 r3   e 7 r1   f 8 r1
g 7 r2   h 5 r4
3 No, 29 is an odd number so you cannot divide into two equal groups without a remainder.
4 8

Answers for Workbook 4 page 86

1 a 6   b 6 r7   c 7 r4   d 10 r4   e 12 r4
2 a 7   b 9   c 14   d 17   e 19
3 a 5   b 6   c 3   d 4   e 7
f 4   g 5   h 5   i 10   j 10
k 5   l 3   m 5   n 2   o 6

Divide by 10 and 100

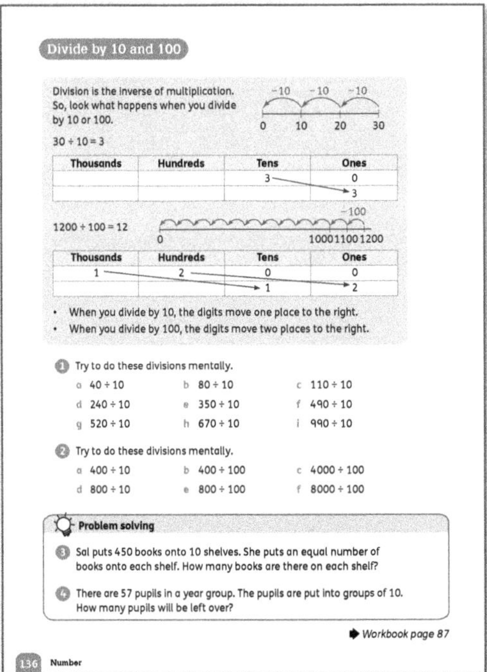

Materials
Place-value tables (see page 22); number cards.

Warm-up
Select suitable activities involving multiplication and division facts from the 'Activity bank' on pages 23–30 or develop your own.

Focus
- Revise multiplication by 10 and 100 using place-value tables as necessary before working through the division calculations shown on **Pupil Book 4 page 136**. Encourage the children to do the calculations mentally as far as they can.
- Ask questions such as:
  - *How many times smaller than 70 is 7?* (10 times smaller)
  - *What number is 10 times greater than 4?* (40)
- <u>Problem solving</u>: The children can work on questions 3 and 4 independently and then share their strategies with the class.
- Have a class discussion about what happens when we divide other numbers by 10 or 100. For example, when we divide: 7 by 10 (0.7); 35 by 10 (3.5); 240 by 100 (2.4); 13 by 100 (0.13).
- Display a place-value table that includes the decimal point, tenths and hundredths (see **Pupil Book 4 page 34**) and model these divisions.

Follow-up
- Use the function machine questions on **Workbook 4 page 87** as additional practice.
- <u>Problem solving</u>: Let the children solve the scaling problems in questions 2 and 3 cooperatively and share how they worked to solve each one.

Challenge
Adapt the activity 'How many ways can you make 100 by multiplying?' from the 'Unit introduction' for Unit 17 (see page 153) to make a division task. For example, ask: *How many different ways can you find to make 10 by dividing?* You could choose different numbers for different groups.

Support
Division is sometimes viewed as a problem area in mathematics, but it really shouldn't be treated as something more difficult than other operations. You can find some useful guides to division online, which have been developed specifically for teachers.

Answers for Pupil Book 4 page 136
1 a 4 b 8 c 11 d 24 e 35
   f 49 g 52 h 67 i 99
2 a 40 b 4 c 40 d 80 e 8
   f 80
3 45 **4** 7

Answers for Workbook 4 page 87
1 a (right, from top) 7 (provided as an example); 21; 35; 40; 48
   b (left, from top) 70; 800; 190; 230; 450
   c (left, from third row) 1800; 3200; 900; (right, from top) 4; 16
2 420 cm **3** 3 cm

Divide or multiply?

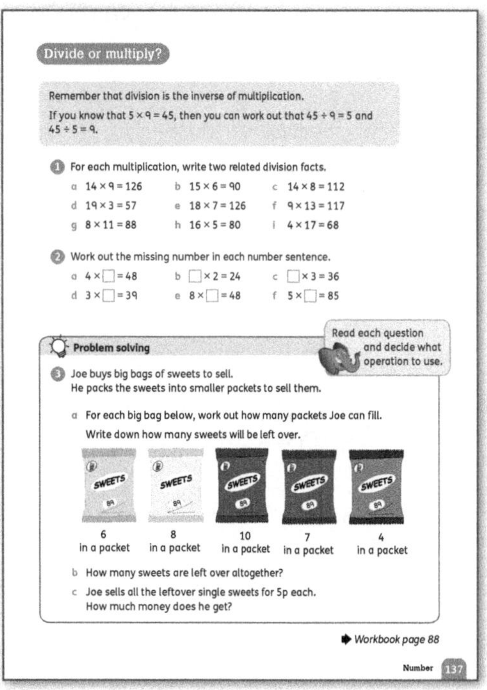

Warm-up
Select suitable activities involving multiplication and division facts from the 'Activity bank' on pages 23–30 or develop your own.

Focus

No new concepts are introduced in this lesson. The children already know about fact families and *inverse operations*. They have also explored words that tell them to multiply and words that tell them to divide.

- The children should work independently through questions 1 and 2 on **Pupil Book 4 page 137**.
- <u>Problem solving:</u> Check that the children read each part of question 3 carefully and decide what to do before estimating the answers and then solving the problems.

Follow-up

- Use **Workbook 4 page 88** to check that the children understand when to use multiplication and when to use division. Let the children check each other's work.
- <u>Problem solving:</u> The children can work independently on questions 2 and 3. Ask different children to share the riddles they have made up. Other children can work out the answers as they listen to them.

Challenge

Give the children examples of divisions with some digits missing. Ask them to work out what the missing digits are and to explain to their groups how they did this. For example:

$$2 \,\square\, 5 \qquad\qquad 8 \,\square$$
$$2\,)\,4\ 5\ \square \qquad 6\,)\,\square\ 2\ 2$$

Support

Ask the children to make up five word problems that can be solved by division. Give them as set of cues, for example: *Use one of these terms in each problem:*

equal share	equal groups	left over
how many	divided among	

The number values they choose are less important than using the words correctly and writing problems that require division.

Answers for Pupil Book 4 page 137

1
- **a** 126 ÷ 9 = 14; 126 ÷ 14 = 9
- **b** 90 ÷ 15 = 6; 90 ÷ 6 = 15
- **c** 112 ÷ 14 = 8; 112 ÷ 8 = 14
- **d** 57 ÷ 19 = 3; 57 ÷ 3 = 19
- **e** 126 ÷ 18 = 7; 126 ÷ 7 = 18
- **f** 117 ÷ 9 = 13; 117 ÷ 13 = 9
- **g** 88 ÷ 8 = 11; 88 ÷ 11 = 8
- **h** 80 ÷ 16 = 5; 80 ÷ 5 = 16
- **i** 68 ÷ 17 = 4; 68 ÷ 4 = 17

2 **a** 12 **b** 12 **c** 12 **d** 13 **e** 6 **f** 17

3 **a** (6 in packet) 14 packets, 5 left over;
(8 in packet) 11 packets, 1 left over;
(10 in packet) 8 packets, 9 left over;
(7 in packet) 12 packets, 5 left over;
(4 in packet) 22 packets, 1 left over
b 21 **c** £1.05 or 105p

Answers for Workbook 4 page 88

1
- **a** ÷ **b** × **c** ÷ **d** × **e** ÷
- **f** × **g** × **h** ÷ **i** ÷ **j** ×
- **k** × **l** ÷

2 **a** 6 **b** 9 **c** 21 **d** 18 **e** 30 **f** 82

3 Individual answers

Think and solve

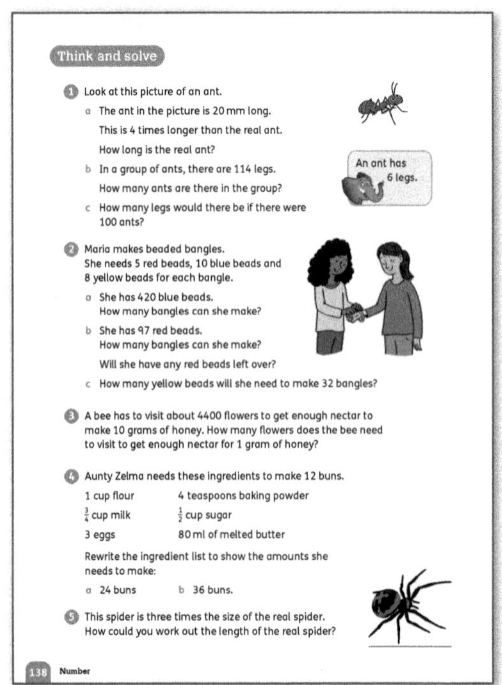

Warm-up

Use one of the 'Mental problem solving' activities on page 27 as a warm-up for this lesson.

Focus

- The problems on this page involve careful reasoning about multiplication and division as well as scaling and proportional thinking.
- Let the children work in groups to discuss each problem in questions 1–5 on **Pupil Book 4 page 138** and then ask them to solve them.
- As a class, discuss the answers and the methods that the children used.

Interesting mistakes

The children may not understand how to increase or decrease amounts in proportion because they don't understand the concept of multiplicative reasoning.

It is important to emphasise that doubling, halving, finding three times as much or a quarter as much, all involve multiplying or dividing.

Many of the ways in which we compare amounts and express change involve multiplicative reasoning. For example, we might say: *This picture of a doll is a third of the height of the real doll.* The children need to understand that this means the real doll is 3 times taller than the one in the picture.

Answers for Pupil Book 4 page 138

1 a 5 mm   b 19 ants   c 600 legs

2 a 42   b 19   c 256

3 440 flowers

4 a 2 cups flour: 1.5 cups milk; 6 eggs; 8 tsp baking powder; 1 cup sugar; 160 ml melted butter

   b 3 cups flour; 2.25 cups milk; 9 eggs; 12 tsp baking powder; 1.5 cups sugar; 240 ml melted butter

5 Measure the length of the spider and divide the length by 3.

End-of-unit check

Ask some or all of these questions to assess how well the children have understood and can apply the concepts from this unit.

* *How can you use the fact that 7 × 8 is 56 to find 56 ÷ 8? (Division is the inverse of multiplication, so 56 ÷ 8 = 7)*
* *What is (give a number) divided by 10 (or 100)?*
* *Explain the method you used to find your answer.*
* Show the children a completed division calculation with a remainder. *In this division problem, there is a remainder of (say the number). What should you do with it?*
* Show some completed divisions, some with mistakes. *Which of these calculations are correct/incorrect? What has this person done wrong? How could you help them to correct it?*
* Show a recipe. *This recipe is for 8 people. What do you have to do to the ingredients to make enough for 2 people (or 4 people)? (Divide the amounts by 4 (or 2).)*
* *Bilal has painted a picture of a monkey. In real life the monkey is 95 cm tall. This is five times taller than the monkey in Bilal's painting. How tall is the monkey in his painting? (19 cm)*

Mixed practice 3

You can use Mixed practice 3 on **Pupil Book 4 pages 139–140** to assess the children's confidence in the concepts from Units 16–20.

Answers for Mixed practice 3, Pupil Book 4 pages 139–140

1 a 135   b 57   c 252

   d 288   e 243   f 144

2 3 × £429 = £1287

3 a 79   b 84   c 32

   d 11 r1   e 10 r3   f 7 r5

4 a True   b True   c False   d True

5 a 150   b 280   c 720

   d 480   e 350   f 800

6 a A (2, 5), C (5, 4), F (2, 1)

   b D   c hexagon

   d rectangle   e G (1, 2), H (1, 4)

   f right-angled scalene   g A (0, 7), F (0, 3), E (2, 3)

7 a 170   b 19

   c 19 × 4 = £76   d yes, 8

8 a add 15   b 85, 100, 115

   c 320, because there is a difference of only 5 between 320 and 325 (which is a term in the sequence)

9 a How much water is left in (or flows out of) a barrel over a 3-hour time period from 8:00 to 11:00.

   b 45 ℓ   c 9:00

   d It stayed the same.   e 5 ℓ

   f At 10:00 – we know this because the amount of water stays the same after this time.